UNIVERSITY LIBRARY
UNIVERSITY OF ILLINOIS AT URBANA-CHAMPAIGN

The person charging this material is responsible for its renewal or return to the library on or before the due date. The minimum fee for a lost item is **$125.00, $300.00** for bound journals.

Theft, mutilation, and underlining of books are reasons for disciplinary action and may result in dismissal from the University. *Please note: self-stick notes may result in torn pages and lift some inks.*

Renew via the Telephone Center at 217-333-8400, 846-262-1510 (toll-free) or circlib@uiuc.edu.

Renew online by choosing the **My Account** option at: http://www.library.uiuc.edu/catalog/ 2/22/11

Agricultural Uses of By-Products and Wastes

ACS SYMPOSIUM SERIES **668**

Agricultural Uses of By-Products and Wastes

Jack E. Rechcigl, EDITOR
University of Florida

Herbert C. MacKinnon, EDITOR
H. J. Baker & Bro., Inc.

Developed from a symposium sponsored
by the Division of Fertilizer and Soil Chemistry

American Chemical Society, Washington, DC

Library of Congress Cataloging-in-Publication Data

Agricultural uses of by-products and wastes / Jack E. Rechcigl, Herbert C. MacKinnon, [editors].

p. cm.—(ACS symposium series, ISSN 0097–6156; 668)

"Developed from a symposium sponsored by the Division of Fertilizer and Soil Chemistry, at the 212th National Meeting of the American Chemical Society, Orlando, Florida, August 25–29, 1996."

Includes bibliographical references and indexes.

ISBN 0–8412–3514–7

1.Fertilizers—Congresses. 2. Waste products as fertilizer—Congresses.

I. Rechcigl, Jack E. II. MacKinnon, Herbert C., 1930– . III. Series.

S633.A275   1997
631.8—dc21   97–16403
                                                              CIP

This book is printed on acid-free, recycled paper.

Copyright © 1997 American Chemical Society

All Rights Reserved. Reprographic copying beyond that permitted by Sections 107 or 108 of the U.S. Copyright Act is allowed for internal use only, provided that a per-chapter fee of $17.00 plus $0.25 per page is paid to the Copyright Clearance Center, Inc., 222 Rosewood Drive, Danvers, MA 01923, USA. Republication or reproduction for sale of pages in this book is permitted only under license from ACS. Direct these and other permissions requests to ACS Copyright Office, Publications Division, 1155 16th Street, N.W., Washington, DC 20036.

The citation of trade names and/or names of manufacturers in this publication is not to be construed as an endorsement or as approval by ACS of the commercial products or services referenced herein; nor should the mere reference herein to any drawing, specification, chemical process, or other data be regarded as a license or as a conveyance of any right or permission to the holder, reader, or any other person or corporation, to manufacture, reproduce, use, or sell any patented invention or copyrighted work that may in any way be related thereto. Registered names, trademarks, etc., used in this publication, even without specific indication thereof, are not to be considered unprotected by law.

PRINTED IN THE UNITED STATES OF AMERICA

Advisory Board

ACS Symposium Series

Mary E. Castellion
ChemEdit Company

Arthur B. Ellis
University of Wisconsin at Madison

Jeffrey S. Gaffney
Argonne National Laboratory

Gunda I. Georg
University of Kansas

Lawrence P. Klemann
Nabisco Foods Group

Richard N. Loeppky
University of Missouri

Cynthia A. Maryanoff
R. W. Johnson Pharmaceutical
  Research Institute

Roger A. Minear
University of Illinois
  at Urbana–Champaign

Omkaram Nalamasu
AT&T Bell Laboratories

Kinam Park
Purdue University

Katherine R. Porter
Duke University

Douglas A. Smith
The DAS Group, Inc.

Martin R. Tant
Eastman Chemical Co.

Michael D. Taylor
Parke-Davis Pharmaceutical
  Research

Leroy B. Townsend
University of Michigan

William C. Walker
DuPont Company

Foreword

THE ACS SYMPOSIUM SERIES was first published in 1974 to provide a mechanism for publishing symposia quickly in book form. The purpose of this series is to publish comprehensive books developed from symposia, which are usually "snapshots in time" of the current research being done on a topic, plus some review material on the topic. For this reason, it is necessary that the papers be published as quickly as possible.

Before a symposium-based book is put under contract, the proposed table of contents is reviewed for appropriateness to the topic and for comprehensiveness of the collection. Some papers are excluded at this point, and others are added to round out the scope of the volume. In addition, a draft of each paper is peer-reviewed prior to final acceptance or rejection. This anonymous review process is supervised by the organizer(s) of the symposium, who become the editor(s) of the book. The authors then revise their papers according to the recommendations of both the reviewers and the editors, prepare camera-ready copy, and submit the final papers to the editors, who check that all necessary revisions have been made.

As a rule, only original research papers and original review papers are included in the volumes. Verbatim reproductions of previously published papers are not accepted.

ACS BOOKS DEPARTMENT

Contents

Preface ix

1. Future Directions of By-Products and Wastes in Agriculture 1
   Rosa M. C. Muchovej and R. S. Pacovsky

   FACTORS AFFECTING USES OF BY-PRODUCTS AND WASTES
   IN AGRICULTURE

2. An Entrepreneurial View of the Future for the Use of Wastes
   and By-Products 22
   Dale F. Galloway and John M. Walker

3. U.S. Environmental Protection Agency Regulations and Other Stakeholder
   Activities Affecting the Agricultural Use of By-Products and Wastes 28
   John M. Walker, Robert M. Southworth, and Alan B. Rubin

   ORGANIC BY-PRODUCTS AND WASTES

4. Biosolids Processing, Products, and Uses 50
   Jane B. Forste

5. Characteristics and Standards for Processed Biosolids in the Manufacture
   and Marketing of Horticultural Fertilizers and Soil Blends 63
   Terry J. Logan, Billie J. Lindsay, and Steve Titko

6. Agricultural and Environmental Issues in the Management of Poultry
   Wastes: Recent Innovations and Long-Term Challenges 72
   J. Thomas Sims

7. Agricultural and Environmental Issues in the Management of Cattle
   Manure 91
   H. H. Van Horn and M. B. Hall

8. Agricultural and Environmental Issues in the Management of Swine
   Waste 110
   Robert L. Mikkelsen

9. Composition and Uses of Organic Fertilizers 120
   E. E. Huntley, Allen V. Barker, M. L. Stratton

10. Composition and Uses of Compost 140
    Allen V. Barker

11. Composition and Uses of Uncomposted Wastepaper and Other
    Organics 163
    J. H. Edwards

12. Composition and Land Application of Paper Manufacturing Residuals 185
    J. J. Camberato, E. D. Vance, and A. V. Someshwar

INORGANIC BY-PRODUCTS AND WASTES

13. Land Application of Boiler Wood Ash in the Southeastern United
    States 204
    C. C. Mitchell and E. D. Black

14. Agricultural and Industrial Uses of By-Product Gypsums 226
    W. P. Miller and M. E. Sumner

15. Ammonium Sulfate Fertilizer as By-Product in Flue Gas Desulfurization:
    The Dakota Gasification Company Experience 240
    Daniel L. Wallach

16. By-Product Usage in Fertilizer Micronutrients 255
    James M. Wyatt

17. Iron Humate Production, Processing, Properties, and Usage 268
    B. E. Rehberg and G. L. Smith

INDEXES

Author Index 278

Affiliation Index 278

Subject Index 279

Preface

THE NATIONAL COMMITMENT TO MANAGEMENT of wastes in a cost-effective, environmentally sound manner leads to the realization that wastes may be viewed as resources for a variety of applications. Many waste materials contain essential plant nutrients vital to successful crop production. Utilization of these materials in agriculture, in combination with commercial fertilizers through incorporation or substitution, may offer economic advantages while ameliorating possible harmful effects due to stockpiling of wastes and overloading of landfills.

This volume was developed from a symposium presented at the 212th National Meeting of the American Chemical Society, titled "Uses of By-Products and Wastes in Agriculture", sponsored by the ACS Division of Fertilizer and Soil Chemistry, in Orlando, Florida, August 25–29, 1996. Leading experts were invited to discuss the various agricultural uses of industrial, municipal, and farm by-products.

In addition to the introductory overview, the book comprises three sections concerning various factors affecting the use of such materials as municipal biosolids, organic by-products and wastes, and inorganic materials. The organic materials include compost, municipal biosolids, poultry wastes, cattle manure, swine wastes, and organic fertilizers. Inorganic materials cover various wastes from paper manufacturing, boiler wood ashes, gypsum, phosphogypsum, by-product ammonium sulfate and micronutrients, and wastewater by-products. Whenever applicable, environmental implications for the use of such materials are discussed, and a special chapter is devoted to government regulations. As is customary with ACS series, each chapter has been evaluated by at least two outside experts.

This book should be useful to researchers, students, government regulators, policy makers, and the general public concerned with agriculture, recycling of wastes, and environmental quality.

Acknowledgments

The editors thank the individual contributors and reviewers for the time and effort spent in preparing and reviewing the chapters. In addition, special thanks are due to the editorial board and the staff of ACS for their advice and assistance in the preparation of this volume.

JACK E. RECHCIGL
Soil and Water Science Department
Range Cattle Research and Education Center
University of Florida
3401 Experiment Station
Ona, FL 33865

HERBERT C. MACKINNON
H. J. Baker & Bro., Inc.
11404 Tullamore Place
Temple Terrace, FL 33617

February 28, 1997

Chapter 1

Future Directions of By-Products and Wastes in Agriculture

Rosa M. C. Muchovej[1] and R. S. Pacovsky[2]

[1]Southwest Florida Research and Education Center, Institute of Food and Agricultural Sciences, University of Florida, P.O. Box 5127, Immokalee, FL 34143-5002
[2]Western Regional Research Center, Agricultural Research Service, U.S. Department of Agriculture, 800 Buchanan Street, Albany, CA 94710

World population increase and increasing urbanization have resulted in various negative effects for society, such as the rising demand for food and the production of large amounts of animal and human wastes. These various organic residuals include livestock wastes, crop debris, biosolids (sewage sludge) and septage, food processing wastes, industrial organic wastes, logging residues, and wood manufacturing wastes, and municipal refuse. Many inorganic by-products and wastes also result from human activities. Bottom ash, gypsum, phosphogypsum, and by-product micronutrient sources are some of the predominant inorganic wastes being generated currently. The disposal of these residues has become an increasing problem and a growing expense for industry, farmers and municipalities, especially in light of the more restrictive environmental regulations that have been enacted. We will review how some of these wastes can be used in agriculture, which is the theme of this symposium. As we shall see, in many cases, current research is designed so that a residue is easier to use, less expensive than alternatives or contains value-added features that make it more desirable to the end user. Individual chapters in this book cover the uses of several by-products and waste products. As a main conclusion, considerable progress has been made regarding proper utilization of "wastes", both organic and inorganic nature. It is imperative, however, that society takes more interest and becomes more educated and involved in effectively re-utilizing materials that have increased in production as a result of population growth and demands for "old and new" products. In this introductory chapter an attempt has been made to highlight some major points that are relevant in this area.

It has been estimated that the world population will increase from 5.4 billion in 1990 to 8.5 billion by 2025 (1). An increase in food production by 60 to 70% will be necessary to feed this growing population; furthermore, societal needs will result in increasing demand for fiber, fuel, and construction materials. To increase agricultural production there must be either an expansion of land area used for crop cultivation and livestock, or production must increase from land already under cultivation (2).

An increase of highly intensive, non-sustainable agriculture or short-term management of our natural resources will result in: 1) nutrient depletion and mining of the soil resource through long-term removal at rates faster than can be replenished; 2) declining soil organic matter due to removal of crop residues; 3) other soil-related problems, such as compaction, decrease in water permeation, and sodicity; 4) build-up of pests, such as weeds, diseases, and insects (1), and increased production of wastes in a more concentrated area. Organic and inorganic by-product production in the United States has been estimated at more than 1 billion tons annually (3). Even if a small portion of these residues could be used in food or fiber production, it would result in enhanced farm efficiency and would minimize that waste problem that society faces.

To protect our natural resource base for future generations, technologies which minimize the on-site and off-site negative environmental impacts of agricultural production are required. Traditional disposal methods are currently perceived as having serious environmental implications, are not sustainable, and are becoming quite costly. In this symposium, most papers have focused on the processes by which municipal biosolids or agricultural wastes have been converted directly into value-added products or into the precursors of such products. The processing options for waste-water biosolids and the appropriate uses for these products were presented (4). The manufacture and marketing of horticultural fertilizers and soil blends have considered the size and influence of the commercial greenhouse industry, commercial nursery production, sod industry and turfgrass maintenance, and home gardening (5).

We need to recognize that we must manage the by-products that result from industrial or societal processes. We can no longer treat these residues as waste, discharging them into the air or water indiscriminately, burying them in a landfill or sinking them at sea. These materials, formerly considered by-products and waste, may be regarded as rich sources of plant nutrients. By themselves, these components may not provide the proper balance of nutrients or growth-promoting substances, but when mixed with one another or amended they can provide a low-cost or specific-use fertilizer to farms, ranches, or nurseries.

This symposium evaluated the current situation regarding uses or potential use of several by-products and waste residues. This overview aims to highlight some of the major points raised about the various materials discussed during the symposium. As a main conclusion, it is clear that considerable progress has been made regarding proper utilization of "wastes", both organic and inorganic in nature. However, it is imperative that society takes more interest, becomes better educated and more involved in effectively re-utilizing materials whose production has increased due to population growth or increased societal demand.

Often the by-products generated from municipal, manufacturing, or barnyard sources (biosolids, sludge, manures) are organic-matter rich. Therefore, these complex fertilizers are normally more beneficial at improving the edaphic characteristics of a soil than are inorganic fertilizers, that provide the same chemical nutrients in a strictly mineral form. Organic by-products have a significantly higher carbon content (approximately 50%) than do inorganic fertilizers (often 1% C or less). In addition, the

N, P, and S present in organic residues is often covalently bound to C as well, so microbial decomposition is required to release ammonia, phosphates, or sulfates from the sources. Therefore, addition of organic residues to a soil results in stimulating soil microbial activities and building soil biomass in a fashion that is not duplicated by inorganic fertilizers. Since microbial decomposition is required for the release of N, P, and S into a plant-available form, organic fertilizers supply nutrients in a slow-release from. Such controlled release of nutrients usually results in significantly less leaching of nitrates and phosphates into the ground water. Controlled-release fertilizer technology has demonstrated that even for inorganic mineral nutrients, resin-encapsulated or polymer coated pelleting may also reduce leaching losses significantly.

Analytical methods for the characterization of these waste materials has not kept up with the wide range of new product formulations. New terminology, analytical methodology, and standards for the manufacture and use of these by-products is now being developed between industry, agricultural, and regulatory personnel. New formulations and additives for controlled-release fertilizers have been developed for newer agriculture practices, such as no-till or low-input sustainable agriculture. Incorporation of enzyme inhibitors into slow-release, urea-based fertilizers has the potential to provide substantial benefits for conservation tillage growers. Similarly, incorporation of surfactants into solid or liquid inorganic fertilizer formulations can enhance nutrient release and subsequent plant assimilation so that markedly less fertilizer per hectare would be required to obtain the same yield.

New materials, such as iron humates, have been recovered from municipal water processing and have a proven ability to act as a matrix for the incorporation of other mineral nutrients (6). Such a product not only demonstrates a decreased rate of nitrogen release, but also provides readily available chelated iron over an extended period of time.

Regulatory requirements govern the processing, distribution, use and disposal of organic and inorganic by-product materials (3). Regulations cover the production of hazardous waste, the practices at solid-waste disposal facilities, and the policies involved in the beneficial use and disposal of these materials (3). However, standards need to be developed for each type of fertilizer, soil amendment, manufactured soil or products containing biosolids (5). These new regulations must address three important areas, since agricultural commodities are produced eventually for human and animal consumption: reduction of disease-causing pathogens, reduction of vector attraction, limits to trace pollutants, and other specific requirements (4).

Many of the by-products and wastes that were discussed in the symposium are very rich in nitrogen and organic matter as they are derived from various manures. Animal wastes, when concentrated in one area, have the potential of being an enriched source for the contamination of ground water by nitrates. Effective means to store, handle and apply such manures to crop production systems is still a problem in modern agriculture (7). In areas where there are insufficient stores of soil-available N and P, poultry waste may be integrated into fertilizer management (7). Most animal manures are P-rich relative to the N if applied to field crops, and such concerns must be taken into account when whole-farm nutrient budgets are calculated (8). Where application of cattle manure to crops for fertilizer recovery is the primary method used to manage animal wastes, additional processing and partitioning alternatives need to be considered, such as anaerobic treatment, solids separation, or export of nutrients off-farm (8). When swine manure is applied to crop land for extended periods of time, accumulation of nutrients in excess of the plant requirement may occur (9).

American consumers have become increasingly concerned about the use of pesticides (insecticides, herbicides, fungicides) and other toxic chemicals in the production of their food. Although the term "organic farming" is vague and lacks scientific precision to the average consumer, many organic farmers would characterize their practices and materials as being derived from naturally-occurring renewable resources. Organic fertilizers are usually derived from living organisms and do not contain petroleum-based fertilizers (10). Organic farming normally uses large quantities of compost. Composts present variable composition, depending upon the raw materials used in the composting process. The materials used for composting may include crop debris, city refuse, manures, biosolids, and food wastes (11). Utilization of certain organic residues, such as waste paper, may actually require composting to prevent long-term N immobilization (12). Additions of inorganic by-products and wastes, such as bottom ash, gypsum, phosphogypsum, and by-product micronutrient sources represent rich sources of calcium, sulfate, and trace elements, and may be added to materials being composted to enrich them.

A holistic approach must be taken to attain solutions or devise alternatives for by-products and waste utilization. Whatever method is employed for the disposal of the "waste", it must not degrade surface and ground water quality, water use, air quality, and must assure protection of the environment.

Public Acceptance

Public perception is the key to the acceptance of by-products and wastes for new uses. Most often, the public has an inaccurate perception about wastes and by-products and considers most of them to be harmful in some way. What can be done? Initially there is need for public education and awareness programs that reach all levels of society. Many residuals are still presently viewed as a disposal problem, rather than as a resource to be utilized.

There is a vast amount of literature on the benefits and potential harmful effects of residuals. The majority of this literature indicates that the benefits outweigh the risks. We should strive to increase public acceptance; better yet, we need to effectively change people's attitudes and behavior. Nordstat (13) stated over two decades ago that we cannot expect to change public attitudes and behavior by telling them something or by coercion. We must have the general public become involved by interests and concerns, and point out the existing economic advantages of utilizing the by-product.

There has been a slight shift in public perceptions about the environment: wastes that were once discarded are now recycled, and soon it is likely that formerly recycled wastes will be turned into their own value-added products. For these changes to take place, it will be necessary to segregate different industrial or societal by-products so that one residue does not contaminate another (14). In addition, we must identify the potential products that can be produced from what was formerly considered waste, determine what markets and opportunities exist for such a product, and find the highest value for the particular residue that has been generated (14). Potential sources for what may be considered wastes at the present, but that will be considered important "raw materials" in the twenty first century, include water-processing residues, municipal sludges, barnyard manures, crop residues, agricultural and industrial composts, boiler ash, inorganic materials from smokestack scrubbers, pulp mill residuals, mining wastes (gypsum, phosphogypsym) and by-product micronutrient sources.

The benefits of recycling have been consistently confirmed by highly respected researchers and institutions over the past decades. Recycling of biosolids on agricultural land, and private vegetable and flower gardens has been performed for thousands of years. Potential health and environmental effects of the various inorganic and organic waste products are discussed in subsequent chapters of this book. There is a large amount of scientific data to prove that there is no human health or environmental risk associated with the beneficial use of biosolids, provided the application it is done properly.

Quality standards for these materials need to go beyond aspects of health and environment. Galloway and Walker (14) indicated that in the future we will tend to produce by-products that are more uniform. Also, more uniform labeling of the by-products will likely be adopted in the U.S., especially for those materials containing biosolids. Another area that deserves attention is periodic analyses of the waste material; manures are rarely analyzed prior to land application. Before application of any by-product or waste onto agricultural land, the user should have a chemical analysis of the material performed and should have knowledge as to what the soil contains and what it can hold.

Some specific information on by-products and wastes was discussed during the symposium, and some of the benefits and problems inherent with their utilization in agriculture now follows.

Organic Wastes

Organic materials are valuable resources when properly applied to land, both as a fertilizer and as a soil amendment. Organic wastes can be applied to agricultural land to help improve soil physical properties (e.g., water retention, infiltration, aggregate stability), as well as the chemical characteristics of soils (e.g., cation exchange capacity and plant nutrient status of soils) (15), and the biological properties (16). Once applied to soils, the major biochemical processes occurring with both animal and municipal wastes include decomposition, transformation of N, S, and P, and modifications of trace metal solubilities. These processes and the rates of nutrient release will influence the potential environment impact (17), specifically that of soil and groundwater degradation.

Organic wastes will not necessarily replace inorganic fertilizers, although many compare favorably with commercial N fertilizers. Usually, organic wastes lack a balance of plant essential nutrients and do not fully meet crop requirements. However, they can be used in conjunction with inorganic fertilizers, thereby reducing commercial fertilizer inputs.

The timing and method of organic waste application to cropland are additional, critical components of a safe and effective management program. Utilization of organic wastes, as well as any other fertilizers, in crop production should be timed so that nutrient availability from the material coincides with the nutrient needs and uptake patterns of a given crop. Improper utilization of organic materials may adversely affect crops through imposition of nutrient imbalance, total salt toxicity, or trace element toxicity. Surface applications usually result in a higher loading of N and P in the runoff, as compared with areas where the organic materials have been incorporated. Amending soil with high rates of organic materials may also result in increased NO_3 and metal concentrations in the soil profile, and these may be subject to leaching. These processes (erosion and leaching) can cause eutrophication of lakes and streams. Eutrophication is generally favored by NO_3 concentration greater than 1 mg per kg (*18*). Health disorders and problems associated with excess nitrates have been discussed (*19, 20, 21*).

The beneficial and possible harmful effects of organic amendments are extensively discussed in current reviews, and the reader is referred those compiled by Raviv et al. (*22*), Eck & Stewart (*23*), Hue (*15*), Mullins & Mitchell (*24*), and Stratton et al. (*16*). Nevertheless, proper management of land-applied wastes will prevent environmental hazards such as groundwater contamination or trace element biomagnification through the food chain.

Health Effects of Metals

In the last 25 years, research on environmental geochemistry and health has focused on the toxicology problems of Pb and Cd. Relatively low blood levels of Pb can cause mental impairment; research on Cd has been less successful. The possible health implications of Hg, I, Cd, Cu, Al, Pt, Se, Cu, As, Zn, and Au have been reviewed by Warren (*25*). Prolonged high levels of Cd in the diet have been shown to cause kidney failure in humans (*26*). Cadmium accumulation has also been determined in kidneys and liver of sludge-fed animals (*26*). Cobalt, Cu and Mo in excess have been shown to disturb the metabolism of ruminant animals (*15*). Lime residues present in many sludges may raise soil pH levels, and this will render Mo more available to plants (*27*). Scant information is available regarding Mo accumulation in soil from sludge additions, but studies have intensified due to more accurate laboratory methods for Mo determination. The primary concern with Mo is the potential for molybdenosis in ruminants from ingesting forages containing high levels of Mo (*28*).

The bioavailability of metal pollutants in by-products and wastes is reduced by organic matrices from biosolids, organic matter and phosphates (*3*), as well as by soil organic matter. Therefore, metal toxicity problems from land application of by-products and wastes are clearly diminished, when applied in an organic form.

Biosolids

The symposium addressed biosolids resulting from waste water treatment plants and from paper manufacturing plants. A brief overview of the main points discussed is given.

Biosolids from Waste Water Treatment

A survey by the U.S. Environmental Protection Agency (EPA) revealed that in 1990 the U.S. generated a total of 160 million tons of solid waste and 8.5 million tons of biosolids. By the year 2000, 193 million tons and 12 million tons per year are projected to accumulate, respectively, because of increased population and more advanced sewage treatment processes (*15*).

Since biosolids are the solids resulting from treatment of wastewater, the degree and type of processing will govern the specific applications for these residuals. Production of biosolids in the United States is in the vicinity of 9 tons per year (*29*). Biosolids contain all nutrients considered essential for plant growth, but their specific quantities vary somewhat depending on the processing and the materials present in the wastewater. The EPA has established concentration standards for 10 trace metals allowable in sludge and compost (*30*), as a protective measure for public health and safety. These maximum concentrations for Exceptional Quality (EQ), in mg per kg, are as follows: As 41, Cd 39, Cr 1200, Cu 1500, Pb 300, Hg 17, Mo 18, Ni 420, Se 36, and Zn 2800. The EPA regulations apply to any waste material containing biosolids.

Technological advances, along with more stringent regulations regarding metal detoxification of biosolids, have contributed to the generation of relatively "clean" products. Currently, most U.S. biosolids are considered fit for land application. A 1988-1989 EPA national survey of sewage treatment facilities determined that 70% of the wastewater residuals met the criteria for EQ with regard to trace metal concentration.

A "typical" nutrient composition of biosolids is 3.2% N, 1.4% P, 0.2% K, 2.7% Ca, and 0.4% Mg (*15*). Based on this composition, these residuals can be used as low grade N and P fertilizer and also as a source of Ca (especially the lime-stabilized biosolids), Mg and Fe. In Fe-deficient calcareous soils, biosolids may be a valuable Fe source, and they may be even more effective than commercial fertilizers (*31*). Application of municipal sludge has been demonstrated to correct Fe deficiencies in sorghum and barley in New Mexico (*32, 33*).

Traditionally application rates for organic wastes have been based on the N requirement of the crop and on the N content of the waste, assuming that a certain percentage of that N would be mineralized in the first year of application. Since the P content of the materials can be as high as that of N (*4*), much attention has been dedicated to phosphates recently.

Instead of basing the application rates on the N concentration, P may have to be the first consideration in areas where soil levels of P are naturally high. Generally, Fe is an abundant element in biosolids, especially when ferric chloride is the chemical conditioner used in the processing. Thus, the residual can be a valuable source of slowly available Fe for plants growing in sandy soils, especially if the pH is alkaline. On the other hand, when lime is used in the stabilization process for vector and pathogen reduction, N content is reduced. However, the pH of the final product is high (in excess of 12), and therefore, it may be used more as a liming material, than as a fertilizer. Nevertheless, biosolids are still good soil conditioners due to the organic matter they contain.

Heat drying of biosolids eliminates pathogens while reducing volume and weight. Although an increase in cost is incurred, considerable savings can be achieved on transport. Processed biosolids products, such as the thermal dried/pelletized, granular forms, and composted biosolids have been receiving more attention, making them more desirable as fertilizers (*5*). Although agriculture has, by far, the greatest potential demand, the horticultural markets for these soil additives is being underestimated, and if cost is low, it is believed that use of biosolids and other organic waste products will intensify rapidly (5).

Sludge from Paper Manufacturing

Paper manufacture produces various residuals, in the form of primary and secondary sludges that have a variable composition but generally have a wide C:N ratio. Addition of N fertilizers to these materials will overcome N immobilization and provide nutrients that would be readily available for plant growth (34). The paper-making industry also produces grits, dregs, and lime mud that have a high lime potential. Concerns with trace concentrations of dioxins (TCDD) and furans (TCDF) from chlorine bleaching processes on wildlife exposed to mill sludges have, so far, been unwarranted (35).

Land application of paper mill sludges and other by-products are currently regulated mainly at the state level, and the regulations are quite variable. Similar to other wastes, specific composition of materials, soil characteristics and crop requirements need to be evaluated so that the material can be adequately utilized as a source for specific nutrients.

Animal Wastes: Poultry, Cattle, and Swine

With the intensification of the livestock and poultry industries to supply the demand for protein and meat products of a growing population, significantly larger quantities of manure are generated each year. This manure must be collected, stored, and utilized efficiently. In certain regions of the country, a minimum amount of land is available for manure utilization, and producers are feeling greater pressure from society and regulators to dispose of the waste in an environmentally sensitive fashion. Swine and poultry industries are concentrated and regionalized, and this results in large quantities of animal waste produced in a relatively small area (7, 9). Animal wastes contain large quantities of water, and transport becomes prohibitively expensive if areas receiving the waste are located outside a certain distance.

Before inorganic fertilizers were used, producers used considerable amounts of manures and green manures in their farm operations on a routine basis (36). Reports on the beneficial effects of manures on plant growth and nutrition are abundant in the literature (23). The development of larger specialized operations has resulted in large commercial feedlots that produce an abundance of manure. Manure from animals represents about 22% of all organic wastes produced in the U.S. (37). Sources of animal wastes are feces and urine excreted by beef and dairy cattle, chickens, swine, sheep, goats, turkeys, ducks, and other types of animals. About 96 000 Gg is excreted on pasture, rangeland, and cropland and thus is automatically returned to the land (37). This manure has the potential to pollute groundwater and cause other detrimental environmental consequences if not used or stored properly. Livestock management has been identified as a major source of agricultural non-point source pollution (38, 39). The method, timing, and rate of manure application are significant factors in determining the likelihood of environmental degradation, particularly of water quality.

The concentration of elements in manures is highly variable, and depends on the animal species, ration, feed, materials used as beds, as well as the type of collection, storage, and handling (23). Manures usually contain substantial amounts of salts of K, P, Ca, Mg, Na, S, and other elements. Unless application rates are excessive, salts do not affect plant growth; however, when leached from the soil, salts may become contaminants of ground and surface water (23). If these materials are applied to land based on N concentration, P and K may be supplied in excess in these materials compared with crop demands. Other potentially adverse effects from excessive rates of manure application are increased soil pH, salt imbalance in the soil profile, and increase in K in soil solution due to release from clay minerals. The total amount of all mineral elements present in the soil, except C and N, should remain constant with time. As manure decomposes, part of the C is given off as CO_2, and some N is mineralized. If nutrient release from manures is slow, growth of crops may be limited, especially in those crops with a large N demand in early spring, when rates of mineralization are low (40). Most of the components of manure are immobile and unreactive in the soil if they remain in the organic form. However, upon mineralization by microbes, they will become more available and mobile.

Fresh manure is mostly water (more than 80%), which encourages high application rates; consequently, high nitrate and salt levels may occur in soil, runoff and leachate. Problems with odor, transport and presence of viable weed seeds and pathogens are magnified in fresh manure. Composting of manures usually reduces these problems, yielding a material with better physical, chemical and biological properties.

After World War II, concentration of livestock production in large-scale, confinement-type operations in the U.S. magnified the problems of handling animal wastes, including health hazards, nuisances with insects and rodents, odors, and aesthetics. Furthermore, substitution of wastes for inorganic fertilizers helped turn animal excreta from a resource to a waste product in the U.S. Many farmers are reluctant to use manure and sludge on their land for reasons such as: a) may contain foreign items (such as concrete from feedlots); b) poor physical condition, which prevents uniform spreading; c) presence of weed seeds; d) odors; e) composition is highly variable and may cause burning; and, f) pathogenic microbes may be present. Some of these problems can be overcome by composting or stabilizing the manure prior to land application. The composted product contains as much N, but relatively more P and K than the uncomposted manure.

Fedkiw (41) estimated that approximately 180 million tons (dry weight) or two-thirds of the total production of manure is available for use on crop land or must be disposed of in a manner that does not cause pollution. A comprehensive national scale analysis of production and usage of composted materials calculated that over 45.4 tons of composts can be produced per year, and the calculated potential usage is 453 tons. All the composted material could be easily used in the agricultural sector alone.

Since many of the animal producing systems are agricultural in nature, they often combine animal and crop production (7). Intensive crop production with double or triple-cropping is often necessary to utilize all the animal waste so that producers are not forced to export manure or fertilizer co-products to other fertilizer users (8). Prior to application, waste is frequently stored in lagoons or pits (9). Animal wastes are also piled, composted in windrows, or simply deposited on land. In these situations, a great percentage of N is lost from the waste as ammonia. Spreading of the manure in a thin layer may reduce this N loss. Since animal wastes are very wet, nitrogen in the nitrate form is easily leached.

Potential manure contamination of surface water and groundwater may result from leaching or runoff, when the manure is improperly handled. This contamination may include nitrogenous compounds, such as ammonia and nitrate, phosphorus, trace elements, pesticides added to feed, bacteria, viruses, hormones, antibiotics, and other nutrients (7). In the soil, excess manure applications may cause mineral nutrient imbalances, accumulations of certain minerals, and salt buildups. In the air, high concentrations of gases and other volatiles can cause odors or can result in airborne bacteria, viruses, fungi, and dusts. Potential treatment methods to overcome these problems are well discussed in the chapters dealing with cattle (8), poultry (7) and swine (9) manures. If integrated manure management systems are used, individual site-specific needs can be met in an economically and ecologically sound manner for each site where manure is generated.

Potential solutions to the problems with animal manures may rely on: 1) incentives given to producers to implement current environmental practices and technologies that result in increased operational costs; 2) animal diets changed to decrease output of certain nutrients, such as P, and odorous compounds; 3) development of new treatment processes or improvement of existing manure treatment or storage processes; 4) controlling nutrients from manures for cropping systems; and 5) conducting economic analysis of manure system alternatives.

It is extremely important to develop positive relationships between producers, regulators, policy makers, scientists, and educators in: " determining the risk/benefit for current and future best management practices (BMPs); encouraging a voluntary approach to meeting water and air quality standards; initiating incentives to adopt integrated best management practices; testing BMPs for a system, site-specific, watershed approach; providing economical mens to process animal manures for feeding and value-added products"(42).
In the educational field, training and educating of environmental specialists, the development of efficient methods of technology transfer to producers, and effective communication with consumers and youth will aid in alleviating some of the problems with public acceptance of animal manures utilization.

Composts

Approximately 9 million tons of compost are produced per year in the U.S. (43) and production equals the use of compost (11). Potential production and demand are estimated at about 50 million tons and 500 million tons, respectively, on an annually basis (11). Farmland alone would be able to accommodate all he composted municipal solid waste (MSW) currently generated in the U.S. (43).

In the last 20 years, a number of reviews on utilization benefits or potential harmful effects of compost utilization have been published. A comprehensive review on compost production processes, benefits or potential adverse effects upon land application was published recently (*16*). The composition of compost is variable and is dependent on the raw materials used and processes employed (similar to biosolids), but typically N ranges from 1 to 3% and P from 0.4 to 1.0 % (*11*). Potassium concentrations vary from 0.5 to 1.5%, and, thus, K values are higher in compost than in biosolids.

When rates of compost application are based on nutrient content, especially on N, often excessive quantities of P but insufficient quantities of K to satisfy the crops needs are applied (*44*). Therefore, the addition of supplemental fertilizers should be made to balance nutrient levels with crop demand. It is always difficult to estimate the quantity of compost that contains long-term slow-release nutrients, to mix with a given quantity of an inorganic fertilizer. To speed up release of nutrients from organic wastes, different processing or storage techniques may be employed. For instance, smaller particles (greater surface area) and frequent mixing (improves aeration) of the material being composted allow for a more rapid mineralization. In static piles of compost, it takes approximately 2 to 3 months to convert NH_4 to NO_3 (*45*).

Composts can be used for agronomic, horticultural, silvicultural, home landscaping, golf courses, and reclamation purposes. However, the largest potential user of composts is the agricultural industry (*46*). Composts can be successfully used as a proportion of the potting medium for horticultural plants. However, when used as the sole potting medium, problems such as heavy metal toxicity (*47*), B toxicity (*48*), poor aeration (*48*) and high soluble salts (*48, 49*) have been reported.

Compost made from biosolids and woodchips or saw-dust may be used to grow many different horticultural crops under field and container conditions (*50*). The concentration of biosolids that can be applied to soils or used in formulating a potting medium is determined by a number of factors, including total soluble salts, particle size, stability of the product, dewatering procedure used in processing the biosolids, nutrient concentrations found in soils, anticipated crop needs and storage conditions (*51*).

Compost plays a critical role in the publicly praised landscaping of a major theme park's beauty; it is used as a mulch in floral beds (*52*). Composted municipal biosolids, combined with lime have been successfully applied to strip mines for remediation of these polluted, degraded areas (*53*).

In the U.S. MSW production 165 million dry tons annually. The EPA (*54*) estimated that by the year 2010 approximately 220 million tons will be produced in the country. Co-composting of MSW and biosolids produces a compost with a higher N content. Enhanced soil physical and chemical properties following application of MSW compost have been reported on several occasions (*55, 56, 57*).

In most states, however, the end user is located more than 50 kilometers from the available sources, and transportation costs greatly reduce the use of these materials. Furthermore, some waste products contain large amounts of water, and if the waste is used to supply a certain nutrient, it becomes less expensive to simply purchase an inorganic fertilizer, containing a higher percentage of that nutrient, so that lower quantities need be transported to the application site. It has been suggested, for instance, that sites for composting facilities should be selected to minimize transportation costs, and that some or all of the transportation and application costs should be covered by tipping fees at MSW composting facilities. Obviously, government subsidies and cost sharing programs could diminish application costs of many waste products.

Uncomposted Waste Paper and Other Organics

Municipal solid waste production in the U.S. has increased tremendously in the last few years, and with the reduced availability of landfill space, disposal problems have intensified (*58*). The waste paper component of MSW is resistant to decomposition by soil microbes and may form a physical barrier preventing penetration of plant roots and water. For these reasons, direct land application is not normally recommended for uncomposted paper. Waste paper can be ground, pelletized, and composted with other organics to improve decomposition in soil.

Newsprint contains Ba, Cd, Cr, Pb, Cu, Co, Mn, and Al in concentrations varying according to the amounts of the various colors used in printing (*12*). The presence of these metals may also be a concern when these wastes are applied to land. Studies in this area are incipient, and research needs are evident from the scarce literature

On the other hand, waste paper materials have been shown to reduce soil erosion and wind erosion (*12*). If composted *in situ* with other organics, such as animal manures, for a period of 4 to 6 weeks, some mineralization from waste paper occurs and some nutrients are made available in the soil. Other applications for waste paper or newsprint include amendments for container medium and mulch, when applied in pelletized form. Currently other uses for waste paper include cat litter from phone books; animal bedding, soil substitutes, hydromulch, construction materials, and feedstock for compost from paper mill sludge. Recycling of used paper materials has also intensified in the last few years.

Specialty Fertilizers

Iron Humates

The clarification process in water treatment was traditionally performed with aluminum sulfate (alum) which produced an aluminum sludge. A newer process is to coagulate the brown humates with iron salt coagulants (*6*). The process yields high-quality drinking water and iron humates, a by-product that contains between 16.5 and 19.5% Fe by weight (*6*). When compared with biosolids, the production of iron humate is small (*6*). These humates are currently being tested for granule integrity, maintenance of high iron availability in the soil, increase of total nutrient concentration in low-cost products, and formation of matrix fertilizers by combining special ingredients with the iron humates.

Iron deficiency is frequently observed in various crops grown in Florida's sandy soils with an alkaline pH. This is especially true for citrus trees. Micronutrient fertilizers are costly, particularly chelated forms that maintain the Fe available for plant uptake during a longer period of time. Iron humates, a waste product, may prove to be very beneficial in these situations. Iron humates are also being tested for use as an Fe supplement for animals. Nevertheless, by determining a special use and obtaining a material that meets the target use, disposal will no longer be a problem, since what may have been considered a "waste" is clearly now a product.

Manufacture By-Products

Land application of non-hazardous waste materials from manufacturing industries may potentially reduce excessive amounts of bioavailable P in the soil. Waste materials that contain hydrous oxides (alum sludge, bauxite red mud) or Ca (cement kiln dust) are readily available (59) and may be used to immobilize or sequester minerals in soils. Application of alum hydrosolids, cement kiln dust, and treated bauxite red mud to soils has been shown to reduce excessive amounts of bioavailable P (59). These authors reported that these by-products may potentially improve drinking water quality and provide savings for municipalities. Although most natural soils do not have problems with excessive levels of P, many manures contain high levels of this element relative to N. Therefore, alum sludge, bauxite red mud or cement kiln dust could be added to manures or to soils that have received high inputs of manure to prevent phosphate from reaching water and result in eutrophication. Such combinations of different residuals can solve a nutrient imbalance in one of the wastes being utilized.

Inorganic By-Products and Wastes

Boiler Wood Ash

In the Southeastern U.S. alone, over 40 million tons of firewood per year are consumed, and the pulp and paper industry generate about 3.6 million tons of boiler wood ash per year (60). Wood ash from industry has a high $CaCO_3$ equivalent, in the order of 38%, and this ash can be used safely as a limestone substitute, as well as a K source. The overall nutrient content is low, and therefore, boiler wood ash would not provide a substantial contribution as a fertilizer (61). No problems have been reported with trace metals, carcinogens, or organics, when broiler wood ash was applied at "reasonable" rates. The major problems with this material appear to be transport cost and spreading technology. Broiler ash use is regulated by state environmental agencies and state departments of agriculture (61).

Fly ash is composed primarily of CaO, gypsum, and $CaCO_3$, while lime kiln dust is primarily CaO, MgO and dolomite. The use of flue gas desulfurization (FGD) by-products of coal-fired boilers as substitutes for agricultural limestone represents a potential beneficial use alternative to landfill disposal of these materials. A greenhouse study conducted with alfalfa (*Medicago sativa* L.) and tall fescue (*Festuca arundinacea* Schreb.) indicated that these materials were able to increase yield, Ca, Mg, and S contents in the plant tissues, without increasing trace elements except for B and Mo (62). Soil pH increased from 4.5 to approximately 7.5 by the end of 9 months.

A recent greenhouse evaluation of pulp and paper mill combined broiler ashes on growth and nutrient uptake by oat (*Avena sativa* L.) and bean (*Phaseolus vulgaris* L.) indicated that plants had higher yields than those amended with lime alone or unamended controls (63). The authors concluded that the ash was as effective as dolomitic lime in raising pH. Plants treated with combined broiler ash had higher tissue concentrations of P, S, and B. Soil Zn, Fe, Mn, and Cu concentrations decreased as ash application rates increased (63).

Ammonia Scrubber

New technological advances in coal gasification plants to treat waste water and gas streams have improved the process of FGD to scrub SO_2 while generating an ammonium sulfate product that meets all fertilizer-grade requirements. The process is described by Wallach (64). Thus, where previously a waste was being generated, a marketable by-product is being produced, containing both N and S in excess of 20%.

By-Product Gypsums

The processing of phosphorus fertilizers yields large quantities of gypsum by-products. This quantity can reach 50 million tons annually in the U.S., and most of it is generated in Florida and North Carolina (65). Beneficial uses of gypsum by-products include manufacturing of wallboard additives in cement and sub-base material for road construction. A considerable portion of this by-product may be used in agriculture as a Ca and S source. Florida's phosphogypsum is considered unfit for use on land due to its radionucleide content, which exceeds the EPA maximum allowable concentration activity of 10 pCi per gram (66). Extensive research done at the University of Florida indicates that the background levels of Rn are actually higher than emanates from gypsum applied to land on pastures. It has been argued that the EPA maximum limit has been set too low (66), based on several systematic studies conducted by these researchers that show little or no detectable effect on crops, soil, groundwater and air from Florida's gypsum, which exceeds the EPA limit by 100%.

Micronutrients from By-Products

Micronutrient deficiencies in crops have been more pronounced since the fertilizer industry has been required to produce "cleaner" products. In addition, intensive cropping over many decades has depleted a number of important agricultural soils of their micronutrients, increasing the scope of this problem. Although many by-products can be utilized as micronutrient sources, their use is somewhat limited due to their physical form, transportation costs and lack of federal regulations (67). Examples of micronutrient-containing by-products include: baghouse dust, mill scale, wastes from cleaning of sheet steel, various oxides and slags from smelting, spent catalysts, scrubber sludge or other products used in pollution control. (67). These by-products can contribute as sources of Fe, Mn, Zn, B, Cu, Co, Mo, and Cl in varying concentrations.

New Technologies

Better and more perfected technologies for the processing, handling, and disposing of by-products and wastes, including equipment for land application of certain residuals are constantly being sought and developed. For instance, preconditioning processes can be performed to provide aerobic thermophilic digestion to disinfect and condition wastewater biosolids prior to anaerobic digestion, and newer anaerobic digestors for dairy and swine manures provide better odor control (one of the most important problems with these residuals). In addition, newer comprehensive approaches to odor control have been developed. These include multi-stage wet chemical scrubbers and better roller presses for dewatering wet biosolids and slurries such as swine and cattle manure, food processing residuals, pulp and other similar fibrous products.

Tailoring By-products and Wastes

Different residuals present different characteristics regarding chemical, physical, and biological make up. Each one of these characteristics will interact with the soil in different ways, therefore affecting the possible uses. Some residuals may be combined in varying proportions to produce, for instance, a more complete fertilizer. We should also realize that it will take resources to dispose of these materials. Most residuals should not be viewed as a complete soil amendment, but rather, as part of an overall nutrient management strategy. Upon consideration of the nutrient supplying capacity of the soil, the nutritional requirements of the plant species, and the management practices used, inorganic residuals may be valuable when applied alone or as a complement to fertilizers or to organic by-products, such as animal manures and sludges.

Blending of one or more wastes with each other or with fertilizers to obtain a more complete fertilizer material or a richer soil additive, along with co-composting, animal feeding and determining new uses, are all possibilities for enhancing use and/or turning a waste into a product that can find use in agriculture or other areas. Estimates of nutrient and metal contents of the various by-products and wastes were given by several authors in this volume. A complete analysis of the by-product or waste material, will allow the determination of possible nutrient sufficiency, deficiency, or excess, that may be supplied or corrected with another by-product or with a commercial fertilizer to make the material adequate for the particular land where it is to be applied. Knowledge of the soil characteristics is also fundamental in this process.

The development of tailor-made products is a research area that can help turn wastes into resources. The reduction of certain trace metal bioavailability in contaminated soils, disease suppression, or a combination of wastes with different desirable characteristics for a determined use would be examples of new areas of research (*68*).

U.S. Environmental Protection Agency Regulations

Several rules, policies, and guidances impact the use of organic and inorganic by-products (*3*). The current regulation strongly encourages beneficial use of residuals such as biosolids. As a result of the new EPA rule, 90% of the biosolids produced in the states of Maryland and Wisconsin are being utilized.

The EPA strategies concerning waste, specifically biosolids are to: 1) promote recycling; 2) provide consistent technical information; 3) form partnerships and teams between producers, consumers, and regulatory personnel; 4) involve stakeholders; 5) become proactive; and, 6) assure compliance. These could possibly be extended to most by-products and wastes.

Tailoring of the waste with respect to certain characteristics may address problems such as reduction of soil Pb availability, disease suppression, obtaining a more balanced nutrient source, use of waste product combinations (for instance, kiln dust and biosolids, fly ash and biosolids, gypsum waste and biosolids).

Conclusions

Considerable progress has been made in research leading to a more environmentally safe manner of waste application on land. However, there are many unanswered questions that warrant further investigation. Some of these issues were recently discussed regarding land application of biosolids (*69*), but the proposed research could also apply to any waste. Among these critical research priorities are: long-term studies to determine the incremental changes in soil and plant composition at a given site and more adequate fundamental knowledge for predicting beneficial effects. Similarly, we need to assess the risks from the introduction of toxic organics, trace metals, and pathogens into the farm ecosystem and possibly into the food chain and water resources. More systematic research is required, along with an appreciation of the ecological effects. Cost of utilization and the sociological-economic considerations with emphasis on public acceptance and awareness are economic and social issues that warrant study.

Other considerations, such as the development of new markets/uses for many wastes or by-products, should be exploited. Examples of this approach are given for by-product gypsum, widely used in wallboard and plaster manufacture, as well as a cement additive. The ultimate target is to minimize waste materials by maximizing their utilization.

With rational use, most types of organic and inorganic by-products and wastes can contribute substantially to the improvement of environmental quality, in the form of soil recuperation, concomitant with the preservation of water resources and production of food for a fast growing population. A systematic holistic approach involving researchers, stakeholders, the fertilizer industry, regulatory agencies, and the public is fundamental to elucidate the problems inherent to each specific by-product or waste material. If this kind of approach is undertaken, undoubtedly the majority of these by-products and waste materials (straw) can be transformed into marketable valuable products (gold).

Acknowledgments

Florida Agricultural Experiment Station Journal Series no. N-01365. The authors wish to express their gratitude to Drs. Thomas Obreza and Fritz Roka (University of Florida, SWFREC, Immokalee) for their critical reviews and to Ms. Pamela Hunter, for assistance with the typing of this manuscript.

Literature Cited

1. Juo, A. S. R.; Russell, D.F., Eds; *Agriculture and Environment: Bridging Food Production and Environmental Protection in Developing Countries;* ASA Special Publication no.60, Madison, WI, 1995.

2. Hossner, L. R.; Dibb, D. W. In *Agriculture and Environment: Bridging Food Production and Environmental Protection in Developing Countries;* A. S. R. Juo; Russell, D.F., Eds; ASA Special Publication no.60, Madison, WI, 1995, pp. 17-32.
3. Walker, J. M.; Southworth, R. M.; Rubin, A. B. *This Vol.* Chapter 3.
4. Forste, J. B. *This Vol.* Chapter 4.
5. Logan, T. J.; Lindsay, B. J.; Titko, S. *This Vol.* Chapter 5.
6. Rehberg, B. E.; Smith, G. L. *This Vol.* Chapter 17.
7. Sims, J. T. *This Vol.* Chapter 6.
8. Van Horn, H. H.; Hall, M. B. *This Vol.* Chapter 7.
9. Mikkelsen, R. L. *This Vol.* Chapter 8.
10. Huntley, E. E.; Barker, A. V.; Stratton, M. L. *This Vol.* Chapter 9.
11. Barker, A. V. *This Vol.* Chapter 10.
12. Edwards, J. H. *This Vol.* Chapter 11.
13. Nordstat, F. A. Human behavioral factors in waste management. In *Proceedings of the 30th Annual Meeting of the Soil Conservation Society of America (Land Use: Food and Living)*, August 10-13, San Antonio, TX, 1975, pp.153-157.
14. Galloway, D. F.; Walker, J. *This Vol.* Chapter 2.
15. Hue, N. V. In *Soil Amendments and Environmental Quality;* J. E. Rechcigl, Ed; Agriculture and Environmental Series. Lewis Publishers, Boca Raton, FL, 1995, pp. 199-247.
16. Stratton, M. J.; Barker, A. V.; Rechcigl, J. E. In *Soil Amendments and Environmental Quality*; J. E. Rechcigl, Ed.; Agriculture and Environmental Series, Lewis Publishers, Boca Raton, FL, 1995, pp. 249-309.
17. Sommers, L. E.; Nelson, D. W.; Silveira, D. J. *J. Environ. Qual.* **1979**, 8,287-293.
18. United States Department of Agriculture (USDA). *Agricultural Statistics*, U.S. Printing Office, Washington, D.C., 1991.
19. Keeney, D. R.; Nelson, D. W. In *Methods of Soil Analysis, Part 2;* ASA-SSSA, Madison, WI, 1982, pp. 643-698.
20. Muchovej, R. M. C.; Rechcigl, J. E. In: *Soil Processes and Water Quality;* B.A. Stewart, Ed; Advances in Soil Science Series, Lewis Publishers, Boca Raton, FL, 1994, pp.91-135.
21. Muchovej, R. M. C.; Rechcigl, J. E. In *Soil Amendments and Environmental Quality;* J. E. Rechcigl, Ed; Agriculture and Environmental Series. Lewis Publishers, Boca Raton, FL, 1995, pp. 1-64.
22. Raviv, M.; Chen, Y.; Inbar, Y. In *The Role of Organic Matter in Modern Agriculture;* Y. Chen; Avnimelech, Y., Eds.; Dordrecht, Martinus Nijhoff, 1986, pp.257-287..
23. Eck, H. V.; Stewart, B.A. In *Soil Amendments and Environmental Quality;* J. E. Rechcigl, Ed; Agriculture and Environmental Series. Lewis Publishers, Boca Raton, FL, 1995, pp. 169-198.
24. Mullins, G. L.; Mitchell, C. C. In *Soil Amendments: Impacts on Biotic Systems;* J. E. Rechcigl, Ed; Agriculture and Environment Series, Lewis Publishers, Boca Raton, FL, pp.1-40.

25. Warren, H. V. *Soc. Sci. Med.* **1989**, 8,923-926.
26. Lisk, D.J.; Boyd, R. D.; Telford, J. N.; Babish, J. C.; Stoewsand, G. S.; Blache, C. A.; Gutenmann, W. H. *J. Anim. Sci.* **1982**, 55,613-616.
27. Pierzynski, G.M.; Jacobs, L. W. *J. Environ. Qual.* **1986**, 15,323-326.
28. CAST Application of sewage sludge to cropland: Appraisal of potential hazards of the heavy metals to plants and animals. Report no. 64, Council for Agricultural Science and Technology, Ames, IA, 1976.
29. Haug, R. T.. *The Practical Book of Compost Engineering*; Lewis Publishers, Boca Raton, FL 1993.
30. United States Environmental Protection Agency (EPA). *Standards for the Use and Disposal of Sewage Sludge;* Environmental Federation, 58 FR 9248, Washington, D.C., 1993.
31. Calcutt, T.; Moss, J. *Water Pollut. Control* **1984**, 83;163.
32. McCaslin, B.D.; O'Connor, G. A. *Potential Fertilizer Value of Gamma Irradiated Sewage Sludge on Calcareous Soils;* New Mexico Agric. Exp. Stn. Bull. 692, 1985.
33. McCaslin, B.D.; Davis, J. G.; Chacek, L. C.; Schulter, L. A. *Agron. J.* **1987**, 79,204-209.
34. Camberato, J. J.; Vance, E. D.; Someshwar, A. V. *This Vol.* Chapter 12.
35. Vera, C. J.; Servello, F. A. *J. Wildl. Manage.* **1994**, 58,719-727.
36. Keely, W. C. *HortScience* **1990**, 25,168-169.
37. Follett, R. F.; Gupta, S. C.; Hunt, P. G. 1987. In *Soil Fertility and Organic Matter as Critical Components of Production Systems;* SSSA Special Publication Number 19, SSSA, ASA, Madison, WI, 1987, pp 19-51.
38. Humenik, F. Ed. *Water Quality: Agriculture's Role;* Task Force Rep.120. CAST, Ames. IA, 1992.
39. National Research Council (NRC). *Opportunities to Improve Soil and Water Quality: an Agenda for Agriculture*; National Academy Press, Washington, D.C., 1993, p.79.
40. Redman, M. H.; Wrigglesworth, S.A.; Vinten, A. J. A. In *Nitrogen in Organic Wastes Applied to Soils;* J. A. Hansen; Henrikson, K.A.J., Eds.; Academic Press, London, 1989, pp. 98-112.
41. Fedkiw, J. *Progress and Status of Livestock and Poultry Waste Management to Protect the Nation's Water 1970-1992*, A working paper of the USDA working group on water quality, USDA, Washington, D.C., 1992.
42. CAST Integrated Animal Waste Management. Interpretive Summary, November, Council for Agricultural Science and Technology, Ames, IA, 1996.
43. Slivka, D.C.; McClure, T. A.; Buhr, A. R.; Albrecht, R. *Biomass and Bioenergy* **1992**, 3,281-299.
44. Sommers, L. E.; Giordano, P. M. In *Nitrogen in Crop Production*; ASA-CSSA-SSSA, Madison, WI, 1984, pp.207-220.
45. Vega-Sanchez, F.; Gouin, F. R.; Willson, G. B. *J. Environ. Hort.* **1987**, 5,66-70.

46. Parr, J. F.; Hornick, S. B. In *Soil Microbial Ecology: Application in Agriculture, Forestry, and Environmental Management;* B. Metting, Ed.; Marcel Dekker, New York, NY, 1992.
47. Chu, L. M.; Wong, M. H. *Plant and Soil* **1987**, 103(2),191-197.
48. Lumis, G. P.; Johnson, A. G. *HortScience* **1982**, 17, 821-822.
49. Sanderson, K. C. *HortScience* **1980**, 15,173-178.
50. Gouin, F. R. *Hort Technology* **1993**, 3(2),161-163.
51. Chaney, R.L; Munns, J. B.; Cathey, H. M. *J. Amer. Soc. Hort. Sci.* **1980**, 105, 485-495.
52. Logsdon, G. *BioCycle* **1995**, 36 (10), 48-50.
53. Garland, G.A.; Grist, T. A.; Green, R. E. *BioCycle* **1995**, 36 (10), 53-56.
54. United States Environmental Protection Agency (EPA) *National Survey of Solid Waste (Municipal) Landfill Facilities: Final Report.* Office of Solid Waste. EPA/530-SW-034. Washington, D.C., 1988.
55. Haan, S. de. *Neth. J. Agric. Sci.* **1981**, 29 (1), 49-61.
56. Gallardo-Laro, F.; Nogales, R. *Biol. Wastes* **1987**, 19,35-62.
57. Chaney, R. L. *BioCycle* **1990**, Oct. 68-73.
58. Finstein, M. S. In *Environmental Microbiology;* R. Mitchell, Ed;. Wiley-Liss, NY, 1992, pp.355-374.
59. Peters, J. M.; Basta, N.T. *J. Environ. Qual.* **1996**, 25,1236-1241.
60. Milner, R.; Unwin, J. *Solid Waste Management and Disposal Practices in the US Paper Industry;* NCASI Tech. Bul. No. 641. National Council of the Paper Industry for Air and Stream Improvement Inc., New York, NY, 1992.
61. Mitchell, C. C.; Black, E. D. *This Vol.* Chapter 13.
62. Stehouwer, R. C.; Sutton, P.; Dick, W. A. *Soil Science* **1996**, 161 (9), 562-574.
63. United States Department of Agriculture (USDA). *Agricultural Statistics,* U.S. Government Printing Office, Washington, D.C., 1991.
63. Krejsl, J.A.; Scanlon, T. M. *J. Environ. Qual.* **1996**, 25,950-954.
64. Wallach, D. L. *This Vol.* Chapter 15.
65. Miller, W. P.; Sumner, M. E. *This Vol.* Chapter 14.
66. Alcordo, I.S.; Rechcigl, J. E. In *Soil Amendments and Environmental Quality*; J. E. Rechcigl, Ed; Agriculture and Environmental Series. Lewis Publishers, Boca Raton, FL, 1995, pp. 365-425.
67. Wyatt, J. M. *This Vol.* Chapter 16.
68. Walker, J. M. In *Sewage sludge: Land Utilization and the Environment*; C. E. Clapp; Larson, W. E.; Dowdy, R. H., Eds.; SSSA Misc. Publication, Madison, WI, 1994, pp. 67-79.
69. Clapp, C. E.; Larson, W. E.; Dowdy, R. H., Eds; *Sewage Sludge: Land Utilization and the Environment,* SSSA Misc. Publication, Madison, WI, 1994, 258p.

FACTORS AFFECTING USES OF BY-PRODUCTS
AND WASTES IN AGRICULTURE

Chapter 2

An Entrepreneurial View of the Future for the Use of Wastes and By-Products

Dale F. Galloway[1] and John M. Walker[2]

[1]RDE, Inc., 101 North Virginia Street, Crystal Lake, IL 60014
[2]Office of Wastewater Management, U.S. Environmental Protection Agency, 401 M Street Southwest, Washington, DC 20460

>Great opportunities exist for creative management of wastes and by-products that result from the centralization, concentration, and intensification of processing and production of animal and plant feedstocks. If these feedstocks are viewed holistically and if animals, humans, factories, and soils are viewed as organic matter and mineral processors, then greater opportunities exist to creatively benefit from the feedstocks at many different times during the various processing and reprocessing sequences.

If societal wastes and by-products are considered holistically, then the idea that these wastes and by-products are too valuable to waste is made paramount. The By-products and wastes left after processing become feedstocks for other uses, some of which may involve recycling to land and some of which may not. It is important to think broadly. At the most basic level, the organic building blocks of life are also the by-products of life, (i.e., oxygen is the by-product of plant photosynthesis and carbon dioxide is the by-product of both animal and plant respiration. Plants use oxygen and carbon dioxide to produce food, feed and fiber and animals use these plant products and oxygen to sustain their life. As we in society learn to make our life more comfortable and pleasant, we create more complex goods and generate more complex wastes or as we should view them - feedstocks for other uses. The real limitation is our ability to see and develop the opportunities for these other uses of wastes and by-products.
If we are able to realize these opportunities, then the initial feedstock users, the "wastes" they generate, the subsequent users of the wastes and by-product and the environment will both profit and benefit.
When life was more simple, the waste problem was more simple. The driving force behind our current interest in creatively managing wastes and by-products is centralization and concentration of animal

22 © 1997 American Chemical Society

production and plant product processing in confined areas - in short the intensification of it all. Visualize a picture that I have of my grandmother and an uncle taken in the early 1900's. This picture shows them driving about 100 hogs 4 miles to a rail siding for transport to market. This was the largest batch of hogs they had ever grown, and they earned a reasonable profit from their sale. Today, large-scale producers raise 15,000 or more hogs at a given location - 150 times more than my grandmother grew. Think also of a 1970's soybean processing plant. This plant would typically crush 200 tons of beans per day. Today, a competitive plant must process or crush 10 times more beans or 2000 tons per day. To supply the soybeans needed to keep this 1990's plant operating, all soybeans grown in a 55 mile radius would have to be gathered. Now think about what a friend of mine told me who has been in the tanning industry all of his life. He said that an average tannery in the early 1900's might tan 100 hides per day. Today an average tannery tans 6,500 hides per day or 65 times greater volume. What once was a 5 gallon per minute stream of waste generated by the 1900's tannery is now a 325 gallons per minute waste stream from the 1990's tannery. (1)

Research is showing us more about how to properly manage wastes and how improperly managed wastes can cause environmental problems. The intensification (confinement, concentration, and corporate like structure) of plant and animal production and management which I have briefly described magnifies the potential for environmental problems. Federal and state governments have enacted several rounds of environmental legislation to help ensure the proper management of the wastes and by-products from society's various enterprises. Hence, the system for managing these wastes is vastly different than before intensification and regulation. Because of large magnitude and concentration of operations, the large-scale hog producers and soybean processors have to ship feedstocks in from great distances to supply their needs. And, lakes or mountains of waste are created from these very confined production and processing facilities. Before rules and regulations, these wastes would often remain in lagoons and piles or be applied at high rates on soils. After regulation these options are being changed. The problems limiting wise agricultural use in accordance with these new rules (both from a pollutant and nutrient management point-of-view) include odors and the high costs for transporting what currently are regarded as low value wastes. Hog producers would normally like to use the hog manure on nearby land. But the problems limiting its use include, but are not limited to, odors, excessive transportation costs and the need to manage nutrients.

Repeated and/or high rate applications of manure on land have caused the build up of nutrients in excess of crop needs and the quality of ground and surface waters can be degraded within watersheds. Degradation of watersheds from the excessive use of manures and the catastrophic accidents that can occur are adding considerable pressure and the need for better management and use of the by-products and wastes. A recent example of such an occurrence was the recent accidental spillage of millions of gallons of waste into the waterways from hog waste storage lagoons in North Carolina.

Watershed degradation from by-products and wastes can be minimized if we change our mind-set. In our earlier example, think of hogs and soybean pressing facilities as processors of grain. And,

rather than viewing the stuff that comes out of the back door of
soybean processing plants and the back side of the hogs as a foul
refuse, think of these wastes and by-products as recycling
profit-making opportunities. In this view of the opportunities for
using these partially processed feedstocks (by-products and wastes)
as feedstocks for other than land recycling the organic matter and
nutrients. If alternative uses are not included, the ability to use
them wisely and sustainably may not be possible.

Minimization of waste is being achieved. Some industries like
the oil refining industry have made great strides in converting most
all of their feedstock into useful products. For example, the oil
industry can extract more than 3,000 products from each barrel of
crude oil (2) with essentially no waste. The oil refining industry
has had several great advantages, (i.e., favorable economics and the
need to avoid generation of wastes which can be environmentally toxic
and costly to dispose. They also have had very smart and creative
people working to find new and better uses for each part of the crude
oil feedstock. On the other hand, residues from the processing of
animal and plant feedstocks have been viewed as major problems.
Careful examinations of these co-products, however, reveals that they
have many of the properties found in the crude oil feedstock.

The task of using municipally generated wastes is made somewhat
easier because of subsidization. This subsidization helps make the
use of municipal solid waste (MSW) and biosolids economically
competitive with other feedstocks currently in use for various
practices. This subsidy is often available from municipalities that
generate by-products and wastes because they must pay for their
disposition by some method anyway. For example with subsidization,
uses of MSW and biosolids may have a competitive advantage over corn
when used as a feedstock for ethanol production. A firm is now
designing a full-scale 600-ton per day facility for separating
recyclable plastics, aluminum, and ferrous components from MSW, mixing
in biosolids, and, after also mixing in biosolids, producing the
ethanol.

This process has been carefully researched and piloted and
offers promise. Tipping fee and recycling revenue offset the $1.75
per gallon cost of producing ethanol from MSW and biosolids mixture
and can compete with the $0.75 per gallon cost of ethanol production
from grain. (3) Society is willing to pay this much differential
because it is equivalent to or greater than the amount that would
have to be paid for disposition by other means.

Most producers and processors of animal and plant feedstocks do
not have the advantage of subsidization. As stated several times
previously, the challenge for these producers and processors of
cattle, hogs and swine and other foods is to change their view to
seeing the wastes and by-products as resources and not refuse. The
current refuse mind-set stems from the corporate paradigm of focusing
energy first and foremost on the production of core products. With
such a focus, the animal producer and food processor mind-set has
been to get rid of by-products in the quickest and cheapest way
possible. (4) In the future the corporate concept should be expanded
to include the identification and development of the use of left over
partly utilized by-products and waste now considered as refuse. The
current president of DuPont is quoted in a recent issue of Chemical
Week (5) as saying, "The goal of our Company is to produce zero

waste. Waste is raw material we paid for that didn't make it into the product we sell." This attitude will probably have more influence on our realizing that great benefits exist for future use of these "wastes and by-products" both inside and outside of agriculture.

The following discussion illustrates other opportunities for using by-products and wastes as feedstocks for other processes which in turn yield by-products and wastes that can be used as feedstocks for still other purposes. By this shift from narrowly focusing on core business there can be more profit as well as being environmental stewards and wasting the left-over by-product will not be tolerated. Consider three examples:

o One company I have worked with extracts pharmaceuticals from the lining of the small intestines of swine. From a 20-ton truck load of this lining they are currently able to harvest only about 8 pounds of the pharmaceutical. The 20-ton minus 8 pound balance of swine ingesting feedstock is certainly not core business but is now being used as a feed for baby pigs.

o Paper companies extract cellulose from wood to make paper - their core business. Cellulose makes up only 50% of wood and as a result the paper industry is based on utilizing only 50% of their feed stock. Over the years, some improvement has been made in converting a small part of this 50% waste to saleable products. The important distinction is that the part of wood that is not cellulose can not be given the developmental muscle needed as long as the core business is making paper.

The press for efficiency will not allow industries of the future to waste 50% of the feedstock going into a plant. This wastage will continue until companies who make paper will change their description of their core business to wood or fiber processors.

o Animal agriculture has existed for many years on the premise that its function was limited to producing meat, milk, or eggs. The drive for ever increasing efficiency has driven animal production units to ever increasing sizes. Throughout the years animal producers have been aware that from 50% to 80% of the feedstock that goes into a unit is not sold as a product. This large "waste" could effectively be used as crop food and actually was not a waste.

The vastly increasing size of animal producing units is eliminating the productive use for the wastes. This evolves as the soil in the proximity of the animal unit reaches its capacity to carry minerals and nitrates. As a result, these animal feeding units have moved towards hauling the manure to ever remote areas or the use of lagoons. In either case, the disposal of the waste has become a production cost.

A solution for this problem can be found by taking a broader view of feeding an animal. When a pound of feed is ingested, the digestive tract makes a fractionation. It absorbs those nutrients which are readily digestible and rejects those which are not. This is precisely what an oil refinery does in each step of refining oil; except that the oil engineers have cleverly devised a sequence of extractions that allows them to sell practically all of the feedstock going into a refinery.

It seems logical that we view the digestive tract as being

the first process in refining grain or feed generally. This first step has extracted the more labile factions and necessarily has concentrated the less labile. One demonstration of this is found as we learned that the early settlers found that buffalo chips were a better fuel than was the grass eaten by the buffalo. The digestive tract of the buffalo had concentrated the carbon and it became a better fuel per unit of volume.

If we adopt this broader view, we must answer the important question, "What have we produced that someone can profitably use?" We can not yet answer this question because research on manure has largely been focused on re-feeding manure to animals and these data are of little use in searching for industrial uses for the component of manure.

We see rapidly increasing interest in industry for proteins that can be used as adhesive resins for building materials and for foundry core binders. The carbohydrate fraction shows promise of finding a home as a low cost adhesive in industry to bind foundry dust, coal fines and limestone. Work has been done in using the fiber fraction in making particleboard and hardboard

The opportunity exists, then, of turning a production cost into a profit center by making animals processing units.

A generalized view of the properties of several animal manures are given in Table I. Indication is made of fractions that are possibilities as feedstock products for industrial uses. To repeat again, if we look at animal agriculture holistically as organic matter processors, then we can begin to look at manure as a partially utilized resource that should not be wasted. The witches' brew wastes that have been characteristic of the past may well disappear. Wastes and by-products will become more uniform and suitable for other uses. There will be less burning, less dumping into the landfill, and less land application in this process. This does not mean that the opportunities for land use of these by-products will disappear, but these co-products will become more refined and appropriate for other uses. As a result of dietary design, genetic engineering, and other creative combinations and structuring of wastes and by-products, profitability will increase to the extent that wasting is not economically feasible.

TABLE I. Potentially Partitionable Fractions of Animal Manures (1)

Fractions	Types of Manure		
	Cattle	Swine	Poultry
	% by dry weight		
Protein	15-30	20-25	30-34
Ash	10-20	10-15	15-22
Ether Extract	5-10	3-4	2-3
All Other*	50-70	56-67	41-53

*Lignin, Cellulose, Hemicellulose

Literature Cited
(1) Bliss, Earl. Private communication. L.H. Lincoln, Currwinsville, PA.
(2) Pampe, M. Petroleum. How It Is Found and Its' Uses p 47.
(3) 1996. Beneficial Use of Solid Waste and Sludge with the CES OxyNol Process. Masada Resource Group, LLC, Birmingham, AL.
(4) Galloway, D.F. 1995. Finding Uses For Industrial Wastes. p 4. Dorrance Publishing Co., 643 Smithfield St., Pittsburg, PA 15272
(5) Roberts, M and Fairley, P. Sustainable Development Is A New Global Agenda. Chemical Week July 3/10, 1996 p 46.

Chapter 3

U.S. Environmental Protection Agency Regulations and Other Stakeholder Activities Affecting the Agricultural Use of By-Products and Wastes

John M. Walker[1], Robert M. Southworth[2], and Alan B. Rubin[2]

[1]Office of Wastewater Management, and [2]Office of Science and Technology, U.S. Environmental Protection Agency, 401 M Street Southwest, Washington, DC 20460

More than one billion tons of agriculturally recyclable waste organic and inorganic by-products are generated each year in the United States. A large number of Federal and state guidances, policies and rules govern the processing, distribution, and agricultural use or disposal of organic and inorganic by-product materials. Important opportunities exist for tailor-making these by-products and wastes to enhance their benefits when used separately and/or together with conventional fertilizer and soil conditioning materials. This paper discusses the need for stewardship and the implementation of a strategy that makes use of by-products and wastes in a holistic and safe manner, minimizes nuisances, and optimizes the opportunities for sustained agricultural use.

Of the more than one billion tons of organic and inorganic agriculturally recyclable by-products generated each year in the United States, about 40 percent are crop residues, 5 percent dairy and beef manures, 3 percent poultry and swine manures, 15 percent municipal solid wastes and less than one percent biosolids (*1,2*). A considerable portion of these by-products are being recycled on land. Some are being used wisely and some are being used in a manner that causes nuisances or adds nutrients in excess of need. Optimum useability of these by-products requires a knowledge of the fate and effects of the by-product constituents, how to minimize odors, and how to tailor their use to need. Minimizing (or in some cases maximizing) the presence of nutrients in by-products will become more important. Ways of tailoring nutrient levels in the by-products include plant breeding or using dietary additives to obtain more digestible nutrients, blending of by-products with other by-products and chemicals, and processing and storage of the by-products.

An important element in being able to maximize the usefulness of these by-products is the availability of scientifically sound and sensible Federal and state rules that apply uniformly to the use of these materials. The risk assessment process and the resultant rules for biosolids use and disposal can serve as a model for development of sound and sensible rules and guidance for safely processing and recycling other by-products and wastes. Another important element in the recycling of these by-products is the views of municipalities, farmers, scientists, and the chemical fertilizer industry. Do these groups view these by-products as agriculturally beneficial and is there a vision of the high potential for developing, improving and marketing specially designed by-products? Will the chemical fertilizer industry join as partners with the producers of these by-products and wastes, universities, and agencies like the U.S. Department of Agriculture (USDA) and the EPA to study, develop, and enhance their usefulness? Will studies be directed at enhancing by-product usefulness as suggested above? This paper encourages a holistic approach for using these by-products.

A brief philosophic discussion of the risk assessment approach used for the development of the 40 CFR Part 503 rule for the use or disposal of biosolids (sewage sludge) is included in this paper. This discussion is given from two perspectives. The first perspective explains the U.S. Environmental Protection Agency's (EPA's) establishment of a user-friendly and protective rule based on a scientific determination of acceptable change. The second perspective is on the applicability of the Part 503 requirements to the use or disposal of other organic and inorganic by-products.

By far the largest factor limiting the use of organic waste materials like biosolids, animal manures and food processing wastes are nuisances, especially odors. Such nuisances, coupled with public perception that the uses of wastes like biosolids are potentially detrimental, are further compounded by the lack of resources to provide careful oversight. With decreased Federal and state resources available for the oversight of the utilization of waste by-products like biosolids, stakeholders will need to fill the gap.

Rules, Policies and Guidances

There are a number of rules, policies and guidances that both directly and indirectly impact the use of various inorganic and organic by-products. Some of these are listed in Table I and discussed below.

Criteria for Classification of Solid Waste Disposal Facilities and Practices (40 CFR Part 257). Part 257 contains the requirements that have to be met when solid waste is disposed on the land. Under Part 257, solid waste includes, but is not limited to, garbage, refuse, sludge from water treatment plants, air pollution control facility solids, and other discarded material. Prior to publication of the Standards for the Use or Disposal of Sewage Sludge

TABLE I. Rules, Policies and Guidances Impacting the Processing and Agricultural Use of Organic By-Products, e.g., Biosolids, Other Municipal Residues, Animal Manures, Food Processing Wastes, and Certain Inorganic Waste Materials[1]

R,P,or G & Act[2]	Citation	Materials Impacted[3]
R, CWA	40 CFR Part 503	Biosolids or materials containing biosolids, technical rules
R, CWA	40 CFR Part 501	Biosolids or materials containing biosolids, programmatic rules
R, CWA	40 CFR Part 122, 124 (NPDES)	Discharge of all wastes and wastewaters into waters of the United States
R, RCRA	40 CFR Part 261-268	Hazardous Wastes
R, RCRA	40 CFR Part 257	Solid Wastes
R, RCRA	40 CFR Part 258	Solid Wastes
R, CERCLA	40 CFR Part	All qualifying wastes
R, TSCA	40 CFR Part 761	All qualifying wastes
R, Air		Radioactivity in fertilizers, biosolids and other by-products

[1]This table only lists generalities about the by-products and wastes covered. Details and exclusions regarding coverage of each by-product and waste must be determined by an examination of the rule. Some additional information regarding coverage is provided in the text of this paper.

[2]R = Rule; P = Policy; G = Guidance; CWA = Clean Water Act; RCRA = Resource Conservation and Recovery Act; CERCLA = Comprehensive Environmental Response, Compensation and Liability Act; TSCA = Toxic Substances Control Act; CZARA = Coastal Zone Act; Farm Bill = The Federal Agriculture Improvement and Reform Act of 1996 Reauthorization Amendments of 1990; Air = EPA Office of Radiation and Indoor Air; EPA = U.S. Environmental Protection Agency; DOI = Department of Interior; USDA = U.S. Department of Agriculture; and FDA = Food and Drug Administration.

[3]The rules listed show those by-products and wastes most directly impacted. These rules also indirectly impact other by-products and wastes as discussed in the paper.

TABLE I. Continued.

P, EPA	49 *FR* 24358 (6-12-84)	Beneficial Use of Biosolids
P, EPA, DOI, USDA	56 *FR* 33186 (7-18-91)	Beneficial Use of Biosolids on Federal lands, Interagency Policy
P,G, EPA, FDA, USDA	1981	Land Application of Municipal Sewage Sludge for the Production of Fruits and Vegetables
R,G, CZARA		Guidance for costal zone states regarding erosion, nutrient management plans, soil testing, feedlots, and waterway protection
R,G, USDA Farm Bill		EQUIP Program; targeted incentives
G, EPA	EPA 625/10-84-003	Use and Disposal of Municipal Wastewater Sludge
G, EPA	EPA 832-B-92-005	Domestic Septage Regulatory Guidance: A Guide to the EPA Part 503 Rule
G, EPA	EPA/625/R-92/013	Control of Pathogens and Vector Attraction in Sewage Sludge
G, EPA	EPA 832-R-94-009	Biosolids Recycling: Beneficial Technology for a Better Environment
G, EPA	EPA 832-R-93-003	A Plain English Guide to the EPA Part 503 Biosolids Rule
G, EPA	EPA 832-B-93-005	A Guide to the Biosolids Risk Assessments for the EPA Part 503 Rule

(Biosolids) (40 CFR Part 503), the requirements in Part 257 had to be met when biosolids were applied to the land or placed on a surface disposal site. Part 257 requirements do not apply to biosolids unless the land-applied biosolids are generated during the treatment of a combination of domestic sewage (i.e., waste from humans and household operations) and industrial wastewater, and if the treatment works are located at an industrial facility. Part 257 was published in 1979 under the authority of both the Resource Conservation and Recovery Act (RCRA) and the Clean Water Act (CWA).

Solid Waste Disposal Facility Criteria (40 CFR Part 258). Part 258 contains the minimum national criteria for all municipal solid waste landfill units, including municipal solid waste landfills that receive biosolids. These criteria were published on October 9, 1991, under the authority of both RCRA and the CWA. A municipal solid waste landfill unit (MSWLF) is a discrete area of land or an excavation that receives household wastes and other types of waste. The approach taken in Part 258 to protect human health and the environment is containment. Part 258 contains requirements such as the need that solid wastes pass a paint filter test, liners, and leachate collection systems that control the release of pollutants from the solid waste into the environment. Human health is also protected by preventing exposure to the solid waste through access restrictions and daily cover requirements which reduce the attraction of vectors to solid waste placed in a MSWLF.

Hazardous Wastes (40 CFR Parts 261-268). The criteria for identifying hazardous waste are in Part 261. A waste can either be listed as a hazardous waste or be determined to be hazardous through testing, e.g., for corrosivity, ignitability, explosiveness, or toxicity. Part 262 through 268 contain standards for generators of hazardous waste; transporters of hazardous waste; treatment, storage, and disposal of hazardous waste, special hazardous waste, and land disposal of hazardous waste. By-products like biosolids are evaluated to determine whether they are hazardous and may be tested using the Toxicity Characteristic Leaching Procedure (TCLP). Almost no biosolids have failed the TCLP (*3*). Most similar organic by-products would probably not fail the TCLP unless they contain toxic constituents at levels that would cause them to exceed the limiting TCLP parameters.

Some by-products may contain listed constituents that could cause them to be considered hazardous no matter what their concentration. However, there are special provisions for delisting such by-products for cause, particularly when by-products are being used beneficially. Biosolids are exempt from being listed, even though they may contain listed hazardous constituents. This is because of the comprehensive Part 503 rule that regulates the use and disposal of biosolids.

Polychlorinated Biphenyls (PCBs) Manufacturing, Processing, Distribution in Commerce, and Use Prohibitions (40 CFR Part 761). Part 761 establishes the requirements for the manufacture, processing, distribution in commerce, storage, use, and disposal of PCBs and PCB items. Part 761 is being revised to establish requirements for the use of PCB remediation waste, which includes any environmental media with any concentration of PCBs. However, biosolids are not subject to the Part 761 disposal requirements for PCB remediation waste if the concentration of PCBs is less than 50 mg/kg and if the requirements in Parts 257, 258, or 503 are met when the biosolids are used or disposed. All other by-products would potentially be impacted by the Part 761 rule if they contain PCBs in any concentration.

Policies on Beneficial Use. The EPA's *Policy on Municipal Sewage Sludge (Biosolids) Management* (49 FR 24358 June 12, 1984) states that *"The U.S. Environmental Protection Agency (EPA) will actively promote those municipal 'biosolids' management practices that provide for the beneficial use of 'biosolids' while maintaining or improving environmental quality and protecting public health. To implement this policy, the EPA will continue to issue regulations that protect public health and other environmental values. . . . Local communities will remain responsible for choosing among alternative programs; for planning, constructing, and operating facilities to meet their needs; and for ensuring the continuing availability of adequate and acceptable disposal or use capacity."*

As noted in the policy statement, the EPA prefers well-managed practices that beneficially use biosolids. Such practices include land application of biosolids as a soil amendment or fertilizer supplement and various procedures that derive energy from biosolids or convert it to useful products. Practices that recycle biosolids reduce the volume of biosolids requiring disposal by either landfilling or incineration. Other benefits derived from recycling biosolids include improved soil fertility and tilth, reduced need for and enhanced response to inorganic fertilizers, better growth and quality of crops, and decreased consumption of energy.

In 1981, some Northwestern food processors, (e.g., Del Monte Corporation) announced that they would no longer accept fruit and vegetables for processing that had been grown on biosolids treated soils. Officials from U.S. Department of Agriculture (USDA), the Food and Drug Administration (FDA) and the EPA met with representatives of the National Food Processors Association to address the food processors' concerns. After very careful analysis of the available health and safety information pertaining to these practices, the USDA, FDA and EPA issued guidance and a joint policy statement in 1981 that was signed by the Administrators of each Agency. By endorsing the utilization of biosolids on land for the production of fruits and vegetables, the Agencies concluded *"that the use of high quality biosolids, coupled with proper management procedures, should safeguard the consumer from contaminated crops, minimize any potential adverse effect on the environment"*, and *"that, with the adherence to the guidance contained in this document, the safety and wholesomeness of the fruit and vegetable crops grown on biosolids-amended soils will be assured."*

The EPA also has policies that encourage recycling of municipal and industrial solid wastes. In a number of instances, there are preferences given to use of by-products and wastes.

Standards for the Use or Disposal of Sewage Sludge (Biosolids) (40 CFR Part 503) and the EPA Risk Assessment Process. Part 503 contains the requirements for the use or disposal of biosolids (*4*), which are the solid, semi-solid, or liquid residues generated during the treatment of domestic sewage in a treatment works. When biosolids are land-applied, surface-disposed, placed on a municipal solid waste landfill, or fired in a biosolids incinerator, the applicable requirements in Part 503 have to be met. The only exception is for biosolids that are land-applied or surface-disposed when the biosolids are generated during the treatment of a combination of domestic sewage and industrial wastewater, and the treatment works is located at an industrial facility. In that case, the requirements in Part 257 have to be met. Part 503 was published under the authority of the CWA.

Part 503 contains both risk-based requirements and technology-based requirements. A multi pathway risk assessment (14 exposure pathways for land application) was conducted to develop the pollutant limits in a Part 503 standard (*5*). Each pathway protects a highly exposed individual against a toxicity endpoint, either a Q_1^* (cancer) or RFD (minimum tolerance dose). The Part 503 numerical limit was derived for each pollutant from the most limiting pathway for that pollutant. For these reasons, the pollutant limits are conservative. The pollutant limits (Table II) together with management practices, (e.g., requirements that have to be met at a land application site), protect public health and the environment from the reasonably anticipated adverse effects of pollutants in biosolids. The rule also contains frequency of monitoring, reporting and recordkeeping requirements.

Part 503 also contains technology-based requirements. Examples include the pathogen and vector attraction reduction requirements in Part 503. These requirements are called operational standards that in the judgement of the EPA Administrator protect public health and the environment. This conclusion is based on years of research and professional judgement even though a risk assessment was not conducted to develop the operational standards.

The EPA risk assessment approach was used to determine a scientifically-evaluated, research-based acceptable level of environmental change from the presence of each pollutant in biosolids that still does not cause unacceptable human health or environmental effects. This contrasts with the zero to minimal risk-oriented policy approach that has used by some European Countries and Canadian provinces in which there has not been an attempt to determine environmental and public health consequences of the pollutant from exposure to the pollutant. The EPA risk assessment process showed that current practices were associated with low risk even before the Part 503 rule was promulgated.

TABLE II. Part 503 Rule Limits for Heavy Metals in Land Applied Biosolids

Pollutant	Ceiling Concentration Limits (mg/kg)	Pollutant Concentration Limits (mg/kg)	CPLR Limits (kg/ha)
As	75	41	41
Cd	85	39	39
Cu	4,300	1,500	1,500
Pb	840	300	300
Hg	57	17	17
Mo	75	to be added	
Ni	420	420	420
Se	100	100	100
Zn	7,500	2,800	2,800

The Part 503 rule is self-implementing. One of the most interesting aspects of the part 503 rule was the establishment of risk-based limits for biosolids (Table II, column 2) that can be used without concern of exceeding the heavy metal cumulative loading limits (Table II, column 3). If the pollutant concentration limits in Table II, column 2 are met, along with pathogen and vector attraction reduction requirements, then those biosolids, also known as exceptional quality (EQ) biosolids, are not subject to the land application general requirements and management practices in the Federal Part 503 rule. Limits also were developed for toxic organic pollutants (Table III). However, these limits were not included in the final part 503 rule because the organic pollutants were either: (i) banned for use in the United States; have restricted use in the United States; or are not manufactured for use in the United States; or (ii) the pollutant had a low percent of detection in biosolids, based on a national survey of biosolids, or (iii) based on data from that National Sewage Sludge (Biosolids) Survey (NSSS), the limit for an organic pollutant is not expected to be exceeded in biosolids that are used or disposed.

Federal rules, such as the Part 503 rule, set minimum standards for states. While state rules can be more restrictive than these minimum standards, the EPA has encouraged states not to make their rules more restrictive than the Federal

TABLE III. Limits for Toxic Organic Pollutants*

Pollutant	Limit, ug/kg
Aldrin/Dieldrin	2.7
Benzo(a)Pyrene	15
Chlordane	86
DDT/DDD/DDE	120
N-Nitrosodimethylamine	2.1
Heptachlor	7.4
Hexachlorobenzene	29
Hexachlorobutadiene	600
Lindane	84
PCBs	4.6
Toxaphene	10
Trichloroethylene	10,000

*From the land application risk assessment for the Part 503 rule.

standards, especially with respect to pollutant limits and pathogen and vector attraction reduction requirements. States do often have additional requirements for biosolids management practices. Different state requirements for EQ biosolids products would pose difficulty for interstate commerce and could also confuse the public about which standards are actually sufficient to protect public health and the environment.

Applicability of the Part 503 Rule to Other By-Products and Wastes. While not specifically developed for wastes other than biosolids, the pollutant concentration limits in Part 503, (the second column in Table II of this paper) have been used as a reference for discussion of the potential safety and usefulness of a large spectrum of other by-products and wastes that are being considered for recycling. In many instances, the pollutant limits should serve as a first approximation as a means of determining the potential safety of using the by-product under consideration. The Composting Council is currently conducting an evaluation of the applicability of the Part 503 provisions to other composted waste materials (Personal Communication, The Composting Council, Alexandria, VA).

The ultimate applicability of the part 503 limits to other by-products and wastes will hinge on their characteristics, especially the presence of oxides or iron, aluminum, and manganese and also phosphates. The ability of these oxides and phosphates to bind and make pollutants less bioavailable in biosolids is strong. The binding effects persist after they have been applied to soils, even after decomposition of the added organic matter. The bioavailability of the added biosolids' remains much less lower than if the pollutants had been added to soil in the form of inorganic salts. An important exception to this strong binding of the added biosolids' metals is when the soil pH is low. At low pHs, added biosolids' metals increase in solubility. Simultaneously, however, there is also an increased solubility of aluminum which is universally present within the mineral fraction of soils and which has a dominating toxic effect on most plants. This is why pH management is a normal farming practice regardless of the type of fertilizing and soil conditioning material being used.

Examination of field data, gathered as many as 60 to 100 years after the use of irrigation wastewater and/or biosolids on soils, supports the concept of the continuity of the binding effect by the oxides and phosphate fractions of the biosolids even after the biosolids' organic matter has had time to degrade. This binding phenomenon and the extent of applicability of the Part 503 rule to other wastes are discussed more fully in papers by Ryan and Chaney (*6*), and Chaney and Ryan (*7*).

State and Local Rules. *It is important to emphasize that the only rules discussed in this paper are Federal rules, and that the many additional state and even local rules which pertain to the management of by-products and wastes have not been included in these discussions.*

Future Federal and State Rule Making. Additional rules and guidances governing the use of these materials may likely be developed at the Federal level. However, under the current streamlined mode of government (*8*), there will likely be increased involvement of stakeholders and negotiations for the development of rules and guidances. For example, several rounds of simplifications and corrections to the Part 503 rule, that governs the use and disposal of biosolids, are underway with the expected addition of criteria for dioxins, furans and coplanar PCBs in the year 2000.

There also will likely be increased development and use by the regulated community of self-governing policies to assure good practice as is discussed below. There also have been increased efforts at the Federal level to simplify existing rules and permitting requirements through simplifying steps and reduced monitoring frequency based upon common sense and previous ability of the regulated entity to consistently meet compliance limits (*9*). Enforcement has begun to place a much greater emphasis on the demonstrated ability to consistently meet compliance limits, (i.e., good performance) rather than meeting bureaucratic requirements. Assistance on meeting necessary compliance limits is becoming of equal importance with sanctions for noncompliance (*10*).

Other Factors Impacting the Beneficial Use of By-Products and Wastes

By far the most important factor limiting the use of organic by-products and waste materials like biosolids, animal manures and food processing wastes are nuisances, especially odors (*10,11*). Concern caused by odor, coupled with the public's perception that the use of many by-products is potentially detrimental, has resulted in lack of acceptance for recycling to land and increased demands for careful oversight of land applied by-products. Unfortunately, resources at the Federal and state levels are often not sufficient to support comprehensive oversight. It is therefore important to develop a strategy to help overcome these problems.

Development and Implementation of the Strategy for Holistic Management of By-Products and Wastes in a Manner That is Publicly Acceptable and Agriculturally Sustainable

Some key components of a strategy for the agricultural use of organic and inorganic by-products and wastes including oversight and gaining public acceptance are first summarized and then discussed in greater detail. This discussion uses, as its primary example, the activities that have been undertaken to foster the beneficial use of biosolids. These components of the strategy are:

- Becoming proactive

- Implementing a sound communication strategy (*13*)

- Providing consistent technical information

- The early formation of partnerships and teams (*14*) and continued active involvement of these stakeholders including environmentalists with planning and assuring oversight and compliance

- Promoting relevant research to provide a sound basis for utilization practices

- Promoting the beneficial use of by-products and wastes and educating the public about the benefits of by-product recycling

- Holistic consideration of all by-products, wastes and more conventional materials that are being applied to land

- Changing the paradigm: Consider now that by-products and wastes are feedstocks for a broad continuum of beneficial agronomic and non agronomic uses.

Proactive. One of the most important considerations is to become proactive. That means the active participation of each individual in those components of this strategy for the use of organic and inorganic by-products and wastes for which he or she has any interest or responsibility. Each person needs to think and talk about his or her role with a colleague and then begin to carry out this role within 24 hours. This means that each individual will assume responsibility for him/herself as well as the colleague.

Communication. Communication must be effective. An excellent communications strategy was prepared for the Water Environment Federation (WEF) for use by wastewater management professionals regarding the beneficial use of biosolids (*13*). The communication objectives included in this strategy are to enhance public perception of biosolids recycling to gain broad public acceptance of its use, to help advance the goals of EPA and WEF in fostering beneficial use, to support municipal programs that use biosolids, and to improve the environment and protect human health. The strategy stated that the best method of accomplishing this goal is to focus communication efforts on gatekeepers and the process of recycling of biosolids rather than on biosolids themselves. It also stated that the key gatekeeper audiences to educate are (i) academics and agricultural scientists, (ii) water quality professionals (iii) farming groups, (iv) environmentalists, (v) regulatory officials and (vi) media. These "gatekeepers" are the individuals and organizations who are asked by the public for their opinion about biosolids utilization because of their expertise, authority or position.

Consistent Technical Information. A sound communication strategy must be supported by valid and consistent technical information. We need to foster the use of skillfully-maintained pages in the Internet. We need to provide funds and support for research and training so that the needed information can be generated. We need to have people with the skills and time to provide the needed information. We need to encourage ourselves and our colleagues to work with others to creatively solve real world problems. In so doing, we will discover how exciting and rewarding this can be. We also need a system to make sure that the information is consistent and technically correct. An important example of the need for consistent and correct technical information is via fact sheets that are now being written under a cooperative agreement with WEF to disseminate the correct technical information about alleged problems from the use of biosolids. The attempt to gain consistency and valid relevant information is by way of a broad spectrum of stakeholder oversight and review.

Stakeholder Partnerships and Teams. Ongoing and additional planned stakeholder activities are underway in many areas. One example is with biosolids stakeholders who met during the latter part of 1995 to develop a vision for a national program to manage the beneficial use as well as disposal of biosolids. These stakeholders recognized that the EPA has decreased resources assigned to biosolids management because of the low risk associated with biosolids utilization. Despite these low risks, the stakeholders also recognized the high level of public concern. Components of the vision are shown in Table IV. A second national biosolids stakeholder meeting will be scheduled early in 1997 to help decide more fully about how to implement the vision and who will assume responsibility for its different components.

The EPA awarded a grant in 1996 to the Northwest Biosolids Management Association to help with regional implementation of the biosolids stakeholder vision. This grant included the challenge to the NBMA to promote and assist others in forming regional technical groups that support the beneficial use of biosolids as well as the establishment of a coalition like the Greenway Trust that use biosolids. The highly successful Greenway Trust uses biosolids as part of a broad program to enhance and protect the natural and historic qualities of the 60-mile corridor along Interstate 90 that leads from the Cascade Mountains to the Puget Sound. The partners in the Greenway Trust include, the King County, WA, The University of WA, Weyerhaeuser Co., the WA Department of Natural Resources, and a large number of community and environmental groups. Countless volunteers are contributing to the successful achievement of the Trust's goals.

Oversight and Compliance. With decreased Federal and state resources available for the oversight of the utilization of by-products and wastes like biosolids, stakeholders need to fill the gap. There is an overriding need for stewardship and good practice to minimize nuisances and optimize opportunities for sustained agricultural use. A key component of an implementation strategy is to provide augmentation to State and Federal oversight. This oversight would include a public component, a Code of Good Practice, independent inspection, and enforcement by regulatory authorities to assure compliance.

The Code of Good Practice should include (i) a pledge by by-product and waste producers and users to follow practices to minimize odors, manage nutrients, and follow applicable State and Federal rules; (ii) procedures for tracking and monitoring to assure the quality of operations; (iii) keeping track of and being in compliance with any required pollutant levels, pathogen and vector attraction reduction requirements, and nutrient needs of the soil and crops that are being grown; (iv) tracking and assuring proper stability of the by-product for the desired end-use and to minimize odor potential when applied or stored; (v) insuring cleanliness and having a spill clean up plan; (vi) keeping accessible records; and (vii) operator and inspector training and certification.

Relevant Research. The EPA has also helped fund a number of research projects about the benefits and risks of using biosolids directly and indirectly through such means as the Water Environment Research Foundation. More recent research has been showing that the organic and inorganic matrices of the singular and blended waste products can have a major impact on the bioavailability of the pollutants in these matrices. Learning about the nature of these matrix impacts leads to by-product management practices that maximize benefit and minimize potential adverse impacts. This knowledge also suggests

TABLE IV. Stakeholder Vision of A Nation Biosolids Management Program

OVERSIGHT AND COMPLIANCE
VISION: To establish an effective monitoring, compliance and enforcement program that will lend credibility to this program to enhance public acceptance of biosolids.
DISCUSSION: Because EPA resources are becoming scarce, and given the low risk associated with the use and disposal of biosolids, extensive oversight by EPA is unjustified. On the other hand, non-compliance with the Part 503 rule and with practices that cause nuisances is of high priority because these could impede public acceptance.

STATE AUTHORIZATION AND PERMIT ISSUANCE
VISION: State acceptance of the responsibility for managing the Federal Part 503 biosolids rules will ensure the most cost-effective and environmentally protective approach for managing the nation's biosolids.
DISCUSSION; EPA should define a simple process for state authorization and a framework so that there is a national consistency of interpretation of administrative requirements for authorization.

PUBLIC ACCEPTANCE
VISION: A program should be designed that will accelerate and enhance public acceptance of for the use/disposal of biosolids. This vision will be a partnership among federal, state, and legal regulatory authorities and other stakeholders. Well designed education and outreach activities are essential. Regional cooperative information transfer groups should be formed and alliances with outside groups developed and strengthened. A communication plan should be updated every 3 years. Messages should be consistent. EPA should have the review capability to endorse peer-reviewed educational materials.

TECHNICAL STANDARDS/TECHNICAL ASSISTANCE
VISION: The Part 503 rule should be kept current, relevant and scientifically based. Maintain the function of the Pathogen Equivalency Committee and extend its function to include Vector Attraction Reduction equivalency. Develop carefully prepared and reviewed appropriate technical guidances. Provide on-going interpretation of the Part 503 rule and the associated risk assessments. Periodically compile updated information on biosolids quality and recycling practices in assessing trends in the quality of biosolids.

RESEARCH AND DEVELOPMENT
VISION: Continue biosolids related research that will support a scientifically based sound regulatory management program and determine priorities for research. Priority areas for research currently identified in the vision are to study the bioavailability of pollutants in biosolids, develop better procedures for analyzing trace organic compounds in biosolids as well as airborne contaminants emanating from various methods for treating biosolids, define biosolids stability, and develop alternative processes for achieving Class A biosolids with respect to pathogens and vector attraction reduction requirements. Define through research the opportunities for tailor-making biosolids together with other by-products and wastes to enhance their agronomic benefits, ability to remediate problems soils and the sustainability of their use.

ways to design and blend and otherwise tailor-make wastes to enhance their benefits including using them together with conventional fertilizer and soil conditioning materials. Knowledge of these by-product matrix impacts also helps establish appropriate limitations on their use. A recent grant to WEF will support the ability to track and promote scientific soundness of municipally funded research. Future research efforts should be more holistically oriented.

Promoting Beneficial Use. The EPA has supported a large number of other activities to promote and support the beneficial use of biosolids. The EPA's Office of Wastewater Management awarded a grant to the U.S. Department of Agriculture (USDA) to establish an Information Sharing Group (ISG) in New York and New Jersey. This group was charged with the task of identifying and then helping overcome factors limiting the beneficial use of biosolids in those two states. As a result of recommendations by the ISG, (i) a large research study was funded in the ecologically sensitive Pinelands of New Jersey; (ii) work was undertaken with bankers to help overcome their concern about lending money to farmers who landspread biosolids (*15*); (iii) a round table was established to determine a position about using biosolids in New York State; and (iv) Rutgers University and stakeholder (SCORE) efforts have been funded in New Jersey to promote sound practice and the beneficial use of biosolids. Another grant was awarded to the National Research Council of the National Academy of Science to determine the appropriateness of the utilization of municipal wastewaters and biosolids. Their report is an independent assessment that supports the EPA's scientific risk assessment approach and confirms the basis for the Federal policy that properly treated municipal wastewater and biosolids can be used safely and effectively for crop production (*16*). Still, another grant went to the WEF to develop the previously mentioned communications strategy. Another grant was awarded to the WEF for the development of an outlook for the beneficial use of biosolids in each state.

Educating the Public. The EPA has sponsored and co-sponsored a number of informational workshops to help educate rulemakers, municipalities, their consultants and the public. The EPA has also produced a series of user-friendly guidances, [e.g., on the benefits of using biosolids (*17*), the part 503 Rule (*4*), the risk assessment for the part 503 rule (*5*), the rules for domestic septage (*18*), permitting (*19*), and land application practices (*20*)]. The EPA is also supporting biosolids management excellence via its annual regional national Beneficial Use of Biosolids Awards Program for operating projects and technology development, research, and public acceptance activities.

Holistic Approach. Many people are now recognizing the need for a holistic approach to manage these by-products. The EPA established an Interagency Agreement with USDA in September 1996 to begin the coordination and organization of a cooperative effort to promote the holistic management of recyclable organic and inorganic resources. These recyclable resources include organic residues such as biosolids, other municipal residues, animal manures, food processing wastes, and certain inorganic waste materials. This Interagency

Agreement was established to provide for a series of activities that will foster and improve the sustainable use of these by-products in agriculture and to maximize the beneficial impacts of their use in watersheds. One of the first activities will be to establish guidance for the field storage of biosolids and other similar by-products.

Changing the Paradigm. Each individual needs to join the effort to creatively turn the environmental problems now being caused by improper use or disposal of inorganic and organic by-products and wastes into benefits to society. The view should now be that biosolids, animal manures, food processing wastes and the recyclable organic fraction of municipal solid wastes are part of the natural cycle of life and are feedstocks for a continuum of uses both agronomically and non agronomically as discussed by Galloway and Walker (*21*). Greater attention needs to be given to the properties and the opportunities for altering and tailor-making these by-products so that they are even better feedstocks for a whole continuum of uses.

The processing of a waste impacts the way a by-product looks, its odor potential and its public acceptability. It also determines stability and nutrient content and will also determine the extent of reduction of any pathogens and attractiveness of the by-products to vectors. The processing also impacts how the product can be stored and used. And, because of all these impacts, the form of processing will lead to varying degrees of restriction on use (*22*). Some agronomic opportunities for enhanced useability are mentioned below.

Metal bioavailability of these by-products can be managed. The bioavailability of metal pollutants in a waste material such as biosolids is reduced due to matrixes' components like oxides of iron, aluminum and manganese, humic acids from the biosolids organic matter, and phosphates (*7,23,24*). Dewatering biosolids with an iron salt such as ferric chloride enhances its ability to reduce bioavailability of metal pollutants both in biosolids and in soils to which biosolids have been added, even when added at an agronomic rate. Ongoing research is showing that because of the matrices impacts, biosolids can reduce soil lead bioavailability. Hence, children who might ingest such biosolids-amended high lead-containing soils, (e.g., contaminated by lead-containing paints or past automotive emissions), would be at reduced risk from lead poisoning (*25*).

Research by Hoitink et. al. (*26*) have demonstrated and the potting industry has adopted the use of specially prepared compost to suppress plant root disease. Research with municipal solid waste compost as a soil amendment has resulted in increased yield and quality of field and fruit crops (*27*) and the use of composted biosolids by themselves and in combination with other feedstocks have been shown to aid in leaf retention and hence more vigorous roots, enhanced plant growth and in the case of cantaloupes more yield and sugar content, because of longer leaf retention so that newly forming fruits could mature (F. Gouin, Dept. of Horticulture, U of MD, personal Communication).

Research by Falahi-Ardakani et. al. (*28*) have shown how composted biosolids can be used in potting media as a substitute for peat and at the same time reduce the need for chemical macro- and micro-nutrient fertilization. Recent cooperative studies at the Rodale Institute involve the comparative root and plant growth responses to inorganic fertilization and organic composts. This research is being conducted to examine how plant root structure and effectiveness are impacted in soils due to the stimulation of fungi and other microorganisms, better physical structure, and other related factors (Schettini, Rodale Inst, Millner, USDA and Lynch, Penn State Univ, personal communication).

Blending biosolids with kiln dust provides a more balanced fertilizer material that also has usefulness as a lime substitute (*29,30*). In addition, certain blends of biosolids with other feed stocks have been shown to make high quality mulches that reduce soil erosion.

Other unique opportunities exist for tailor making by-products such as alterations made prior to feeding animals. For example, the breeding of low phytate corn seed results in corn grain whose phosphorus is much more bioavailable than phosphorus in normal high-phytate containing corn grain. Thus, the litter from poultry that has consumed low phytate corn grain will contain reduced levels of phosphorus -- of value where this material is being applied to soils whose phosphate levels are already very high. Another example is the development of phytase overexpressing bacterial strains to enhance phosphorus availability to animals and thereby reduce phosphorus in manures (Personal communication, Xingen Lei, Department of Animal Science, Cornell University). Still another example is the design of feed rations which more closely match nutrient contents with animal nutritional needs (*31,32,33*).

As can be seen from the examples, greater mechanistic knowledge of the beneficial properties of biosolids leads directly to improved methods of benefiting society including remediating other environmental problems. The proper use of these waste products helps achieve the goal of sustainable farming. The production of these tailor-made by-products creates opportunities for new jobs in both rural and urban areas.

Conclusion. Change from the paradigm that the use and disposal of by-products and wastes of food and fiber production often result in insurmountable environmental problems. The new paradigm is that by-products and wastes are valuable feedstocks for a series of agronomic and non agronomic purposes that will enhance and sustain society. Knowledge of the fate and effects of the by-product constituents, how to minimize odors, and how to tailor their use to need is required to make optimum use of these by-products. Implementing the strategy described in this paper will help achieve the goal of sustainable use in society with very little ultimate waste within less than 10 years. This is indeed possible, (e.g., currently, there virtually is no waste from processing crude oil). Individuals and their organizations need to be proactive, take a holistic view of the many by-product and waste feedstocks that can be used, and work to obtain resources to support needed research and educational programs to fully reap the benefits from these valuable recyclable resources.

Literature Cited

1) Wright, R., Ed. 1994 Preprint. Agricultural utilization of industrial byproducts, agricultural, and municipal wastes, USDA, ARS, Beltsville, MD.

2) USDA. 1978. Improving soils with organic wastes, Agricultural Research Service, Beltsville, MD.

3) USEPA. 1991. Cooperative testing of municipal sewage sludges by the toxicity characteristic leaching procedure and compositional analysis. EPA 430/09-91-007. Office of Water, Washington, DC.

4) USEPA. 1994. A plain English guide to the EPA part 503 biosolids rule. EPA 832-R-93-003, Office of Water, Washington, DC.

5) USEPA. 1995. A guide to the biosolids risk assessments for the EPA part 503 rule. EPA 832-B-93-005. Office of Water, Washington, DC.

6) Ryan J. and R. Chaney. 1993. Regulation of municipal sewage sludge under the Clean Water Act Section 503: a model for exposure and risk assessment for MSW-compost. In Science and Engineering of Composting. Worthington, OH: Renaissance Publications, 1993.

7) Chaney, R. and J. Ryan. 1993. Heavy metals and toxic organic pollutants in MSW-composts: research results on phytoavailability, bioavailability, fate, etc. In Science and Engineering of Composting. Worthington, OH: Renaissance Publications, 1993.

8) Gore, Al. 1996. Reaching public goals: Managing government for results. Resource guide. National Performance Review, B. Godwin. Washington, DC.

9) USEPA. 1996. Interim guidance for performance-based reductions of NPDES permit monitoring frequencies. EPA-833-B-96-001.

10) USEPA, 1993. Office of Wastewater and Enforcement, Strategic Plan. EPA 833-R-93-001.

11) Walker, J. M. 1991. Fundamentals of odor control. Biocycle 32(9):50-55.

12) Walker, J. M. 1992. Control of composting odors. In H. A. J. Hoitink et. al. Science and Engineering of Composting. Design, Environmental, Microbiological and Utilization Aspects. Renaissance Publ. Worthington, OH. pp 185-218.

13) Powell Tate, 1993. Communications plan on biosolids. Prepared for the Water Environment Federation, Alexandria, VA.

14) Walker, J.M. 1995. U.S. EPA's roles in waste utilization: The need for partnership. In Agricultural Utilization of Urban and Industrial By-Products. Soil Sci. Soc. Amer. Special Pub. No. 59. pp. 11-18.

15) Walker, J. M. and R. N. Allbee. Fall 1994. Yes, but what about the liability? Water Connection. 11#2:10-11. New England Interstate Water Pollution Control Commission, Wilmington, MA.

16) National Research Council/National Academy of Science. 1996. Use of reclaimed water and sludge in food crop production. National Academy of Science. Washington, DC.

17) USEPA. 1994. Biosolids recycling: Beneficial technology of a better environment, Office of Wastewater Management, Washington, DC. EPA 832-R-94-009.

18) USEPA. 1993. Domestic septage regulatory guidance: A guide to the EPA 503 rule. Office of Wastewater Management, Washington, DC. EPA 832-B-92-005.

19) USEPA. 1995. Part 503 implementation guidance. EPA 833-R-95-001.

20) USEPA. 1995. Process design manual: Land application of sewage sludge and domestic septage. EPA/625/R-95/001.

21) Galloway, D. F. and J. M. Walker. 1996. An entrepreneurial view of the future for the use of wastes and by-products? In "Use of By-Products and Wastes in Agriculture". American Chemical Society, Washington, DC.

22) Walker, J. M. 1994. Production, use, and creative design sewage sludge. In Land utilization and the environment. Soil Sci. Soc. Amer. Proc. Miscel. Pub. Madison, WI.

23) Chaney R. L and J. A. Ryan. 1995. Risk based standards for arsenic, lead, and cadmium. in DECNMA. ISBN 3-926959-63-0.

24) Corey, R., L. King, C. Lue Hing, S. Fanning J. Street, and J. Walker. 1987. Effects of sludge properties on accumulation of trace elements by crops. In Land Application of Municipal Sewage sludges: Food Chain Implications. Chelesa, MI: Lewis Publishers, pp 25-51.

25) Heneghan, J. B., D. M. Smith, Jr., H. W. Mielke, R. L. Chaney and J. M. Walker. 1994. Composts/biosolids reduced soil lead bioavailability in rat feeding studies (Poster Abstract). National Conference of the Composting Council. Composting Council, Alexandria, VA, November 1994.

26) Hoitink, H. A. J., M. J. Boehm, and Y. Hadar. 1992. Mechanisms of suppression of soilborne plant pathogens in compost-amended substrates. In H. A. J. Hoitink, et. al. Ed., Science and Engineering of Composting. Design, environmental, microbiological and utilization aspects. Renaissance Publications, Worthington, OH. pp. 601-621.

27) Peverly, J. H. and P. B. Gates. 1994 Utilization of MSW and sludge composts in crop production systems. In Sewage Sludge: Land Utilization and the Environment. Soil Sci. Soc. Amer. Miscel. Pub. Chpt 26 pp. 209-224.

28) Falahi-Ardakani, A. F. R. Gouin, J. C. Bouwkamp, and R. L. Chaney. 1987. Growth response and mineral uptake of vegetable transplants grown in a composted sewage sludge amended medium. Part II. As influenced by time of application of nitrogen and potassium. J Environ. Hort. 5: 112-116.

29) Logan, T. J, and J. C. Burnham. 1994. Alkaline stabilization with accelerated drying process (N-Viro): An advanced technology to convert sewage sludge into a soil product. In D. L. Karlen, Ed. Wastes as Resources, Soil Sci. Soc. Amer. Special Pub. In Press.

30) Logan, T. J, , B. J. Harrison, and S. 1996. Uses of processed biosolids in the manufacture and marketing of horticultural fertilizers and soil blends. In "Use of By-Products and Wastes in Agriculture". American Chemical Society, Washington, DC.

31) Simms, T. 1996. Agricultural and environmental issues in the management of poultry manure. In "Use of By-Products and Wastes in Agriculture". American Chemical Society, Washington, DC.

32) Van Horn, H. and M. B. Hall. 1996. Agricultural and environmental issues in the management of cattle manure. In "Use of By-Products and Wastes in Agriculture". American Chemical Society, Washington, DC.

33) Mikkelsen, R, L. 1996. Agricultural and environmental issues in the management of swine manure. In "Use of By-Products and Wastes in Agriculture". American Chemical Society, Washington, DC.

Organic By-Products and Wastes

Chapter 4

Biosolids Processing, Products, and Uses

Jane B. Forste

Bio Gro Division, Wheelabrator Water Technologies Inc., 180 Admiral Cochrane Drive, Suite 305, Annapolis, MD 21401

Biosolids, the primarily organic product of wastewater treatment, are frequently recycled as a fertilizer substitute. The degree and type of processing will usually determine the specific applications for biosolids products and the way(s) in which they are marketed. A number of specific regulatory requirements (federal and state-by-state) are imposed for biosolids processing prior to use under site-specific or non-site-specific approvals from regulatory agencies.

Wastewater treatment results in two products: treated effluent and biosolids (in the form of slurries from the physical and biological treatment processes). Following wastewater treatment, the effluent must be disinfected (usually by chemicals such as chlorine) before it can be discharged into the receiving water bodies and released to the environment. Similarly, biosolids must also be treated if they are to be returned to the environment through land application. Because these biosolids contain significant amounts of organic matter and inorganic nutrients, which are a valuable resource for crop production, there is increasing interest in biosolids management options that enable treatment facilities to recycle such materials.

Land application is used to describe various means by which biosolids can be recycled onto the land to take advantage of their soil conditioning and nutrient properties. Land application includes agricultural, horticultural and forest uses, disturbed land reclamation applications, cover material for landfills, and a number of other beneficial uses.

USEPA developed policies to encourage the beneficial use of biosolids, and the Agency continues to prefer, wherever possible, well-managed beneficial uses to disposal options (*1*). The key to implementing such beneficial uses is to develop land application systems and products that enable users to take advantage of the beneficial properties of biosolids while ensuring that environmental and health considerations are also addressed. Under a statutory requirement of Section 405(d) of the Clean Water Act, EPA developed a comprehensive risk-based pathway analysis for regulating the concentrations of pollutants from biosolids application which would be protective of public health and the environment. The 40 CFR Part 503 Regulations are the result of an extensive risk analysis by EPA and are based on research efforts of agricultural scientists throughout the US who have the most detailed and complete understanding of the chemistry, exposure pathways and other factors included in risk assessment. The final rule, which took effect in February, 1993, remains a landmark regulation for EPA because of the comprehensive risk assessment and the extensive scientific peer review (2).

Properties of Biosolids for Beneficial Uses

Improved soil properties through land application of biosolids result primarily from organic matter and plant nutrients. Both of these are a significant resource for agricultural and other uses of biosolids. The 14 mineral elements established as essential micro or macronutrients are present in biosolids. These elements, combined with biosolids' approximately 50 percent organic matter (typical of digested biosolids) are of considerable value when used in a soil/crop system.

Generally, biosolids are used at rates based on the primary fertility element required by farmers for crop production (i.e., nitrogen), liming potential or need for organic matter, depending on the specific product characteristics and circumstances. The nitrogen content of biosolids varies somewhat, depending on sources of wastewater and wastewater treatment processing. Variability within a particular source of biosolids can be addressed by frequent analyses to establish a database for each specific material. Once that database is available, each biosolids material can be evaluated in terms of its nutrient content. Typically, each source will remain relatively consistent over time within a consistent processing method. Table I shows the nutrient composition of representative anaerobically digested biosolids.

Table I. Composition of Representative Anaerobic Biosolids (2–4)

Component	Range[a]
Organic nitrogen	1% - 6%
Ammonium nitrogen	1% - 3%
Total phosphorus	1.5% - 5%
Total potassium	0.2% - 0.8%

[a]Within each range, a single source of biosolids will generally have a much narrower range.

The organic nitrogen in biosolids (TKN – NO_3 and NH_4N) represents a unique form of nitrogen for crop production; it is mineralized to more available inorganic forms through biological decomposition processes in soil. Environmental conditions determine various N transformations in soils, and the resultant plant available nitrogen (PAN) is calculated to be available for a crop during the first growing season following the application of biosolids. PAN can be estimated as follows (5):

AGRICULTURAL USES OF BY-PRODUCTS AND WASTES

$$PAN = N_{NO_3} + XN_{NH_4} + YN_{ORG}$$

Where:
X is fraction of NH_4-N that does not volatilize (generally 100 percent if incorporated into soil, 50 percent if not)
Y is fraction of N_{ORG} that is expected to mineralize based on the representative values shown in Table II

In order to calculate plant available nitrogen, estimates of organic nitrogen mineralization rates must be made for each biosolids source (Table II).

Table II. Estimated Percentages of Organic N Mineralized Following Application to Soils

		% N_o		
Time After Application (Years)	Unstabilized Primary and Waste Activated	Aerobically Digested and Lime Stabilized	Anaerobically Digested	Composted
0 - 1	40	30	20	10
1 - 2	20	15	10	5
2 - 3	10	8	5	3

Table II can be used to calculate nitrogen-based agronomic application rates for biosolids. There is some variability in these estimates; plus, the amount of N absorbed by plants varies considerably with weather, moisture, pH and other soil conditions. Therefore, for agricultural production, a reasonable database with appropriate analytical information on nitrogen forms coupled with knowledge of biosolids processing can provide a reasonable estimate of how much available nitrogen a biosolids source will supply. It may also be appropriate, particularly for commercial biosolids, to develop more sophisticated field data on mineralization of these products under specific local or regional conditions.

An ongoing field study in Florida is evaluating heat-dried biosolids pellets as a potential source of nutrients (particularly N and Fe) for pasture, as well as characterizing the mineralization and nitrification rates of heat dried biosolids in a typical Florida soil. The rate and extent of mineralization of the heat-dried biosolids are also being determined in a laboratory study. This research is a good example of the type of evaluation needed to assist in determining more accurately biosolids' product value in specific applications. Initial results show significant increases in dry matter production and crude protein content of bahiagrass grown using biosolids pellets--total dry matter yields of 17 Mg/ha (comparable to fertilized controls) were found at a biosolids application rate of 4.4 Mg/ha. The highest application rate (17.6 Mg/ha) yielded 22 Mg/ha dry matter (6).

Based on the percentages of total N recovered, higher mineralization rates (70%+) appear to have resulted from lower biosolids application rates and vice versa, as shown in Table III.

Table III. Total N Recovered by Bahiagrass Fertilized With Biosolids[a]

Treatment	N Recovery (%)
0.55 Mg/ha	77
4.4 Mg/ha	50
17.6 Mg/ha	30
Fertilized control	62

[a]Adapted from ref. (6)

No environmentally significant changes in plant and soil metal concentrations were found with any of the biosolids application rates.

The laboratory study indicated that released N was primarily in the NH_4 form in the initial phases of incubation, with increasing NO_3 concentrations over time. Mineralization rates were greatest four weeks after adding biosolids pellets to soils, and remained substantial between six and eight weeks into the incubation period (7).

Organic matter from biosolids can significantly change the physical, chemical and biological properties of soils. Organic matter -- the non-living, heterogeneous mixture of organic compounds from microbial and chemical transformations of organic debris (as shown in Figure 1) -- has long been recognized as greatly contributing to soil productivity. Organic matter can improve soil structure (porosity, aggregation and bulk density), as well as alter the content and transmission of water, air and heat, and contribute to improved soil strength. Mineralization of nitrogen during decomposition of organic matter is an important component in improving soil productivity. Other chemical properties of soils, such as pH, electrical conductivity and redox potential, also improve with the addition of organic matter (8). Since organic matter is an energy source for microorganisms, improvements in biological diversity of soil microbial population will also occur with organic matter additions. In general, the desired effect of lower bulk density values occurs to a greater degree with increasing biosolids use. The resultant well-structured soil can better resist erosion and demonstrates improved air/water relationships, thus providing increased hydraulic conductivity, infiltration rate, air diffusivity, surface drainage and ease of root penetration. While these profound and complex interactions of soil organic matter are by no means fully understood, it is commonly recognized that organic matter (particularly if applied at relatively high rates or repeatedly over time) imparts to soils a desirable physical condition as well as modifying various chemical and biological relationships (9).

Biosolids Digestion

Biosolids are typically digested biologically, either through an aerobic or anaerobic process (i.e., with or without oxygen). This method of stabilizing, reducing volume and pathogens has been used for many years in treating wastewater solids. These solids from municipal wastewater are typically generated from primary clarifiers or settling basins and secondary biological processes, such as trickling filters, activated sludge and others. Advanced wastewater treatment may also contribute to the volume of biosolids generated by a particular wastewater treatment facility.

The primary goal of digestion processes is to reduce pathogen content and vector attractiveness of the material (generally measured by reduction in readily decomposable organic matter). Improvements in digestion processes can result in increasing efficiency, lower maintenance and the most rigorous pathogen and vector attraction reduction criteria. Pasteurization, as an additional step, can also enhance pathogen reduction but is not biological digestion. Digestion processes also allow microorganisms to decompose organic compounds into simpler compounds. The control of oxygen, temperature, feed rate of solids and a number of other factors are significant in operating these processes in a municipal wastewater treatment facility (*10*).

Biosolids Dewatering

Conditioning and dewatering of biosolids are commonly employed and are essential for many beneficial use options. The conditioning and dewatering steps prepare biosolids for drying, composting, alkaline stabilization or for direct beneficial use as a regulated soil amendment/fertilizer material. Chemical conditioning with either organic polyelectrolytes (polymers) or inorganic conditioning chemicals (e.g., lime, ferric chloride) is common, with increasing emphasis on the polymer technology. Chemical conditioning can be eliminated by using high pressure thermal treatment prior to dewatering. Dewatering typically reduces volume and weight of liquid biosolids by about ten-fold. Dewatered biosolids also have improved handling characteristics as compared to the liquid slurry form. In general, mechanical dewatering is more effective than passive dewatering (e.g., drying beds). This is accomplished through a number of processes, including belt filter presses, centrifuges, plate and frame presses, and vacuum filters. Biosolids thickening prior to dewatering can be accomplished by gravity belt centrifuges, flotation devices, or gravity.

Factors which affect the solids concentration in thickened and dewatered biosolids include: biosolids characteristics, and type of conditioning, thickening or dewatering used. Thickening reduces the volume of material handled in other processing steps (e.g., digestion) and, when employed prior to dewatering, may produce greater throughput and higher solids following the mechanical dewatering process. Mechanical dewatering processes apply pressure through a porous medium (e.g., a belt), a rotating bowl (centrifuge) which separates solids from liquids, or a rotating drum covered by a porous cloth to which a vacuum is applied (vacuum filter). All these processes result in a drier material, but the primary component in dewatered biosolids is still water, up to approximately 80%.

Regulatory Requirements

Increasing popularity of beneficial uses of biosolids during the last 20 years has resulted in complex and detailed technical standards at the federal level (40 CFR Part 503). In addition, specific management practices and site requirements are often imposed by state regulatory agencies. In general, biosolids must meet, or be treated for, the following before beneficial use:

4. FORSTE *Biosolids Processing, Products, and Uses*   55

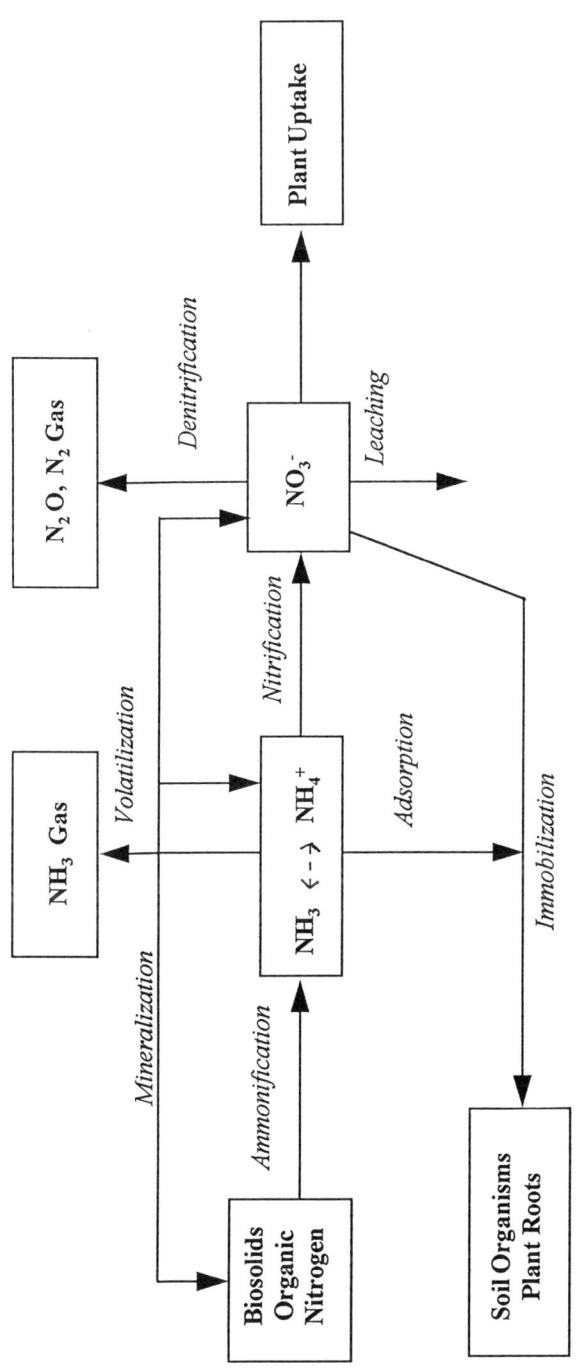

Figure 1. Biosolids N transformation in soils.

- Pathogen reduction: elimination, or reduction to an acceptable level, of potential disease-causing organisms.
- Vector attraction reduction: elimination, or reduction to an acceptable level, of attraction to vectors (e.g., flies, other insects).
- Pollutant limits: biosolids products or materials may not exceed limits established by federal, state and local regulations. For the most, part these regulations apply to the trace metal content of biosolids.
- Other specific requirements (such as monitoring and recordkeeping, and--for biosolids which have received only the basic level of treatment for pathogen reduction--waiting periods before human food crops are harvested).

The federal (503 Rule) requirements form the basis for the following discussion of regulatory requirements for biosolids. As noted above, there may be additional state requirements for specific situations.

Metals Limits

Current regulations establish numerical limits (ceiling concentrations) for trace metals that must be met if biosolids are used on the land (pollutant concentrations). Within those limits, additional numerical limits provide a means of either limiting the cumulative loading of metals on specific sites or limiting the metal concentrations in the biosolids themselves such that no cumulative limit is necessary. From the perspective of commercial biosolids products, this "pollutant concentration" standard is seminally important.

Table IV. 40 CFR §503.13 Metal Limits Summary

Metal	Ceiling Conc. (mg/kg)	CPLR lbs/ac[a]	(kg/ha)	Pollutant Conc. (mg/kg)[b]	APLR lbs/ac/yr[a]	(kg/ha/yr)
Arsenic	75	36.6	41	41	1.8	2.0
Cadmium	85	34.8	39	39	1.7	1.9
Copper	4,300	1,340	1,500	1,500	66.9	75
Lead	840	268	300	300	13.3	15
Mercury	57	15	17	17	0.8	0.85
Molybdenum	75	c	c	c	c	c
Nickel	420	375	420	420	18.8	21
Selenium	100	89.3	100	100	4.5	5.0
Zinc	7,500	2,500	2,800	2,800	125	140

[a]Calculated from the metric units in the 503 Rule
[b]Maximum monthly average concentrations
[c]Molybdenum limits deleted on February 18, 1994-- reproposal anticipated in 1997

CPLR = Cumulative Pollutant Loading Rates; APLR = Annual Pollutant Loading Rates -- These limits only apply to biosolids sold or given away in a bag or other container which do not meet the Table III "pollutant concentration" limits.

Composting

The biological decomposition, or composting, of organic matter is in many ways an extension of the natural biological processes in wastewater treatment. During composting, biosolids are processed aerobically at thermophilic temperatures to stabilize, reduce odors and pathogens, and create a product that can be used beneficially for horticultural and agricultural purposes (*11*). Typically, composting processes biosolids to a level of pathogen destruction which allows the product to be sold commercially; a lesser degree of treatment is allowable if appropriate site restrictions and management practices for compost use are imposed.

Composting relies on a diverse microbial community to accomplish the biological decomposition of organic matter to a humus-like material. Control of moisture, temperature and food source are all important to achieving the regulatory requirements for pathogen reduction, as well as an aesthetically acceptable and marketable final product. Optimum composting temperatures range from 35-65°C. These higher-than-ambient temperatures are achieved by the shift in microbial population within the composting process from mesophilic to thermophilic organisms. In general, this aerobic microbial population will be sustained with approximately 5-10 percent oxygen in the mixture. Good mixing and maintaining moisture levels of about 40-50 percent are also essential for the biological activity of composting. Regulatory requirements for temperature to assure pathogen control are necessary to produce a commercial compost; however, too long a period of high temperature can destroy beneficial organisms needed for the composting process. Control of air circulation, by turning or forced air, can maintain the temperature at a level to meet regulatory requirements but not impede the composting process. Unlike other biosolids processes, composting generally entails the addition of one or more additional feed stocks to optimize the physical and nutrient components of the process. Such materials increase the solids concentration as well as the carbon to nitrogen ratio (C:N) of the compost. Besides these two primary factors, porosity, biodegradability and other characteristics may also be controlled by the use of appropriate amendments as specified below.

Finished compost has a variety of uses. The composting process, as well as the amendments, will determine the product characteristics. Amendments include wood chips, shredded brush and leaves, shredded paper, wood ash, ground corn cobs, rice hulls, papermill residuals, and numerous other types of by-products which can be composted with dewatered biosolids. Most of these materials do not present a problem with respect to contamination with trace metals, but this factor must also be addressed in composting operations.

Finished compost contains the same macro and micronutrient elements (identified previously in the biosolids) as well as more stable organic matter. Compost is used extensively in establishing and maintaining turf grass for horticultural and garden uses, nursery crops and in bedding plant mixes. Successfully marketed compost must have consistent composition and appropriate physical and chemical properties. Careful management of the composting process will ensure that these factors are addressed.

Alkaline Treatment

Alkaline chemicals (particularly lime) have been used for many years to process and treat both wastewater and wastewater solids. Lime, or lime containing materials, are used to raise pH to inactivate or destroy pathogenic microorganisms.
The addition of lime slurry, often for conditioning, prior to dewatering (e.g., vacuum filtration) became an accepted method of solids treatment decades ago (*10*). In more recent years, post-dewatering application of lime has achieved even greater popularity. This process uses lime primarily as a treatment technology for pathogen and vector reduction rather than a chemical conditioner prior to the dewatering process.
For centuries, farmers have recognized the agronomic benefits of liming materials on acid soils to enhance crop yields. Since the value of any liming material used as a soil amendment depends upon the calcium and magnesium oxide content, this becomes the standard measure of the liming potential and therefore the value of alkaline treated materials. For agricultural applications, liming potential is commonly expressed as calcium carbonate equivalency (CCE), based on the ground limestone used by farmers to adjust soil pH.
Lime or other alkaline material may be added to achieve one of two biosolids pathogen reduction processing requirements (referred to as Class A or Class B). The combination of alkalinity (pH), temperature, and in some cases dryness, are the elements of alkaline treatment which achieve the regulatory requirements for pathogen reduction (*12*).
Alkaline treatment, like composting, can be implemented to a lesser level than that required for marketing a commercial biosolids product when treatment is combined with the site restrictions and management practices required by regulations. Unlike composting, this lesser level of treatment is commonly employed in alkaline processing of biosolids. Depending upon the level of treatment, the resultant products are quite different: Class B represents, primarily, a regulated nutrient source to which lime has been added, and Class A processing results in a liming agent which contains some organic nutrient content from the original biosolids. Varying process parameters (e.g., the type and dosage of the alkaline additive, the addition of heat, windrow turning) results in products with different physical and chemical characteristics. Some of the specific components important for various uses include dryness, granularity, spreadability, compactibility and nutrient content.
As with other agricultural lime products, the acid-neutralizing capability of lime-treated biosolids is expressed as calcium carbonate equivalence (CCE) which compares the biosolids product to $CaCO_3$, agricultural limestone. The increasing popularity of post-dewatering lime applications to biosolids has resulted in greater use of quicklime (CaO) than the hydrated form ($CaOH_2$) which was widely used in the past. The ability of quicklime to increase temperature (reactivity) is an important factor in using this material to produce the Class A type alkaline products. From the perspective of the treatment facility, using quicklime eliminates the need to hydrate (slake) the lime, as practiced in the past, and provides a significant advantage.

A number of proprietary processes have been developed for alkaline treatment of biosolids using commercial lime and by-products such as kiln dust and fly ash. As is the case with the amendments used for composting processes, these by-product materials should be evaluated with respect to their trace metal and other contaminant concentrations to ensure compliance with regulatory standards for land application and/or as a commercial product. Some typical characteristics of the BIO*FIX (a commercial process) products made using commercial quicklime (shown in Table V) illustrate some of the relevant properties for different types of products and their potential uses.

Biosolids Drying

Municipal biosolids have been processed for many years using open-air solar drying if favorable climatic conditions exist. In most cases such an option is not feasible in much of the US, and during the last decade heat treatment has grown as the method of choice to meet regulatory requirements and to maximize volume reduction. Heat-drying of biosolids offers a number of advantages: it eliminates pathogens from the end product, reduces the product weight significantly, results in a nutrient-rich product and improves transportability and storability of the material. Heat drying typically produces a material which is at least 90% total solids, often in the form of pellets which are readily integrated into the agricultural and specialty fertilizer markets.

Heat drying of biosolids eliminates pathogens and produces a sterile end product in accordance with the US EPA Part 503 standards (Class A). It also reduces the volume and weight of the biosolids while preserving valuable organic and inorganic nutrients found therein. Heat-dried biosolids represent a valuable end product with revenue potential as a fertilizer, soil amendment or fuel source (13).

Typically, dried biosolids pellets contain over 90% total solids when marketed as a fertilizer. Product moisture content above 10% may encourage microbiological growth and consequent heating of the product. A Class A product which meets the pollutant concentration metal limits contained in Table III can be used as a fertilizer and organic soil amendment product without the management practice and site restrictions imposed for Class B land-applied biosolids. This is particularly important for pelletized material which is intended for use in the broader fertilizer marketplace. Nitrogen content of biosolids pellets depends on the organic nitrogen fraction in the original biosolids. The loss of inorganic nitrogen (NH_3) during the drying process results in a somewhat lower PAN than contained in the original biosolids cake.

Producing a fertilizer product from municipal biosolids through heat drying and pelletizing will increase from the current (1995) 0.3 million tons to over 0.5 million tons by the end of the 1990's. Such a production level still comprises only a small portion of the total US chemical fertilizer consumption which approximates 15 million tons annually. Pelletized biosolids comply with the most stringent federal, state and local standards governing distribution and marketing of biosolids as a fertilizer. In general, heat drying with pelletization, while it is the most expensive of the various biosolids processing options, also provides the greatest opportunity for

Table V. Major BIO*FIX Product Data

Item	Class A	Class B
Total Solids, %	50 - 60	20 - 30
Total Volatile Solids, %	5 - 15	40 - 60
Total Nitrogen (%, dry)	0.3 - 1.0	2 - 5
Phosphorus, P, % Potassium, K, %	As in biosolids feed, diluted by the addition of lime	As in biosolids feed
Calcium (%, dry basis)	30 - 40	10 - 15
Trace Metals mg/kg, dry	Concentrations are reduced by a factor of 2 - 2.5	As in feed
pH	11.5 for several months	11.5+ for 24 hours or longer
$CaCO_3$ Equivalent, (CCE), %	70 - 95	15 - 30
Bulk Density, lb/cf	52 - 58	58 - 60
Physical characteristics	Soil-like, near odorless, crumbly material with very good spreadability and storability	moist, dark brown; good spreadability
Application	Liming agent, landfill cover	Soil amendment and fertilizer

generating revenues from product sale. Biosolids pellet characteristics which represent consumer value for various applications include:

- slow release of organic N content, micronutrient content, organic matter content (soil conditioning properties)
- environmental benefit from reduced likelihood of nitrogen movement to groundwater compared to inorganic N sources

Biosolids fertilizers must be consistent with other fertilizer materials in terms of physical properties. These properties include the following:

- Particle size range, generally from 1 - 4 mm. Biosolids pellets with narrow particle size distribution and minimum dust are considered most valuable.
- Bulk density. Most fertilizer materials range from 40 - 80 pounds/cubic foot; biosolids bulk density values typically fall at the lower end of this range or somewhat less.
- Durability. Pellets must be able to withstand transportation, handling and mixing of bulk materials without producing excessive amounts of dust.
- Pellets are hydroscopic and act as a conditioner when blended with fertilizers such as urea.

Biosolids pellets which meet regulatory standards can be used in a variety of fertilizer applications ranging from use as a slow-release N source and filler in blended materials for specific applications to turf grass and ornamental plant production. There is a substantial demand for heat dried biosolids fertilizers in the US, particularly in the Florida market including a number of specific custom-blended fertilizers. Such applications include citrus, other fruit crops, and a wide variety of vegetable crops. Non-agricultural applications include turf production, golf course use, residential and commercial lawn fertilization and landscape/plant nursery production. New markets are currently developing outside the historically strong Florida area and will be available for the increased production of heat-dried biosolids pellets.

Literature Cited

1. *Municipal Sludge Management Policy*; Federal Register; U.S. Environmental Protection Agency: Washington, DC, 1984.
2. *40 CFR Parts 257, 403 and 503, Standards for the Use or Disposal of Sewage Sludge*; U.S. Environmental Protection Agency: Washington, DC, 1993.
3. *Municipal Sewage Sludge Management: Processing, Utilization and Disposal*; Lue-Hing, C., et al., Eds.; Water Quality Management Library; Technomic Publishing Co., Inc.: Lancaster, PA, 1992; Vol. 4.

4. *Design of Municipal Wastewater Treatment Plants*; WEF Manual of Practice No. 8, ASCE Manual and Report on Engineering Practice No. 76; Water Environment Federation and American Society of Civil Engineers; 1992; Vol. 2.
5. Forste, J.B.; In: *Biosolids Treatment and Management: Processes for Beneficial Use*; Girovich, M.J. Ed.; Marcel Dekker Inc.: New York, NY, 1996; pp 389-448.
6. Muchovej, R.M.C.; Rechcigl J.E.; *Effects of Municipal Biosolids on Forage and Soil Quality*; Proceedings of the International Conference on Livestock in the Tropics, Institute of Food and Agricultural Sciences, University of Florida; 1996.
7. Muchovej, R.M.C.; Rechcigl, J.E.; *Temporal Release of Nitrogen Forms from Soil Amended with Biosolids*; Agronomy Abstracts; American Society of Agronomy, Crop Science Society of America, Soil Science Society of America; 1996.
8. *The Role of Organic Matter in Modern Agriculture;* Chen, Y.; Avnimelech, Y., Eds.; Martinus Nijhoff Publishers: Dordrecht, Netherlands, 1986.
9. Stevenson, F.J.; *Cycles of Soil: Carbon, Nitrogen, Phosphorus, Sulfur, Micronutrients*; John Wiley & Sons: 1986.
10. *Process Design Manual for Sludge Treatment and Disposal*; EPA-625/1-79-011; U.S. Environmental Protection Agency: Washington, DC, 1979.
11. Naylor, L.M.; In: *Biosolids Treatment and Management: Processes for Beneficial Use*; Girovich, M.J., Ed.; Marcel Dekker Inc.: New York, NY, 1996; pp 193-269
12. Girovich, M.J.; In: *Biosolids Treatment and Management: Processes for Beneficial Use*; Girovich, M.J., Ed.; Marcel Dekker Inc.: New York, NY, 1996; pp 343-388
13. Maestri, T.T.; Gratton, P.F.; *The Case for Pelletization*; 10th Annual Residuals and Biosolids Management Conference, Water Environment Federation, Alexandria, VA, 1996.

Chapter 5

Characteristics and Standards for Processed Biosolids in the Manufacture and Marketing of Horticultural Fertilizers and Soil Blends

Terry J. Logan[1], Billie J. Lindsay[1], and Steve Titko[2]

[1]School of Natural Resources, Ohio State University, 2021 Coffey Road, Columbus, OH 43210
[2]Scotts Company, 14310 Scottslawn Road, Marysville, OH 43041

Beneficial use of treated biosolids is not a new practice, but it has increased in the last two decades, primarily in the U.S., and to a lesser extent in other countries. Biosolids at small treatment plants were often dried on sand beds and made available to farmers and gardeners at no cost, and with virtually no regulation. Application to dedicated sites for agricultural production was practiced in the U.S. and Europe. In none of these cases, however, were biosolids viewed as products in the sense that fertilizer, limestone or pesticides were viewed. A few well known exceptions to this are the heat-dried biosolids marketed by the City of Milwaukee for at least six decades as Milorganite, and the composted biosolids marketed by the Kelloggs Company in Southern California. There are no similar early experiences in Europe and other countries. This paper considers characteristics (e.g., solids, nutrient, organic matter and lime contents) and suitability of EQ biosolids (compost, advanced alkaline stabilized biosolids, heat dried pellets) for horticultural markets (greenhouse, comercial nurserys, turfgrass, home gardening), and uniform national product quality standards and labels are proposed.

Beneficial use of treated biosolids is not a new practice, but it has increased in the last two decades, primarily in the U.S., and to a lesser extent in other countries. Biosolids at small treatment plants were often dried on sand beds and made available to farmers and gardeners at no cost, and with virtually no regulation. Application to dedicated sites for agricultural production was practiced in the U.S. and Europe. In none of these cases, however, were biosolids viewed as products in the sense that fertilizer, limestone or pesticides were viewed. A few well known exceptions to this are the heat-dried biosolids marketed by the City of Milwaukee for at least six decades as Milorganite, and the composted biosolids marketed by the Kelloggs Company in Southern California. There are no similar early experiences in Europe and other countries.

© 1997 American Chemical Society

Current Conditions

By the 1990s, a number of factors had led to the development of true potential market opportunities for biosolids derived products. One was the publication of the USEPA 503 biosolids rule (EPA, 1993) that established the Exceptional Quality (EQ) biosolids concept: a class of highly treated biosolids that could meet high, quantifiable standards for pathogen destruction, trace elements and vector attraction reduction; having met these standards, these biosolids would be deregulated under federal law. It is interesting to note that the EQ biosolids status has been de facto awarded to Milorganite for many years in most states. Secondly, there has been a ban or schedule for phasing out ocean dumping of biosolids in the U.S., Canada, the EC, and Australia, thereby increasing pressure on land-based options, including beneficial use. Large cities are commonly located on the coast, and shifting biosolids from cities like New York, Boston, Los Angeles, Dublin, and Sydney from the ocean to the land has had, or will have, enormous impact on beneficial use of biosolids. Since many of these communities have opted to produce EQ biosolids materials, the opportunity to penetrate agricultural and horticultural markets with these products is great. The third factor affecting the potential to bring biosolids-derived products to the marketplace is the increased number of technologies available to produce EQ biosolids: composting (static pile, aerated static pile, envessel), advanced alkaline stabilization, and thermal drying/pelletization.

Horticultural Markets for Soil Additives

Horticultural markets are distinctly different in their use of soil additives (these include fertilizer, limestone, organic matter and pesticides) than agriculture. In a study to determine national potential markets for compost, Slivka et al. (1992) identified horticultural markets that included landscaping, delivered topsoil, bagged/retail products (unidentified), container nurseries and sod production (Table 1).

The horticultural markets are small compared to the potential use in agriculture. Taken together, the horticultural markets amount to 38.6 million cubic yards, and most of this (20 million cubic yards) is in sod production (Table 1). We feel that the actual and potential horticultural markets are underestimated and that there will be significant increases in demand if low-cost materials like biosolids or other organic waste-derived products enter the market.

It is difficult to compare market demand against biosolids production. Biosolids production is given as dry tons (Table 2) and the potential to shift from incineration and landfilling to beneficial use in general, and product development in particular, is complicated by socio-economic forces that are regional and even local. If we use an approximate density of biosolids products of 0.5 metric tons/m^3, and assume that a yard is approximately equal to a meter, then 38.6 million cubic yards is approximately 20 million metric tons. If we assume that all biosolids that were land applied or marketed in 1988 were converted to EQ products and used in horticulture, this would be about 2 million metric tons, or 10 % of the potential market. We do not have accurate national statistics on current biosolids use and disposal, but we suggest that the trend is in the direction of beneficial use and towards EQ biosolids production. For example, a significant portion of the 4 million metric tons ocean dumped in 1988 (Table 2) are going into pellets (New

Table 1. Potential demand for compost in the U.S. (Slivka et al., 1992)

Market Segment	Potential Demand (10^6 cubic yards)	Current Saturation (Percent)
Landscaping	2.0	< 20
Delivered topsoil	3.7	< 5
Bagged/retail	8.0	80
Landfill final cover	0.6	< 5
Surface mine reclamation	0.2	< 5
Container nurseries	0.9	50
Field nurseries	4.0	< 1
Sod production	20	< 1
Silviculture	104	< 1
Agriculture	895	< 1
Total (rounded)	1040	< 2

York, Boston), alkaline stabilized EQ biosolids (Middlesex County, NJ, Weschester, NY, Bergen County, NJ) or compost. Even without an increase in the size of the horticultural market, as we have suggested will occur, there appears to be the potential to use a large part of the EQ biosolids in horticulture. It should be kept in mind, however, that other organic waste-derived products (yard waste compost, composted or dried manure) are also increasing in volume (Logan et al., 1995).

Horticultural markets for soil additives are more specialized and diverse than those in agriculture, and the requirements are different (Table 3). Horticulture can be divided into commercial greenhouse production, commercial nursery production, turfgrass production and maintenance, and home gardening. The soil additives used by these segments differ considerably, and product requirements are very different.

The commercial greenhouse industry uses artificial soil media for growing plants in pots and flats. The industry prefers materials that are light and have low bulk density. They are less concerned with water holding capacity or nutrient supply than container or nursery producers because these are normally provided by watering and fertilization systems. A critical factor to the greenhouse producers is wettability of the media. Additionally, materials have to be of high quality, free of disease and weed seeds, and uniform. Unit cost is not critical. Popular commercial products are mixes of sphagnum peat and other materials like perlite.

Commercial nursery production is soil-based, with native soil at the nursery supplemented with organic matter and fertilizer. Commonly used materials include yard waste compost and other composts; unit cost is a consideration because of the large volumes of soil used.

Turfgrass production and maintenance involves turf establishment, maintenance and renovation. Major consideration is for visual turf quality, and primary soil additive requirements are fertilizers, particularly with controlled nitrogen release. Home gardening demands two types of soil additive: specialty fertilizer/pesticide or soil/fertilizer mixes for growth of specific trees, shrubs, turf and flowers; and bagged soil products like peat, "topsoil", or compost.

Table 2. Production and disposal/beneficial use of biosolids in the U.S. in 1988 (EPA, 1993)

Disposal or Beneficial Use	POTWs		Biosolids Production	
	No.	Percent	Volume (x 10^3 mt/yr)	Percent
Land application	2,623	17.1	1,202.2	15.6
Distribution and marketing	106	0.7	705.5	9.1
Municipal landfills	6,664	43.5	3,162.3	41.0
Surface disposal	2,395	15.6	196.4	2.5
Monofills	49	0.3	101.4	1.3
Incineration	169	1.1	1,651.4	21.4
Ocean disposal[†]	25	0.2	424.4	5.5
Other	3,274	21.4	270.0	3.5
Total	15,305	100.0	7,713.6	100.0

Table 3. Horticultural additive requirements and characteristics

Needs	Products	Product Characteristics and Constraints
	Greenhouse Industry	
Potting Media	Synthetic soil blends (peat, perlite)	Low density; high porosity; low soluble salts; pest free; uniform quality
	Commercial Nursery	
Soil	Native soil; native soil blended with composts	Inexpensive local supply; costs
	Turfgrass Industry	
Soil Fertilizer/pesticides	Native soil; native soil blended with composts; synthetic fertilizer/pesticide combinations	Proven performance for turf quality; slow release nitrogen
	Home Gardening	
Soil Fertilizer/pesticides Mulch Organic matter	Synthetic fertilizer/pesticide combinations; fertilizer enriched organic matter; "topsoil"; compost	Proven performance for fertilizer/pesticide products; quality control and cost for soil products

Product Requirements and Regulations

Product requirements for both agriculture and horticulture include: guaranteed analysis (fertilizer and limestone), uniform size and good flowability, and product stability (composts and other organics). In the case of consumer horticultural products, appearance (dark colors) and acceptable odor are additionally important. Horticultural soil products should also meet minimum standards for pH (~ 4-8) and soluble salts (~ < 10 dS/m). Some species like poinsettias and rhododendrons have highly specific standards. Materials must also be storable in bulk or in bags. While fertilizers and other soil additives are not subject to federal regulations on heavy metals, pathogens, or toxic organics, and there are few state standards, products containing biosolids must meet EPA waste requirements (EPA, 1993). While this is not usually a problem where biosolids is only a component of the product and the contaminant levels are diluted in the final product, present labeling requirements are onerous and limit national or regional marketing of regulated products (see section below on labeling).

Soil additives are normally regulated by state departments of agriculture. Regulations include ingredients, uses for the product, guaranteed analysis (calcium carbonate equivalent (CCE) or effective neutralizing value (ENV) for limestone, N-P_2O_5-K_2O for fertilizer), and levels of heavy metals (usually only for biosolids). Products are registered or certified by the agency, and bagged products may be required to carry a label that lists ingredients, gives guaranteed analysis, and instructions for use. There are considerable differences in state _interpretations_ of allowable label claims, and this is likely to be even more the case for a material like biosolids.

The industry also regulates itself. Associations like the U.S. Golf Association (USGA) and the National Bark and Soil Producers Association (NBSPA) set standards for product terminology and provide guidelines and recommendations on product characteristics. The organic growers in the U.S. are attempting to identify products that can be used for organic crop production, and the Compost Council is attempting to set quality standards for finished compost. Scotts Company is working with the American Association of Plant Food Control Officials (AAPFCO) to develop a model label for EQ biosolids.

How Can Biosolids Derived Products Compete in Horticultural Markets?

There is great potential to use treated biosolids in horticulture, and there are successes like Milorganite and Kelloggs as examples. The challenges, however, are formidable. First, the attitudes of most biosolids producers are not consistent with a product industry - the wastewater treatment plant wants the biosolids to go away at the lowest cost, wants it to go away every day, and with the least liability or negative public reaction. Limited attention is paid to product quality control; meeting the regulatory standards for metals, disinfection and vector attraction is the primary goal. Second, different biosolids products are compatible with only certain markets (Table 4).

Third, there is a reluctance on the part of established product marketers to assume the added regulatory burden and potential negative public opinion of biosolids. In this case, a major concern is the possibility of having to have different product labels for each state to reflect state differences in regulation of biosolids. For national product marketers, this is will require costly add-ons to labeling equipment and more complex inventory control.

For biosolids to be fully accepted and integrated into agricultural and horticultural product markets, the following must be met:

1. Biosolids must meet the 503 EQ standards - this is essential for unrestricted use of the products (EPA, 1993).

2. Product standards for each type of biosolids (pellets, advanced alkaline stabilization, compost) must be developed. These will go beyond EQ requirements to deal with issues of use and value, stability, odor and physical characteristics.

3. National, generic labels must be developed for bagged biosolids derived products - this will require cooperation by the states.

4. There must be better interaction between biosolids producers and product marketers. If biosolids marketing is truly the goal, then product and market analysis must start with the consumer. What products does the consumer want and what is the competition? How must the product be priced? Can the product be integrated into the consumer's production system? Although there is limited flexibility in EQ biosolids production, there are some options. For example, the N-Viro process uses an array of waste alkaline materials that include cement kiln dust, lime kiln dust, wood ash, and various coal combustion ashes. The chemical characteristics of these materials vary in content of essential nutrients (particularly Ca, Mg, K and S), and some N-Viro facilities blend different alkaline materials in response to agricultural and horticultural needs.

Uniform Standards For Fertilizers, Soil Amendments and Manufactured Soil

One of the inequities and confusions in attempting to fully utilize recyclable wastes in agricultural and horticultural products is the great discrepancy in how these materials are regulated. Commercial fertilizers and agricultural limestone must provide guaranteed analyses for macronutrients (nitrogen, phosphorus and potassium), acid neutralizing value (given as ENV or CCE), and, for some materials, particle size. These materials are not regulated with respect to content of trace elements, and trace element analyses do not have to be provided. Yet some phosphate fertilizers contain cadmium concentrations that would not meet EQ biosolids limits (39 mg/kg), and some organic fertilizers that contain tannery waste or poultry manure can contain high concentrations of chromium, copper, arsenic and zinc. Livestock manures do not have to meet trace element or pathogen limits for beneficial use, and yard waste composts are virtually unregulated for contaminants in all but a few states (Illinois, California) even though they routinely contain measurable levels of fecal organisms and pathogens. Municipal solid waste composts are not regulated for content of inert plastic, glass or metal. Conversely, present hazardous waste laws may force some relatively benign materials like foundry sands and combustion residuals to be classified as hazardous even when the contents of regulated contaminants are low.

From a consumer standpoint, there is the further problem of a lack of quality standards that go beyond the issues of health and the environment. This is less of a problem for completely manufactured products like synthetic fertilizer than it is with processed wastes where physical, chemical, and biological compositions are variable. Product standards differ with the type of product (Table 5), but groups like the organic producers, the Compost Council, or individual states (like Ohio) are only now beginning to develop uniform standards for particular products like compost. Standards are needed for final products that address health, environmental and product use needs. Once product standards are set, it will be relatively easy to determine the standards for product ingredients to insure that product standards are met.

From the standpoint of product performance and environmental impact of organic products like biosolids, it is important to know the forms, behavior in soil, and plant availability of nutrient elements. Since most of the N in these materials is organic, the rate and extent of N mineralization must be known with some degree of precision and accuracy. Progress has been made in the development of N availability tests, but these are not routine, quick or inexpensive, and they tend to be soil and climate specific (Gilmour et al., 1985). Better prediction of N availability will enhance the fertilizer value of the product and avoid the water quality problems associated with over application. Likewise, the P fertilizer value of these materials is poorly known and no tests are currently available to determine P availability of the materials themselves.

Uniform State Labels For Bagged Products Containing Biosolids

The 503 federal biosolids rule permits unrestricted use of bagged products containing biosolids if the biosolids are EQ (EPA, 1993). The product must be identified as containing biosolids, and general directions for proper product use must be provided on the label. The intent of this provision of the rule was to encourage as widespread use as possible of biosolids that meet the EQ standards. National product marketers like the City of Milwaukee (produces and sells Milorganite, a thermally dried biosolids), or Scotts Company (markets manufactured soil under the Hyponex trademark and is considering the use of EQ biosolids in its soil products), were encouraged when 503 became law. This optimism has been tempered, however, by the realization that many states have decided to impose requirements for bagged biosolids over and above those in 503.

Surveys conducted by the City of Milwaukee, the senior authors of this paper, and by Scotts Company, reveal that states vary widely in their present requirements for labeling of products containing biosolids. Many do not recognize the term biosolids and require instead the term "activated sewage sludge" which has been in effect since 1950 and was instituted to accommodate the marketing of Milorganite. Activated sewage products "...are those made from sewage freed from grit and coarse solids and aerated after being inoculated with microorganisms. The resulting flocculated organic matter is withdrawn from the tanks, filtered with or without the aid of coagulants, dried, ground and screened". This definition was clearly made to encompass the Milorganite process, but is inadequate for other EQ biosolids processes like advanced alkaline stabilization and composting. Scotts has recently proposed to the Association of American Plant Food Control Officials (AAPFCO), whose members are the state plant food regulators, that a national label for plant food products containing biosolids be approved. They have proposed a definition for biosolids that is essentially that developed by the Water Environment Federation

Table 4. Suitability of EQ biosolids for agricultural and horticultural markets

Product Characteristic	Horticultural Market
Biosolids Compost	
Stable organic matter Slow release nutrients	Bagged compost; topsoil blend with native soil
Advanced Alkaline Stabilized Biosolids	
Limestone Slow release nutrients Organic matter "Soil-Like" physical properties	Aged for use as topsoil; Blended with native soil for topsoil; blended with yard waste compost as topsoil
Heat Dried Pellets	
Slow release nutrients; Dry, granular material	Turfgrass; turf fertilizer filler

Table 5. Proposed quality standards for soil additives and manufactured soil

Product Type	Material	Product Standard
Fertilizer	Synthetic	Guaranteed NPK analysis Ingredients Trace element content
	Organic	Guaranteed NPK analysis N mineralization Trace element content Pathogen content Biological stability
Liming agent	Limestone	CCE or ENV Total Fineness Efficiency (TFE)
	Waste liming materials	CCE or ENV Total Fineness Efficiency (TFE) Lime speciation
Soil amendment	Compost	Organic matter content NPK analysis C:N ratio/N mineralization Trace element content Pathogen content Biological stability Content of inerts
Manufactured soil	Blended soil	Bulk density, total porosity pH, CEC, soluble salts Available nutrients CCE or ENV Trace element content Pathogen content

(WEF): "Biosolids are a primarily organic material produced by wastewater treatment processes that can be beneficially recycled for its nutritional content and/or soil amending characteristics".

While states have varying label requirements for products with biosolids, most would permit such products to be registered and sold. This is heartening, and suggests that there is room to work with state regulators, perhaps through AAPFCO, WEF and USEPA, to produce a national label for biosolids derived products that would be acceptable to all.

National and State Uniform Regulations For Waste Derived Soil Additives and Manufactured Soil Products

As the nation implements its commitment to waste recycling, utilization of treated wastes in products for agricultural, horticultural and silvicultural production, in land reclamation and in soil remediation will increase. Extensive research in the last three decades has identified important contaminants in these wastes, environmental fate of these contaminants when the wastes are land applied, and pathways for human exposure. This information was incorporated into the comprehensive risk assessment conducted as the basis for the 503 biosolids rules (EPA, 1993). While that exercise was specific to biosolids and to selected beneficial uses, the risk assessment methodology is general and flexible enough to be applied to any waste and any anticipated use, provided that a sufficient data base exists. We suggest that, if human exposure to a contaminant like cadmium (Cd) is of sufficient concern to require Cd regulation in biosolids, Cd should also be regulated in other products whose use could cause increased human Cd exposure. Because of differences in the chemistry of Cd in diverse wastes like biosolids, fly ashes, phosphate fertilizer, manures, etc., and in the use patterns of these materials (e.g., application rates), different regulatory limits might be developed for each waste, and for products with commingled wastes. The basis for regulation should be acceptable contaminant levels in the products as used with the assumption of unregulated use, i.e., human exposure would be limited by controlling the quality of the product rather than trying to control product use. Guidelines (labels) for product use would be required as would state registration. This approach is consonant with quality control, labeling and registration in other commercial products. By moving to a uniform omnibus approach to product regulations, the present confusion that exists with differential regulations (or lack of them) will be removed and the opportunity to beneficially use our recyclable wastes will be enhanced.

References

Gilmour, J.T., M.D. Clark and G.C. Sigua. 1985. Estimating net nitrogen mineralization from carbon dioxide evolution. Soil Sci. Soc. J. 49:1398-1402.

Logan, T.J., G.M. Pierzynski and R. Pepperman. 1995. National markets for organic products including biosolids: Opportunities, competition and constraints. Conf. Paper. The 4th Joint WEF & AWWA Residuals/Biosolids Conference. Water Environ. Fed.

Slivka, D.C., T.A. McClure, A.R. Buhr and R. Albrecht. 1992. Compost: United States supply and demand potential. Biomass and Energy. 3:281-299.

U.S. Environmental Protection Agency. 1993. 40 CFR part 503. Standards for the use or disposal of sewage sludge. Federal Register. 58:9387-9404.

Chapter 6

Agricultural and Environmental Issues in the Management of Poultry Wastes: Recent Innovations and Long-Term Challenges

J. Thomas Sims

Department of Plant and Soil Sciences, College of Agricultural Sciences, University of Delaware, 147 Townsend Hall, Newark, DE 19717–1303

Modern poultry production systems face a number of complex environmental challenges. Most poultry operations are agricultural in nature, combining animal and crop production. Unfortunately, the inputs of feed and fertilizer required by concentrated animal operations are greater than the outputs in animal products and harvested crops. This often results in large excesses of nutrients on individual farms and in regions where poultry-based agriculture predominates. Many studies have shown that this can result in losses of nitrogen to groundwaters and phosphorus to surface waters, negatively affecting water quality. Other environmental concerns include the fate of trace elements, hormones, antibiotics, and pesticides added to poultry feed. This paper summarizes recent information on the environmental impact of poultry wastes in the U.S., with a particular emphasis on water quality. It also addresses some recent advances in poultry waste management and existing or proposed measures designed to minimize the environmental impacts of poultry based agriculture.

Environmentally sound management of animal wastes is one of the greatest challenges facing modern production agriculture. Clearly, it has been known for centuries that animal wastes have many beneficial properties that make them useful as soil amendments for agricultural crop production. Among the more important of these are their nutrient value and general contributions to soil quality by adding organic matter which can improve soil structure, tilth, and water relations. Consequently, numerous publications are available that describe the most effective means to handle, store, and apply animal wastes for agricultural production systems (1, 2, 3, 4). This is not to say that all questions about the agricultural value of animal wastes as a soil amendment have been answered; indeed many important questions remain about the basic principles controlling the cycling of nutrients in waste-amended soils. Similarly, research on application technologies for animal wastes that can be more cost-efficient and adapted to a wider range of crops is needed. It has, however, become increasingly apparent that developing management programs that not only maximize the agricultural value of animal wastes, in terms of crop growth and yield, but also minimize animal waste impacts on soil, air, water, and food chain quality, is a more complex challenge. Meeting this challenge will require extensive basic and applied research and comprehensive, integrated planning efforts between a variety of scientific disciplines.

It can be argued that the fundamental cause of many of the environmental problems facing modern animal-based agriculture has been the marked intensification of animal production that has occurred since World War II and particularly in the past 25 years. The driving forces behind intensification of production have been an increased demand for meat products as the global per capita income increases and the economies of scale that are realized in production, processing and transportation by "clustering" large numbers of animals in rather small geographic areas (5). In many cases intensification of animal production has markedly stimulated local economies, providing a stronger tax base, continuing employment opportunities, and generally raising the standard of living in rural areas. Indeed, in some situations, animal-based agriculture has become a major contributor to regional, state, and even national economies. Unfortunately, in many areas rapid intensification of animal production has also strained the assimilative capacity of receiving watersheds and airsheds, creating both short and long-term environmental problems (6). Large inputs of nutrients from other states or regions, as feed grain or feed concentrates, often without any reductions in fertilizer use, commonly create significant nutrient surpluses in areas with dense animal populations. This has usually resulted in serious concerns about the contamination of ground and surface waters with nutrients (nitrate-N and phosphorus) (7, 8, 9) and, more recently, by organic chemicals and growth hormones found in animal wastes (10, 11). Questions have also arisen about the effects on soil quality and food chain safety of some trace elements found in or added to animal feeds, especially arsenic (As), copper (Cu) and zinc (Zn). Not surprisingly, sharp conflicts between "stakeholders" with investments in animal production and those primarily concerned about environmental quality have often occurred. To quote Purvis and Abdallah (6), "..When there was plenty of assimilative capacity, conflicts among stakeholders were rare.. Once overburdening is a problem however, negotiations and compromise become increasingly difficult". Resolving the conflicts between economic value and environmental quality in modern, intensified animal production will not be easy, nor can simple, short-term solutions be expected to prevail in the long-term. A fundamental re-thinking of the management of animal wastes is required, one that moves beyond the farm scale to develop innovative regional, or even national, solutions. Such an effort must involve all "stakeholders" (farmers, integrating agribusinesses, fertilizer industries, advisory and regulatory agencies, environmental advocacy groups, and legislatures) if it is to be successful in the long-term.

Modern poultry production systems face many of the agricultural and environmental challenges described above. Today's poultry operations are highly intensified, "vertically integrated", and profitable components of the agriculture of many states and countries. At the same time a rather extensive body of research suggests that intensification of poultry production is contributing to some serious environmental problems in many areas (*12, 13*). Much has been learned about the most effective means to manage poultry wastes for crop production purposes in the past 30 years. Ongoing research efforts are now identifying advances that make agricultural use of poultry wastes even more cost-effective, in terms of crop production. Unfortunately, many of these advances have only served to exacerbate the problem of nutrient excesses common to areas where poultry operations are often concentrated in small geographic areas. For example, recent improvements in our understanding of nutrient availability in soils amended with poultry wastes, in combination with advances in soil and plant testing and the development of more efficient application technologies, have shown that optimum crop yields can often be obtained with lower application rates of poultry wastes than are now commonly used by most farmers. Obviously, lower application rates to cropland equate readily to larger surpluses on the farm. The long-term success of modern, concentrated poultry production systems requires that environmental quality issues be addressed and that permanent solutions be developed. This paper first addresses the nature of the environmental problems facing poultry-based agriculture and then proposes some fundamental principles that must underlie agriculturally and environmentally sound poultry waste management strategies for agricultural systems.

Environmental Issues in the Management of Poultry Wastes

Nutrients and Water Quality. Most of the environmental problems reported to be associated with poultry waste management have centered around the contamination of ground and/or surface waters with nitrogen (N) and phosphorus (P). While poultry wastes contain appreciable amounts of some other nutrients, such as calcium (Ca), magnesium (Mg), potassium (K), and sulfur (S), and trace quantities of most micronutrients, these elements rarely, if ever, create problems with water quality. Geographic excesses of N and P are common in farms, counties, states, and regions dominated by poultry-based agriculture. Hence, understanding the principles that control the cycling of N and P in soils receiving poultry wastes and that affect their transport to ground or surface waters is critical for both crop production and water quality.

Nutrient Budgets for Poultry-Based Agriculture. Regardless of the nutrient source being used, nutrient management can be a challenging proposition for agricultural crop production. However, the physical and chemical properties of animal wastes and the geographic excesses present significantly intensify this problem for poultry based agriculture. Poultry wastes include a wide variety of materials including manures, litters (mixture of feces, urine, and woodchips or sawdust), dead birds and dead bird composts, and sludges and wastewaters at processing plants. Factors affecting the production and composition of these materials has been reviewed before (*11, 14*) and will not be repeated here. In general, however, poultry wastes applied to agricultural lands (manures, litters, composts) will be heterogeneous, bulky materials of variable chemical composition. For most farmers these wastes are difficult to store, handle, and apply uniformly, although cost-sharing programs for storage facilities and new advances in application equipment are improving this situation somewhat. Analyses of most poultry wastes are conducted infrequently at best, hence accurate knowledge of the waste's nutrient content is usually not available at the time of application. Further complicating the matter is the fact that, due to time and labor constraints, farmers must often apply poultry wastes to fields during the fall and winter months, well in advance of crop requirements, and under soil and climatic conditions that promote nutrient losses (e.g. no crop uptake, frozen ground, during periods of maximum ground water recharge and surface water runoff). Logistical problems, then, are serious constraints to more efficient management of poultry wastes in the short-term.

Of far greater significance to the long-term sustainability of poultry-based agriculture is the question of excess nutrients at the farm, state, and even regional levels *(11, 12)*. Simply put, the input of nutrients to a poultry farm, in animal feed alone, often greatly exceeds the output as farm products (harvested crops, animals, and animal products). The intensification of poultry-grain agriculture on the Delmarva (Delaware-Maryland-Virginia) peninsula, one of the more concentrated poultry production areas in the U.S., illustrates the nature and scale of this problem (Figure 1). Consider a typical, medium-sized poultry farm in Delaware, with 100 ha of cropland and three poultry houses (20,000 bird capacity), producing six flocks of broiler chickens per house per year. Major crop rotations used on a farm such as this would include field corn (*Zea mays* L.), soybeans (*Glycine max*, L., Merr.), double-cropped wheat (*Triticum aestivum* L.)-soybeans, and some vegetable crops for fresh market or processing. It is possible, without great difficulty, to construct a mass nutrient balance for N and P on this farm simply by computing the quantities of nutrients entering the farm as feed, fertilizers, and biological N fixation by leguminous crops (soybeans) and subtracting the quantities leaving the farm in grain and animal products (Table I). Other assumptions required are crop yield goals and estimated soil test levels (available from University of Delaware Soil Test Summaries; *15*), both needed for fertilizer recommendations, estimated hectares of each crop, and animal mortality. Results of such an analysis (Table I) show that annual excesses of N and P, without fertilizer purchases, are about 220 and 80 kg/ha/year. Any purchased fertilizer would increase farm-wide surpluses of N and P proportionately.

Table I. Mass nutrient balance for a typical poultry-grain farm in Delaware

Type of Farm Input or Output	Nitrogen	Phosphorus
Inputs	--------------Mg/farm------------	
Broiler feed	50	11
Fertilizers	5	2
Biological N fixation	6	0
Outputs		
Animals sold	24	2
Harvested crops	10	1
Nutrient balance (Inputs - outputs)	+27	+10
Nutrient balance w/o fertilizers	+22	+ 8

The fate of the excess N and P on this "typical" farm is at the crux of the environmental issue. Certainly not all of the excess will be lost to ground and surface waters by leaching, erosion, and runoff. For instance, as much as 20-30% of the N input as feed has been shown to be lost to the atmosphere via volatilization directly from the poultry production houses (*16*) and more would likely be lost by this mechanism during storage, handling, and field application (*17*). Unfortunately, given ongoing concerns about the effect of ammonia on poultry health in the production house, air quality in this region, and the potential for deposition of volatilized ammonia in rainfall falling upon the Chesapeake or Delaware Bays, this form of N loss cannot be regarded as desirable or beneficial. Nitrogen remaining on the farm in excess of gaseous losses and crop removal will enter the soil N cycle and, depending on farming practices and soil conditions, be immobilized in organic forms or gradually runoff or leach from the soil as nitrate-N. Phosphorus, because it is not lost from soils as a gas, will accumulate with time and enhance the potential for surface water degradation if it is lost from fields via erosion, runoff, and leaching/lateral subsurface flow in drainage waters. Recent soil test summaries for Delaware confirm the build-up of soil P to high levels in areas dominated by poultry-grain agriculture. Summaries from Delaware in 1994-1995 (n=2780 samples) showed that 29% of the soils tested by the University of Delaware for Sussex County, DE (site of the poultry industry) were rated as "optimum" (no fertilizer P required) and 61% were rated as "excessive" (soil test P ≥ twice the optimum value for agronomic crops). Given the annual excesses of N and P present in many poultry operations, prudence would dictate that the fate and transport of these nutrients be carefully studied and documented and that management practices for poultry wastes be developed to minimize losses of N and P to air and water.

Nitrogen in Poultry Wastes and Water Quality. Efficient N management is a critical component of an effective nutrient management plan for poultry wastes. Today, most land application programs for poultry wastes are based on N because of the importance of N to crop growth and because contamination of ground and surface waters by nitrate-N is a major environmental concern. As will be discussed below, the need for P-based management of poultry wastes is an emerging issue, one that would drastically change the nature of most land application programs. A large body of research exists on N cycling in soils amended with poultry wastes, most of which has focused on the factors controlling the rate and timing of N *mineralization* (conversion of organic N to ammonium-N), *nitrification* (transformation of ammonium-N to nitrate-N), the gaseous loss of ammonia-N from soils by *volatilization* or nitrate-N by *denitrification*, and the downward movement of nitrate-N through soils in percolating waters (*leaching*). Quantifying the amount of mineralizable N in poultry wastes, and the timing of the mineralization process, is essential to determine the correct application rates and timing of these wastes needed to minimize gaseous losses and nitrate leaching. Laboratory, greenhouse, and field studies have examined the influence of soil properties, soil management, and treatment of poultry wastes on N cycling and loss. Since a number of review articles provide considerable detail on these studies (*11, 18, 19*), only a few will be discussed here, primarily to illustrate the principles that must underlie environmentally sound N management of poultry wastes.

The approach most commonly recommended for N-based management of poultry wastes includes the following steps: (i) establish N requirement of crop to be grown based on realistic yield potential for the area; (ii) estimate N that will be provided during the growing season of the crop by mineralization of organic N in poultry waste and by recovery of inorganic N already present in the waste (NH_3-N and NO_3-N); (iii) estimate loss of N that is likely to occur due to inefficiency of soil-crop system (e.g. leaching, volatilization, denitrification); (iv) estimate N that will be provided from previous applications of poultry wastes (or other N sources), using soil and plant N tests; (v) select poultry waste application rate required to provide crop N requirement, factoring in residual N available from soil and potential N losses; and (vi) apply waste at proper time of year, in synchrony with crop N uptake patterns, using well-calibrated application equipment. Nitrogen requirements of most crops at realistic yield goals are well known and are usually updated regularly by land grant universities. Additionally, N cycling in soils amended with poultry wastes has been studied sufficiently to allow for reasonable predictions of the amount of plant available N provided. These studies have shown that many poultry wastes contain large percentages of rapidly mineralizable organic N and that as much as 20-30% of the total N can be found in ammoniacal forms that are easily lost by volatilization (Figure 2a and 2b). These two factors create the potential for appreciable losses of N and must be considered in the selection of an agronomically and environmentally effective N rate. High concentrations of nitrate-N are commonly found in soils shortly after waste application as the labile organic forms quickly mineralize (Figure 2a; *20*). If the waste application rate is in excess of crop needs or if the waste is applied at the incorrect time of year this can result in NO_3-N leaching losses and increase the potential for ground water contamination (Figure 3; *22*). Similarly, significant (as much as 30-50% of total N applied) losses of NH_3-N can occur during application of poultry wastes or if the waste is left on the soil surface for extended periods of time (Figure 2b; *21*). Nitrogen volatilization not only equates to a loss of plant available N but causes environmental concerns about air and water quality as well. Some recent studies have shown that chemical amendments, such as alum (aluminum sulfate) can markedly reduce NH_3-N losses from poultry litters by volatilization, improving animal health and increasing the fertilizer N value of these wastes (Figure 4; *16*).

Recent advances in soil and plant N testing have increased the efficiency of N use in soils amended with poultry wastes. New tests such as the *presidedress soil nitrate test*, the *leaf chlorophyll meter*, and the *stalk nitrate test* have enabled farmers to more accurately identify the rate of poultry waste needed for economically optimum crop yields (*23*). Similarly, new equipment has been developed to apply poultry wastes in more efficient manners and at lower rates, such as that designed to allow for sidedressing of poultry litters to row crops during the growing season rather than broadcasting them in advance of planting (*24*). Sidedressing N sources is well known to be a more efficient application practice because it more closely matches the timing of N availability in the soil with the period of maximum uptake for many annual crops. Combining new soil and plant tests with advances in equipment design definitely improves the N use efficiency of poultry wastes However, as noted earlier, this also usually means that farmers need lower application rates of poultry wastes than they have traditionally applied to attain optimum yields, further exacerbating the problem of nutrient excesses on some farms.

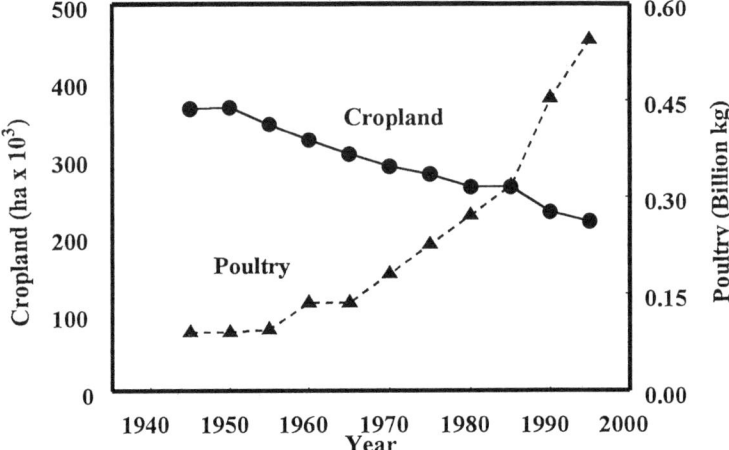

Figure 1. Trends in poultry production and amount of cropland in the state of Delaware (1945-1994).

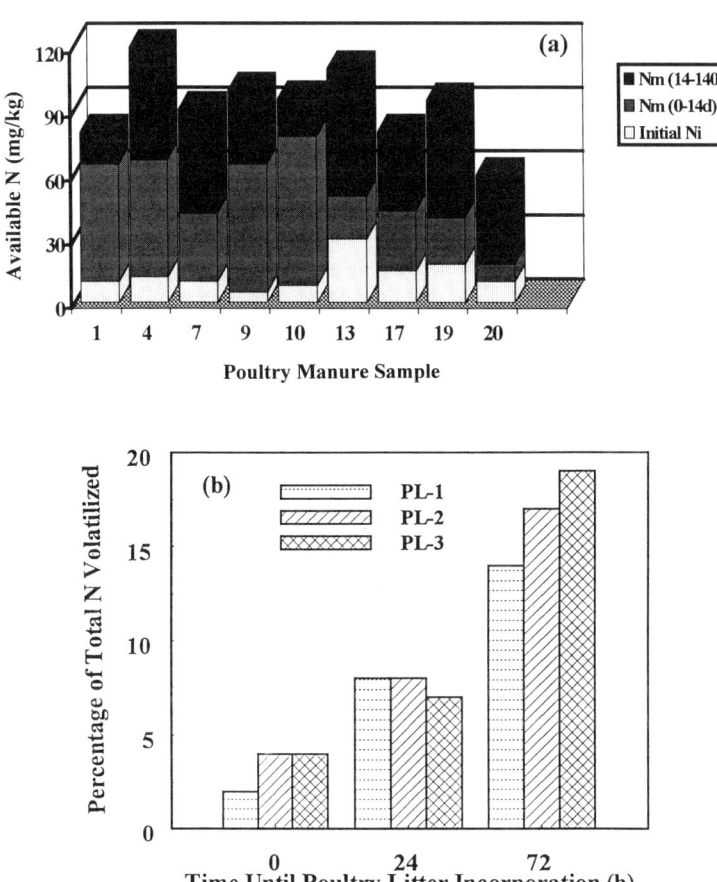

Figure 2. Nitrogen transformations in agricultural soils amended with poultry litters. (a) mineralization patterns for poultry litters (20) and (b) influence of time until incorporation of poultry litters on NH_3 volatilization losses (21).

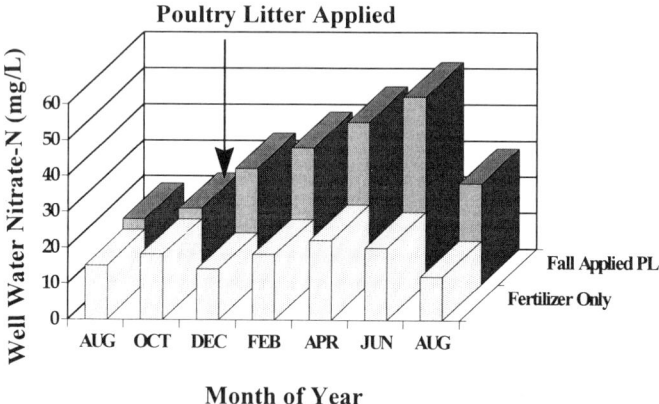

Figure 3. Influence of fall applications of poultry litter on well water nitrate-N concentrations in the Atlantic Coastal Plain of the U.S.(*22*)

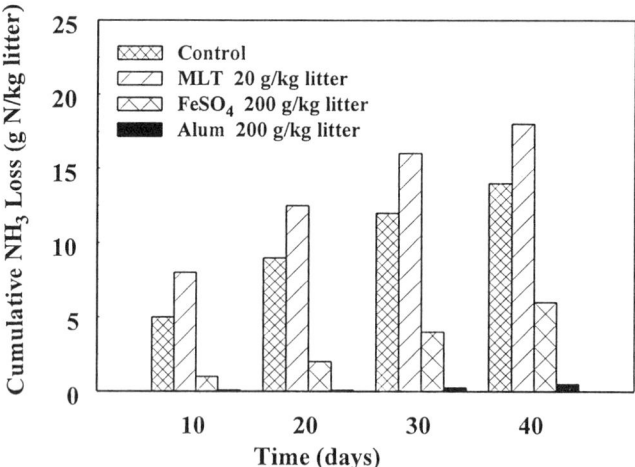

Figure 4. Effect of chemical amendment of poultry litter on NH_3 volatilization losses under laboratory conditions (*16*).

Phosphorus in Poultry Wastes and Water Quality. Phosphorus is well-known to be a major contributing factor to the *eutrophication* (nutrient enrichment) of surface waters. A number of undesirable effects on water quality are associated with eutrophication, including increased growth of algae and undesirable aquatic weeds, depletion in dissolved oxygen, and fishkills when the algal biomass decomposes, foul odors, sedimentation and restriction of navigation, and surface scums that impair recreational uses. Nonpoint source pollution of surface waters by P occurs when soluble or soil-bound P is transported to these waters by erosion (soil-P) or runoff (soluble-P). Minimizing the accumulation of excessive P in soils and reducing P transport are thus essential soil management practices in areas with surface waters that are sensitive to eutrophication. Unfortunately, as mentioned above, a number of studies have shown that long-term applications of poultry wastes to agricultural cropland based on crop N requirements inevitably builds soil P to levels well beyond crop requirements. The unfavorable P:N ratio in poultry wastes, relative to the P:N ratio in the harvested portion of most agronomic crops, almost always results in the application of excess P to soils. For example, application of poultry litter at the rate typically recommended for corn production in the Atlantic Coastal Plain of the U.S. (5 Mg/ha, dry weight basis), adds ~135 kg P/ha to the soil, relative to P removal of ~25 kg P/ha in harvested corn grain. Since few alternatives to land application of poultry wastes are now economically viable, understanding and controlling the loss of P from soils amended with these wastes is of considerable importance if we are to avoid eutrophication of surface waters.

Several recent studies illustrate the complex nature of the P problem in areas dominated by poultry-based agriculture. Poultry production in Arkansas is primarily concentrated in a mountainous agricultural area dominated by pasture land and used for beef production. Continued applications of broiler litter to pastures have markedly increased soil test P concentrations and, as recent studies have shown, the potential for P losses to surface waters in erosion and runoff (Figure 5a; *25*), raising concerns about the eutrophication of streams, rivers, and reservoirs in the area. Encouragingly, studies using alum to reduce ammonia losses from poultry litters (*16*) have also shown that alum reduced P loss in runoff from pastures (Figure 5b; *26*). In a similar, but slightly different manner, increases in soil test P from long-term applications of broiler litter to cropland in the coastal plain of southern Delaware are emerging as a water quality issue. While the flat topography of southern Delaware creates minimal risks for surface erosion and runoff, an extensive ditch drainage system exists in some poultry-grain areas with ditch waters discharging into streams and a nearby national estuary. Although P losses in leaching and drainage has usually been assumed to be a minor issue, recent studies in Western Europe (*27, 28*) have shown that as soil profiles become progressively "saturated" with P the downward movement and lateral flow of soluble P into drainage waters can occur. A "degree of P saturation" of 25% of the total P sorption capacity has been proposed as an upper limit for Dutch soils. Recent studies in Delaware (Sims, unpublished data) have shown that the upper portions of soil profiles in areas with high P, manured soils, can exceed this limit (Table II). The question now being asked is whether P transport via these drainage ditches is a significant environmental problem and, if so, what should be done to minimize P loss in agricultural drainage.

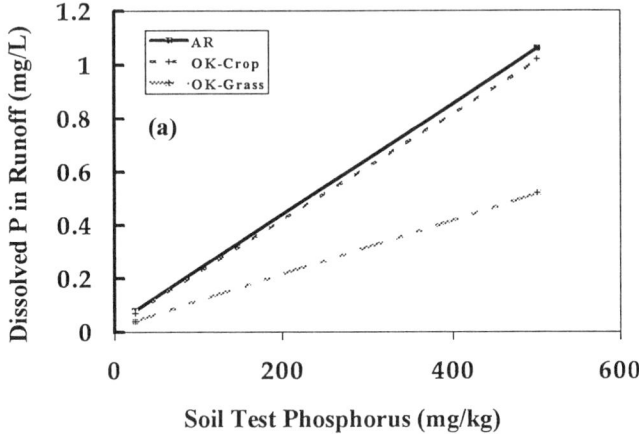

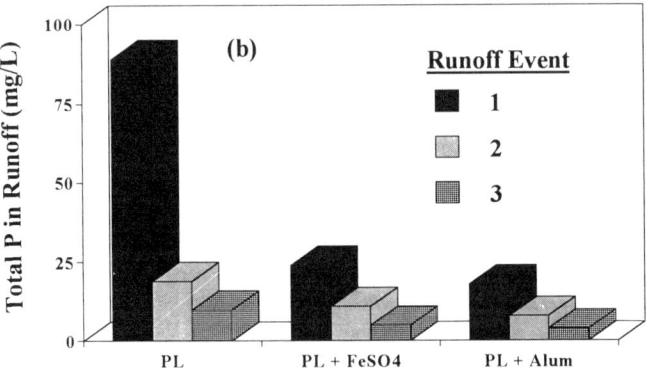

Figure 5. Phosphorus losses from poultry litter amended soils. (a) Relationship between soil test P and P losses in runoff from soils amended with poultry wastes (25) and (b) effect of chemical amendments of poultry litter on P losses in runoff from pastures (26)

Table II. Effects of long-term fertilization and manuring on the degree of P saturation (DPS) of two Delaware soils (Sims, unpublished data)

Moderately Drained Soil			Poorly Drained Soil		
Depth	DPS-OX[†]	DPS-ST[‡]	Depth	DPS-OX	DPS-ST
---cm---	---------------%--------------		----cm--	---------------%--------------	
0-20	85	65	0-15	32	9
60-90	22	17	30-45	8	1
125-150	14	12	55-85	5	1
165-190	4	1	90-120	1	1

[†]DPS-Ox= Degree of P saturation estimated as oxalate extractable P divided by one-half of oxalate extractable (Fe + Al).
[‡]DPS-ST= Degree of P saturation estimated as soil test P divided by P sorption maximum calculated from Langmuir sorption isotherms.

The larger problem facing both areas is the fact that annual applications of broiler litter continue to build soil P values even higher. Restricting land application of poultry wastes to soils that need P for crop production, however, would be extremely problematic and economically devastating to the poultry industry. For example, in southern Delaware, much of the land base would be considered unsuitable for poultry waste application, based strictly on crop P requirements, because >80-90% of the agricultural soils are rated as "optimum" or "excessive" -- i.e. little, if any, P is required for adequate crop growth and yields. Since other alternatives to land application do not exist at this point, the use of P as a "land limiting constituent" for poultry wastes would create enormous problems for poultry-based agriculture in this and many other areas. Even if P applications in poultry wastes were to cease immediately, long-term studies have shown that it may be decades before some high P soils are depleted by crop uptake from "excessive" to "optimum" levels (29). Thus, controlling P losses from waste-amended soils in the short-term, while important, cannot ignore the need for long-term remediation strategies for high P soils, nor the importance of developing other alternatives to land application in areas with large excesses of P.

Environmental Concerns with Other Constituents in Poultry Wastes. Poultry-based agriculture has as its primary goal the profitable production of poultry products (e.g. animals, eggs). Optimizing animal nutrition and overall health and minimizing the incidence of disease are thus of vital importance. As a result, a number of feed additives are used to promote animal growth and prevent the outbreak of disease. Among these are growth hormones, antibiotics, organic chemicals (e.g. insecticides), and feed additives containing trace elements such as arsenic (As), copper (Cu), and zinc (Zn). Many of these materials pass through the animal in the feces or urine and are present at varying concentrations in poultry wastes that are applied to agricultural cropland. Particular concerns exist with As, which has been reported to be present in broiler litters at concentrations that exceed regulatory limits established for land application of municipal sewage sludges (41 mg/kg; *11*). Poultry wastes also contain a very large and diverse population of pathogenic microorganisms (viruses, bacteria, fungi, protozoa), some of which can create serious health problems for humans if they enter ground or surface waters (e.g. fecal coliforms, *E. Coli*). Very little research has been conducted to quantify the effect of poultry waste management on the amount and form of organics, trace elements, and pathogenic microorganisms, and even less on their fate and mobility in soils. Reviews of the limited research available suggests that losses of some of these waste constituents from soils may occur, albeit usually at very low concentrations (*11, 19*). However, at this time there is little evidence of serious environmental problems caused by the feed additives used in poultry production and the pathogenic microorganisms present in poultry wastes when the wastes are applied following accepted agricultural practices. Given their potential environmental impacts, continued research into the effects of low concentrations of these waste constituents on aquatic biota and human health is probably justified, as are studies on their long-term fate in soils that are consistently and continuously amended with poultry wastes.

Environmentally Sound Management of Poultry Wastes in Agriculture

The complexity of modern poultry-based agriculture requires the development of comprehensive, multi-disciplinary plans to ensure that the wastes generated by this industry are used in a safe and beneficial manner. Although the vast majority of poultry wastes produced today are applied to agricultural cropland, there are other options available to re-use these wastes. Given the large excesses of nutrients present in some areas, these other options merit serious consideration as long-term alternatives to traditional land application for crop production. Since the planning efforts required for these other options will be fundamentally different than those needed for land application programs, they will be discussed separately.

Agricultural Crop Production: Land Application Programs. A successful, environmentally sound land application program for poultry wastes should have, as a minimum, the following components; others may be required in geographic areas with site specific environmental concerns.

State or Regional Assessment of Major Environmental Issues. All too often complex, highly detailed land management plans have been developed for poultry-based agriculture without a serious discussion of the environmental issues by all concerned parties. This not only neglects the input of some rather important stakeholders, but often alienates some and divides the efforts of others. Clearly the first step required for a successful, long-term plan is to convene a meeting of all groups and individuals involved to discuss the problems and potential solutions. Given the fact that there will undoubtedly be intense disagreements as to how serious some problems are and the feasibility of many proposed solutions, this must be an ongoing debate, led by a neutral party respected by all participants. The goal should be to prioritize the nature of the environmental problems that must be resolved to ensure the sustainability of poultry-based agriculture in a region. From this consensus can arise the research, technology transfer, economic support programs, and public education needed to sustain the industry.

Comprehensive Nutrient Management Plans, at Appropriate Scales. Previous studies have clearly shown that the first, and perhaps most important, environmental issue that must be resolved will be the impact of nutrients in poultry wastes on water quality. Accordingly, it is imperative to develop farm, county, state, and regional nutrient balances that identify the nature and scale of the problem of nutrient excess. This is an important first step because it immediately identifies areas where land application is unlikely to be the only option required (because of the presence of large nutrient excesses) and other areas (usually remote from the poultry production area) where nutrient deficits may exist and poultry wastes could be used more efficiently. It can also identify the magnitude of excess poultry wastes that may need to be disposed of by means other than land application. Field and farm-scale nutrient management plans for agricultural crop production should also be developed and implemented as much as possible. However, farm-scale nutrient management planning cannot resolve the issue of nutrient excess; it can only provide a systematic approach to efficiently use nutrients given the assimilative capacity of available cropland. County, state, and regional nutrient management plans are also required to find cost-effective alternative uses for the excess nutrients on a farm, or in a region.

Effective Educational and Technology Transfer Programs. The importance of nutrient and animal waste management to production agriculture and environmental quality has resulted in the development of a large body of educational information on the proper use of animal wastes. Fact sheets, extension bulletins on "manure management", videotapes illustrating field application of wastes, and computer programs that calculate proper application rates for a wide range of cropping systems are now readily available to most farmers. Public workshops, live demonstrations of equipment calibration and composting practices, and the increased involvement of agricultural consultants in animal waste management are all signs that an effective network now exists to transfer new and existing information to end-users. Probably the greatest need in this area at the present time is to find out if the intended users (i.e. the farmers) are following the waste management advice provided and, if not, what constrains them from doing so.

Ongoing Basic and Applied Waste Management Research. It can certainly be argued that the advances in agricultural productivity seen in the past few decades have largely resulted because of an intensive basic and applied research effort. It is absolutely critical to continue and even expand upon this research effort, given the very difficult environmental challenges faced by poultry-based agriculture. For example, one of the most difficult questions now being asked is the need for regulating land application of poultry wastes based on P, not N. If this proves to be necessary significant and costly changes will be required in land application programs. More research is needed to determine the fate and transport of P in soils considered "excessive" due to long-term application of animal wastes. Do all soils rated as agronomically excessive in P create environmental risk? If not, how do we identify those that do? Similar questions can be asked about the environmental impact of other waste constituents, such as growth hormones, antibiotics, some trace elements (As, Cu, Zn) and pathogenic microorganisms. Many other research areas exist, ranging from the need for better waste and soil tests that can predict environmental impacts of animal wastes (30) to the need for improved application equipment that can more efficiently apply the wastes to cropland. Beyond the normal realm of agricultural research there is also a need for socio-economic studies that can help resolve conflicts among stakeholders in the animal-agriculture industry and those concerned about the impact of this industry on the environment. Continued support for research is essential to the long-term viability of animal-based agriculture.

State or Regional Environmental Coordinating and Review Committee. If we are truly to be successful in developing environmentally sound land application programs for poultry-based agriculture, there must be an ongoing, critical review of the steps taken. This should be conducted by an established committee that includes representatives from all concerned parties, one that should meet regularly and have clearly defined criteria for success. This group should take the lead in identifying promising alternatives, failing programs, and conflicts that create major stumbling blocks to success. By bringing together all stakeholders an ongoing consensus can be developed and used to provide the influence needed to institute significant changes rather than continuing to work around the edges of these problems.

Beneficial Use of Poultry Wastes other than Land Application. It has become increasingly recognized in some areas where highly unfavorable nutrient balances exist (i.e. large excesses of N and P) that the agricultural land base will be inadequate for proper, environmentally acceptable use of all the poultry waste generated. In scenarios such as these what is needed is a frank and realistic assessment of the economic costs and environmental gains of the alternative options now available for poultry wastes. Since there are few options available that do not require a rather large infrastructural investment, it should probably be expected that land application will predominate for the near future. However, this does not obviate the need to aggressively pursue other options that can ensure the long-term sustainability and economic contributions of poultry-based agriculture. Some of these options are presented below.

Integration of Poultry Wastes into Fertilizer Manufacture. One of the seemingly most logical alternatives for many poultry wastes (e.g. manures and litters) is to blend them into the production stream for existing nutrient sources (i.e. the fertilizer industry). In this process the low nutrient analysis poultry wastes are enriched by blending them with inorganic fertilizers and thus converted into "value-added" fertilizer materials that have better physical properties, more uniform chemical compositions, and higher nutrient contents than the original wastes. These materials would be designed for both large-scale agricultural crop production and specialty fertilizer markets. This approach has been tried in some countries and U.S. states but with only limited success. Very little research is available on the agronomic value of "pelletized" or "organically-enriched" fertilizers made using poultry wastes (31). The chief obstacles to this alternative are economic: farmers are reluctant to part with nutrient sources that are proven to be successful (manures, litters) without monetary compensation, even if they are known to be present in excess quantities on the farm. Similarly, fertilizer manufacturers are reluctant to pay for non-traditional "raw materials" that require changes in storage, handling, and processing infrastructures. Other concerns include the possible transmission of animal pathogens (e.g. poultry viruses) from one farm to the next in poultry-based fertilizers and the need for an educational program to persuade farmers and their advisors to pay a premium for fertilizers manufactured in this manner to ensure that the fertilizer industry recoups initial investment costs. For this approach to be successful it seems likely that state or local governments will need to provide economic incentives at least to farmers and perhaps to the fertilizer industry as well.

Composting. Many state and local governments have constructed composting operations to deal with organic wastes such as sewage sludges, municipal solid wastes, and yard wastes (leaves, lawn clippings). Since, from a nitrogen availability perspective, composts are highly stable materials used primarily to provide organic matter to soils, higher application rates can be applied and land uses other than production agriculture considered. Build-up of P to excessive levels, however, can still be an environmental concern in compost-amended soils. Composting of agricultural wastes has received little attention with the exception of "dead bird composting" that has become a standard practice in some areas to dispose of poultry mortality (*32*). Major constraints to the use of composting to deal with excess poultry wastes are the transportation costs required to deliver the wastes to a centralized composting and marketing facility, concerns about disease transmission (especially of viruses which may survive temperatures associated with most composting processes), and the highly competitive marketplace where agricultural composts would need to shown as superior to established urban-type composts to provide a profitable economic return on the composting investment. Nevertheless because of the intense interest in composting in the U.S. and other countries, this process may provide a viable end-use for at least some poultry wastes.

Animal Feeding Programs. The use of poultry wastes as supplements to feed for cattle (beef and dairy), poultry (broilers and layers), sheep, and swine has been successfully reported (*19, 33, 34*). The wastes are usually dehydrated prior to feeding and often mixed with other grains or silage to improve their nutritional value. For this approach to be economically successful the animals to be fed must be in reasonably close proximity to the poultry operation and the feed mill must be prepared to incorporate the wastes into the existing feed production process. Logistics and cost, again, will be the primary determinants of success. Other concerns reported include the effects of pathogenic organisms, growth hormones, medications, and some trace elements (especially As and Cu) on the health of the animals ingesting the feed. Given these problems it seems unlikely that refeeding will be a large scale end use for poultry wastes, except in localized areas with large and diverse animal production operations.

Bio-energy. Increased interest in renewable sources of energy has raised the question of the use of poultry wastes to produce energy sources, either methane or electricity. Despite reports of the successful use of poultry wastes in the production of biogases that can be used for direct burning as a fuel source, few large-scale operations that use poultry wastes in this manner exist today. Many logistical problems have been reported that preclude farmers from investing the time and money into biogas facilities, most associated with the difficulty of maintaining an anaerobic environment in the biogas digester. Economic analyses of this end-use of poultry wastes have usually been rather unfavorable as well (*19, 35*). In some European countries the drier poultry wastes (e.g. litters) have been used to produce fuel pellets that can be burned in coal-fired electrical plants. Location of a centralized facility to produce fuel pellets for electric power plants seems more likely to be successful in the long run than constructing complex and difficult to maintain biogas generators on individual farms. As with fertilizer manufacture, however, monetary compensation of farmers for poultry wastes will be required as will economic incentives to those desiring to construct and operate a fuel pelletization facility.

Conclusions and Future Directions

The environmental problems faced by modern poultry-based agriculture are formidable. The most immediate challenge is the need to minimize the impacts of nutrients, especially nitrogen and phosphorus, in poultry wastes on ground and surface water quality. At the same time it must be recognized that the large geographic excesses of nutrients present in areas of intensified poultry production represent a serious, long-term problem, one that cannot be resolved by focusing solely on developing "best management practices" to improve the efficiency of land application of poultry wastes to agricultural cropland. Alternative uses, such as incorporation of waste nutrients into fertilizer materials, composting, re-feeding, and energy production should be investigated more thoroughly for areas of nutrient excess. Assessing the environmental impacts of growth hormones, pesticides, trace elements, medications, and pathogenic microorganisms in poultry wastes should receive greater research efforts in the future to determine the short and long-term environmental risks posed by these waste constituents.

Literature Cited

1. Overcash, M. R., Humenik, F. J. and Miner, J. R. Introduction to livestock waste management. In: *CRC Livestock Waste Management*; M. R. Overcash et al., Eds.; CRC Press, New York, New York, **1983**, Vol. 2; pp. 114-182.
2. U.S. Department of Agriculture Soil Conservation Service. *Agricultural Waste Management Field Handbook;* Part 651. U. S. Govt. Printing Office, Washington, D.C. **1992**.
3. U.S. Department of Agriculture. *Animal Waste Utilization on Cropland and Pastureland;* Utilization Research Report No. 6; Science and Education Administration, Washington, D. C. **1979**.
4. Sims, J. T. Animal Waste Management; In: *Encyclopedia of Agricultural Science;* Academic Press, New York, New York, **1995**, Vol 1., pp. 185-201.
5. Narrod, C. and Pray, C. Technology transfer in the poultry industry: Factors associated with increased production. In: *Animal Waste and the Land-Water Interface;* K. L. Steele, Ed.; Lewis Publishers, Boca Raton, FL, pp. 523-532, **1995**.

6. Purvis, A. and Abdallah, C. W. Analyzing manure management policy: Toward improved communication and cross-disciplinary research. In: *Animal Waste and the Land-Water Interface*; K. L. Steele, Ed.; Lewis Publishers, Boca Raton, FL, pp. 533-544, **1995**.
7. Correll, D. L., Jordan, T. E., and Weller, D. E. Livestock and pasture effects on the water quality of Chesapeake Bay watershed streams. In: *Animal Waste and the Land-Water Interface*; K. L. Steele, Ed.; Lewis Publishers, Boca Raton, FL, pp. 107-118, **1995**.
8. Ritter, W. F. and Chirnside, A. E. M. Influence of agricultural management practices on nitrates in the water table aquifer. *Biol. Wastes.* **1987**, 19, 165-178.
9. Sharpley, A. N., Chapra, S. C., Wedepohl, R., Sims, J. T., Daniel, T. C., and Reddy, K. R. Managing agricultural phosphorus for protection of surface waters: Issues and options. *J. Environ. Qual.* **1995**, 23:437-451.
10. Shore, L. S. and Correll, D. L. Relationship of fertilization with chicken manure and concentrations of estrogens in streams. In: *Animal Waste and the Land-Water Interface*; K. L. Steele, Ed.; Lewis Publishers, Boca Raton, FL, pp. 155-162, **1995**.
11. Sims, J. T. and Wolf, D. C. Poultry waste management: Agricultural and Environmental Issues. In: *Advances in Agronomy*; D. L. Sparks, Ed.; Academic Press, New York, New York, **1993**, Vol. 52, pp. 1-83.
12. Narrod, C., Reynnells, R. and Wells, H. Potential options for poultry waste utilization: A case study of the Delmarva Peninsula. *USDA/EPA White Paper, Washington, D.C.*, **1993**.
13. Gassman, P. W. and Bouzaher, A. Livestock pollution: Lessons from the European Union. In: *Animal Waste and the Land-Water Interface*; K. L. Steele, Ed.; Lewis Publishers, Boca Raton, FL, pp. 515-522, **1995**.
14. Simpson, T. W. Agronomic use of poultry industry waste. *Poultry Sci.* **1990**, 70, 1126-1131.
15. Sims, J. T. and Schilke-Gartley, K. L. *The University of Delaware Soil Testing Program: History, Philosophy, and Value. A Thirty-Seven Year Summary (1957-1993).* Coop. Bull. No. 45, **1993**. 120 pp., Univ. of Delaware, Newark, DE.
16. Moore, P. A., Daniel, T. C., Edward, D. R., and Miller, D. M. Effect of chemical amendments on ammonia volatilization from poultry litter. *J. Environ. Qual.*, **1995**, 24:293-300.
17. Wolf, D. C., Gilmour, J. T., and Gale, P. M. **1988**. *Estimating potential ground and surface water pollution from land application of poultry litter-II.* Publ. No. 137. Arkansas Water Resources Research Center, Fayetteville, AR.
18. Sims, J. T. Organic wastes as alternative nitrogen sources. In: *Nitrogen Fertilization in the Environment*; P. E. Bacon, Ed.; Marcel Dekker, Inc., New York, New York, 13, 487-535, **1995**.
19. Edwards, D. R. and Daniel, T. C. Environmental impacts of on-farm poultry waste disposal -- A review. *Bioresourc. Technol.* **1992**, 41, 9-33.
20. Bitzer, C. C. and Sims, J. T. Estimating the availability of nitrogen in poultry manure through laboratory and field studies. *J. Environ. Qual.*, **1988**, 17, 47-54.

21. Schilke-Gartley, K. L. and Sims, J. T. Ammonia volatilization from poultry manure-amended soil. *Biol. Fert. Soils.* **1993**, 16, 5-10.
22. Weil, R. R., Weismiller, R. A., and Turner, R. S. Nitrate contamination of groundwater under irrigated coastal plain soils. *J. Environ. Qual.* **1990**, 19, 441-448.
23. Sims, J. T., Vasilas, B. L., Gartley, K. L., Milliken, B. and Green, V. Evaluation of soil and plant nitrate tests for maize on manured soils of the Atlantic Coastal Plain. *Agron. J.*, **1995**, 87, 213-222.
24. Glancey, J. L. and Adams, R. K. An applicator for side-dressing row crops with solid wastes. *Trans. ASAE,* **1996**, 39(3):829-835.
25. Sharpley, A. N. Dependence of runoff phosphorus on extractable soil phosphorus. *J. Environ. Qual.* **1995**, 24, 920-926.
26. Shreve, B. R., Moore, P. A., Daniel, T. C., Edward, D. R., and Miller, D. M. Reduction of phosphorus in runoff from field-applied poultry litter using chemical amendments. *J. Environ. Qual.* **1995**, 24, 106-111.
27. Breeuswma, A., Rekjerink, J. G. A.and Schoumanns, O. Impact of manure on accumulation and leaching of phosphate in areas of intensive livestock farming. In: *Animal Waste and the Land-Water Interface*; K. L. Steele, Ed.; Lewis Publishers, Boca Raton, FL, pp. 239-250, **1995**.
28. Lookman, R., Jansen, K., Merckx, R. and Vlassak, K. Relationship between soil properties and phosphate saturation parameters: A transect study in northern Belgium. *Geoderma,* **1996**, 69, 265-274.
29. McCollum, R. E. Buildup and decline in soil phosphorus: 30 year trends on a Typic Umprabult. *Agron. J.,* **1991**, 83, 77-85.
30. Sims, J. T. Characterization of animal wastes and waste-amended soils: An overview of the agricultural and environmental issues. In: *Animal Waste and the Land-Water Interface*; K. L. Steele, Ed.; Lewis Publishers, Boca Raton, FL, pp. 1-14, **1995**.
31. Hamilton, C. M. and Sims, J. T. Nitrogen and phosphorus availability in enriched, pelletized poultry litters. *J. Sust. Agric.* **1995**, 5(3), 115-132.
32. Carr, L., Grover, R., Smith, B., Richard, T., and Halbach, T. Commercial and on-farm production and marketing of animal waste compost products. In: *Animal Waste and the Land-Water Interface*; K. L. Steele, Ed.; Lewis Publishers, Boca Raton, FL, pp. 485-492, **1995**.
33. McCaskey, T. A. Feeding broiler litter as an alternative waste management strategy. In: *Animal Waste and the Land-Water Interface*; K. L. Steele, Ed.; Lewis Publishers, Boca Raton, FL, pp. 493-502, **1995**.
34. Fontenot, J. P., Smith, L. W., and Sutton, A. L. Alternative utilization of animal wastes. *J. Animal Sci.* **1983**, 57, 222-333.
35. Badger, P. C., Lindsey, J. K., and Veitch, J. D. Energy production from animal wastes. In: *Animal Waste and the Land-Water Interface*; K. L. Steele, Ed.; Lewis Publishers, Boca Raton, FL, pp. 475-484, **1995**.

Chapter 7

Agricultural and Environmental Issues in the Management of Cattle Manure

H. H. Van Horn and M. B. Hall

Department of Dairy and Poultry Sciences, Institute of Food and Agricultural Sciences, University of Florida, P.O. Box 110920, Gainesville, FL 32611–0920

Manure management systems should account for the fate of excreted nutrients that are of environmental concern. Currently, N and P are most monitored to assure water quality. Land application of manure at acceptable levels to fertilize crops is the basis of most systems. Cattle-producing farms with insufficient crop production potential to recycle manure nutrients need to reduce excretions if possible and develop affordable systems to concentrate manure nutrients to reduce hauling costs and possibly produce a salable product. Composting selected manure solids oxidizes most odorous volatiles, reduces volume, and encourages combination with other organic wastes. Solutions to odor problems are needed. Generating energy from manure organic matter via anaerobic digestion reduces atmospheric emissions of methane and odors but, thus far, has not been economical.

Manure nutrients and other decaying organic matter are natural components of the environment that ultimately are recycled in the production of more plant and animal tissue. Although often called wastes, they are in fact resources. Historically, manure resources were in short supply, valued, and used for fertilizer or fuel. However, in many large, food-animal production units manure nutrients are in excess to fertilizer needs on the farm and, when they are unused, potentially could result in detrimental environmental effects through 1) nutrient losses to ground or surface waters or 2) gaseous losses that reduce air quality or are nuisance odors affecting people living near to animal production units.

Most regions of the world with intensive, domestic livestock production have begun monitoring farms to ensure that losses of nutrients to the environ-

NOTE: Florida Agricultural Experiment Station Journal Series Number R–05350

© 1997 American Chemical Society

ment is avoided (1). Emissions of odorous compounds are regulated in all US states through nuisance legislation and, in several states, through odor measurements taken at the property line (2). Additional regulations sometimes include standards for volatile emissions of ammonia (e.g., in The Netherlands), and studies of methane emissions (3) may lead to regulatory oversight in the future.

Animal manures are unavoidable if we choose to use animal products and to utilize some of the world's fibrous feed resources that are indigestible by humans. The objective of this paper is to review many of the manure-related issues and some of the environmentally accountable manure management alternatives that cattle producers can utilize to produce animal foods. The outline chosen is:
1. Nutrient management to ensure water quality
2. Air quality issues
3. Partitioning and processing alternatives
4. Discussion: Debits and credits associated with cattle manures

Nutrient Management to Ensure Water Quality

Currently, N and P are the primary plant nutrients of concern with regard to water quality. Applying manure to supply nutrients to plants is an ancient method of recycling. To avoid excessive applications of N and P that become risks with regard to contamination of groundwater or surface runoff, food-animal producers must budget nutrient applications to fields to match expected crop production needs. Budgeting requires accurate estimates of manure nutrient excretion and losses during storage, nutrient removals by crops to be produced with manure used as a fertilizer, and expected, acceptable losses and carryover soil reserves. Whole-farm budgets must be developed in order to determine if all of the manure nutrients can be utilized on-farm or, alternatively, plans must be made to export some or all of the nutrients from the farm.

Predicting Manure Amount and Composition. From an environmental perspective, it is critical to have accurate estimates of the amounts of manure nutrients originally excreted and not limit nutrient measures to amounts later recovered from the animal pens. This is because it is important to account for losses that may occur (or may not occur) on-site. Another important factor is to be able to estimate nutrient excretion for cattle under different dietary and performance situations. High producing cattle, on a per animal basis, will eat more than cattle consuming low-performance diets. However, excretion per unit of milk or meat produced decreases as production efficiency increases.

Utilizing measured dairy cow nutrient intakes, nutrient excretions in feces and urine, and nutrient outputs in milk, Van Horn et al. (4) concluded that a simple input, output model accurately predicted P and N excretion and Watts et al. (5) suggested the same approach for beef cattle. Accurate nutrient intake is the most important single source of information needed to estimate original nutrient excretions. Nutrition managers of large food-animal production units, who have access to computerized records of feed nutrient deliveries to animals, are key consultants in developing nutrient budgets. Records of food-product sales off-farm along with measured or estimated nutrient content of the products provide the other component needed to accurately estimate manure nutrient excretions. Nutritionists also are skilled in balancing nutrients in diets so that animal nutrient requirements (6, 7) can be met with as little excess of environmentally sensitive nutrients as possible.

Eliminating dietary excesses, where they exist, is the easiest and first step to take to reduce on-farm nutrient surpluses. It is well documented that many, perhaps most, producers overfeed P. Reducing P to NRC (7) recommendations would reduce P excretion per cow by at least 10 kg/yr (20% or more) compared with feeding levels used by most dairymen (4). Similarly, beef cattle excretions of P often were as high as 20 kg P/yr per steer (5) whereas feeding NRC-recommended levels (6) would have limited P excretion to about 7 kg/steer per year.

Most animal manures are P-rich relative to N needs when applied as fertilizer to crops (see later section on budgets). If P applications to crops become regulated down to agronomic levels, reducing P excretions through optimum diet formulation will become even a higher priority. Rumen microbes provide ample phytase to make organic P available to ruminant animals (8). Thus, feedstuffs high in P should not be discounted to force higher supplementation of inorganic P supplements as is done with nonruminants.

Losses in Storage and from Soil. Nutrient losses vary tremendously, especially for N, from farm to farm depending upon the type of storage and handling systems used. Losses of N due to volatilization of ammonia from manure are likely and variable. Two primary forms of N exist in manure, ammonia and organic N. For cattle, about half the original manure N is urea (from urine) or other easily degraded N compounds in feces that rapidly yield ammonia. Use of anaerobic lagoons and any form of long-term storage add further to atmospheric N losses. In most cattle management systems, more than 50% of the excreted N is lost to the atmosphere before it can be recycled for fertilizer use by plants (9). In addition to ammonia volatilization, airborne losses include denitrification, an anaerobic process in which soil bacteria convert nitrate to nitrous oxide and then to N gas. Denitrification is a major process of N removal from soils. However, enhancing N loss by denitrification on farms is not feasible, and nitrous oxide, an intermediary gas that may be emitted in the process, has been implicated as a significant contributor to global warming. Moore and Gamroth (10) found denitrification losses from 7 to 28% of applied N in wet or poorly drained soils and Newton et al. (11) found losses of 20 to 43%. Losses of other nutrients should be small.

Potential Nutrient Removal by Plants. It is generally accepted that nutrients can be applied to land slightly above the level of the nutrients removed by the crops harvested. When animal numbers are high in relation to the amount of land readily available, we need to maximize crop production on available land and determine the maximum application rates for given soil types and crops that can be efficiently utilized.

A long-term research project at Tifton, Georgia used a triple-cropping system (*11, 12*). Flushed dairy manure nutrients were applied through centerpivot irrigation. The cropping system included Tifton 44 bermudagrass in which corn was sod-planted for silage in spring and abruzzi rye was sod-seeded in fall. Harvests included rye for grazing from about Dec 1 until Feb 15, rye for silage about Mar 20 (corn planted the day following), corn for silage in mid-July, low-quality bermudagrass hay about 10 days later, and high quality bermudagrass hay or grazing until rye was planted again about Nov 1. Manure nutrient application rates were the same year after year so that the amount of available N, which for manure is influenced by the organic matter decay rate (*13*), should have reached a steady state. Using decay series estimates to discount organic N sources the first year or more after application and carrying over N credits to subsequent years is an important consideration in using manure efficiently, especially in colder climates.

Harvests of all crops yielded 13.9 tonnes or more of dry matter (**DM**) per hectare with N-deficient application of 240 kg N/ha and yields plateaued at 28.0 to 29.1 tonnes of DM/ha with manure wastewater applications of 493 kg N/ha or more. Crop removals of N, however, continued to increase after DM yields plateaued because of luxury consumption of N which increased N concentrations of crops harvested and suggested that N removals of over 670 kg/ha were possible without environmental losses occurring.

Phosphorus removals in the Georgia and other experiments suggest that P removals of 100 kg/ha or more are possible. Phosphorus removals were 55, 88, 100, and 104 kg/ha in the 13.9, 22.0, 28.7, and 29.1 tonnes of DM harvested; application rates of P were 51, 92, 144, and 202 kg/ha (*11*). Luxury consumption of P is relatively small compared to the extent that it occurs with increasing fertilization of N and K.

Other forage crops, even legumes, like alfalfa and perennial peanut, have been proposed as being good crops for consuming large quantities of manure nutrients since legumes take up soil N in preference to fixing N from the air when free N is available in the soil to "scavenge." Giant elephantgrass which has been used in field studies in Okeechobee County, Florida (*4*) gives the highest estimated P uptake. Although there may be potential for greater recovery of P in the enormous quantity of biomass harvested in giant elephant grass than from other crops, the estimated digestible energy value of the harvested forage would be low and thus it is not as suitable for recycling nutrients in ruminant systems as the crops used in the Georgia, multicropping program.

Whole-Farm Nutrient Budgeting. After determining total manure nutrient excretion, crop production, nutrient removal potential by crops (production plus potential for luxury consumption), and expected losses of nutrients, a nutrient budget can be developed by combining those elements and making adjustments to balance needs and excesses. As an example, Van Horn et al. (*9*) chose data from Georgia experiments (*11, 12*) to develop the N-use cycle in Figure 1. Crop production from the triple-crop program was estimated to be 29 metric tonnes of dry matter containing 572 kg N which calculates to an average crude protein (**CP**) of 12.3% (572 kg N × 6.25 kg CP/kg DM ÷ 29000 kg DM). To maximize

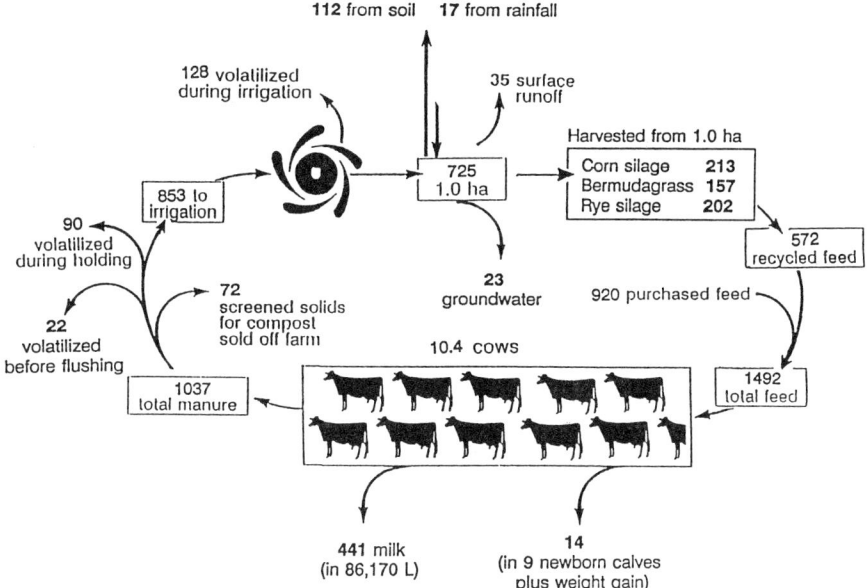

Figure 1. Example of N budget for dairy manure sprayfield system. (Reproduced with permission from ref. 9. Copyright 1996, *J. Animal Sci.*). Bold numbers represent kilograms of N. Crop N harvested and losses of N from soil by denitrification were derived from other data in this paper. Losses of N to groundwater was the amount calculated to equal 10 ppm of NO_3-N added to estimated throughput groundwater. No commercial fertilizer was applied to crops and all forage harvested was fed to cows; forage harvest was 29 metric tonnes (Mg) DM, cows average 22.7 kg milk/d and consumed 17.8 kg DM/d (7).

use of manure N to fertilize crop production in this system, it was determined that manure from 10.4 cows could be utilized if those 10.4 cows were fed the harvested forages as part of their required 17.8 kg DM/day while producing 22.7 kg milk per day (year-around averages). Additionally, 38.6 metric tonnes of DM containing 920 kg N were required to be imported to meet the total dietary N required of 1492 kg N. Thus, the cows had no problem utilizing all of the DM produced in this system.

The estimated acceptable yearly losses of N (Figure 1) were 240 kg to volatilization of ammonia, 58 kg to surface plus groundwater, and 112 kg to denitrification. Screened solids, containing 72 kg N, were sold off-farm. Recovery of 725 kg N for fertilizer application to the field (70% of the 1037 kg N excreted) probably is a best case scenario with regards to the percentage of N recovered from manure for fertilizer use. Most management systems lose more than 50% of excreted N to volatilization and losses during decomposition of organic matter.

Budgets for P and K (9) for the presumed 29 metric tonnes/ha of DM harvested from the triple crop system contained 101 kg P and utilized manure P from 5.6 cows and manure K plus 46 kg K from commercial fertilizer. If lower, assumed average P contents of forages (7) had been used, only 67 kg/ha P removal would have occurred which would have reduced cow carrying potential to 3.7 cows/ha to produce the expected yields and removal of P. The estimated N that would need to be provided from commercial fertilizers would be 334 and 461 kg/ha with the two P scenarios (5.6 and 3.7 cows/ha with 101 and 67 kg P removals). The forage removal of K of 477 kg/ha was based on Newton et al. (11).

Air Quality Issues

Odors emanating from animal manures and their perceived presence are the source of much friction between non-farm rural residents and food-animal producers. Additionally, ammonia and methane emissions are being scrutinized carefully to determine if regulatory oversight of these units will become necessary.

Odor Control. Volatile odorous compounds emitted from manure during transport, storage, treatment, and disposal have become an acute public relations problem for animal agriculture. Odorous compounds usually are present at such low levels (parts per million or parts per billion) that they are not toxic at the concentrations found downwind of livestock production facilities. Thus, the problem depends largely on subjective factors, how much the smell bothers people or the "nuisance value" of the odor. Often, flies add to an odor nuisance, and the two problems may be difficult to separate in the minds of complainants. Odor complaints range from casual comments, indicating displeasure, to major lawsuits and court orders which have the potential to terminate the affected food-animal enterprises (14).

The US Environmental Protection Agency does not regulate odors. However, odor is regulated as a nuisance in every state in the US. States regulate odors as a public nuisance through air pollution control and public health protection statutes. Definitive measures of odor are needed to evaluate the extent of an odor nuisance. One problem is in defining what to measure. Manure odors are caused principally by intermediate metabolites of anaerobic decomposition. Over 75 odorous compounds, in varying proportions, were long ago identified around manure storage areas (e.g., 2) and the number of identified compounds continues to expand as methodology permits identification of more metabolic intermediates. Phenols, volatile fatty acids, and sulfides are thought to be the major odor-causing compounds with total phenols the most important factor associated with odor intensity, as determined by human panelists, of dairy manure products removed from an anaerobic environment (15). However, typical chemical analyses measure concentrations of only a small number of constituents in the complex mixture that contribute to the odor people identify by smell. Correlation of concentrations of individual or multiple analytes with human panelist evaluations have been too low to accept as an accurate assessment of odor or degree of nuisance. Consequently, current odor measurement technology has turned to sensory methods (i.e., using the human nose) using on-site scentometers or to employ sampling methods to bring odorous air or fabric samples, placed on-site to absorb odors, back to a laboratory for panel evaluation.

Odor control methods fall into three broad categories: 1) control of odor dispersion, 2) odor capture and treatment, and 3) treatment of manure. Control of odor dispersion is primarily a function of site selection, system design and construction, and manure handling methods (e.g., sprayfield application may provoke more odor drift than soil incorporation). Odor capture and treatment methods include containment, wet scrubbing, packed-bed adsorption, and soil filter fields (2). These methods are not well suited for open housing conditions on most cattle facilities. Manure treatment methods include anaerobic digestion, aeration, and biochemical treatment. Commercial chemical and biological additives available to date have not reliably reduced odors.

Ammonia Emissions. The major source of ammonia is urea from urine, or uric acid in the case of birds, which can be easily converted to gaseous ammonia (NH_3) by bacterial urease, which is ubiquitous. Urea plus ammonia N from urine usually accounts for 40 to 50% of total N excreted in manure, i.e., feces plus urine (4). In aqueous solution, NH_3 reacts with acid (H^+) to form an ion (NH_4^+), which is not gaseous. Thus, the chemical equilibrium in an acid environment promotes rapid conversion of NH_3 to NH_4^+ with little loss of NH_3 to the atmosphere. However, most animal manures, lagoons, and feedlot surfaces have a pH >7.0, making H^+ scarce and, thus, permitting rapid loss of ammonia to the atmosphere. As a consequence, N losses from animal manures can easily reach 50 to 75%, most as NH_3 before NH_3 is converted to NO_3^- through nitrification.

An important question to be answered is whether it is important to minimize low level emissions of ammonia to the atmosphere, and, if not, should livestock producers be encouraged to use manure management procedures to volatilize more ammonia? In Europe, atmospheric ammonia concentrations have become a public concern through their perceived contribution to acid rain and the destruction of forests (e.g., 16). Consequently, European livestock and poultry operations are being required to utilize practices to minimize ammonia losses to the atmosphere.

Ammonia can be toxic to cells, and the potential exists for plant damage if excessive ammonia is released after manure application. Also, excessive ammonia concentrations in closed buildings used to house large numbers of animals may lower animal performance and may be a potential health hazard for workers. Atmospheric ammonia has been known to cause blindness in chicks and turkey poults. Thus, it is important to avoid ammonia buildup where animals are confined and people work.

Most volatilized ammonia is dissolved in water vapor in the lower atmosphere and washed back to earth by rainfall. During this process, ammonia neutralizes the acidity of the rainwater. In industrial regions with somewhat acid rainfall, e.g., Pennsylvania, neutralization is one potential benefit of ammonia release (17). If techniques were used to promote ammonia volatilization, a portion would be redeposited from the atmosphere to nonagricultural, N-poor areas such as forests. The resulting increase in soil fertility would be a potential benefit of increased volatilization. However, soil pH would begin to drop over time, just as continued application of ammonia-containing fertilizers acidifies agricultural soils. Current data do not prompt concern in North America about negative effects on the environment caused by diffuse ammonia emissions from animal manures. However, local concern about animal, human, and plant health is warranted when ammonia concentrations are high (*17*).

Methane Emissions. Methane emissions from animal production systems do not present an odor-control problem because methane is odorless. The concern with methane relates to its role as a greenhouse gas and as a potential contributor to global warming (*3, 18, 19*). Carbon dioxide is the most abundant greenhouse gas and is being added in the greatest quantity; carbon dioxide is expected to cause about 50% of the global warming occurring in the next half century. Methane is generally held to be the second most important greenhouse gas and is expected to contribute 18% of future warming (*18*). Indeed, molecule for molecule, methane traps 25 times as much of the sun's heat in the atmosphere as carbon dioxide. Thus, methane is estimated to contribute 18% of future warming from <1% of the total greenhouse gas emissions. In addition to warming effects, increased atmospheric methane will likely be detrimental by increasing ozone pollution near the earth's surface and, conversely, by decreasing ozone in the stratosphere, which shields the earth from harmful solar ultraviolet radiation.

The origin of methane produced by animals is microbial action in the gastrointestinal tract, which occurs to varying degrees in all animals. Major fermentative digestion, allowing utilization of fibrous dietary components, occurs in ruminants. This, coupled with large body sizes, dry matter intakes, and animal numbers, results in 95% of animal methane emissions arising from ruminants, about 80% from the Bovidae family alone. Sheep and goats account for another 12%, and horses and pigs contribute about 2 and 1%, respectively (*20*).

Energy losses through methane produced in the rumen are usually 6 to 8% of gross energy intake in cattle consuming high forage diets; greater losses occur when forage is of low digestibility (*3*). Dairy cows fed moderately high concentrate diets convert about 5% of their gross energy intake into methane and belch this methane into the atmosphere (e.g., Figure 2).

The methane produced by animals and animal manures constitutes about 16.4% of estimated annual methane emissions [from (*18*)], which translates roughly to 2.9% of the estimated contribution of all greenhouse gases to global warming (i.e., 16.4% of 18%, the projected contribution of all methane sources). Although an extremely small part of the total, the feasibility of reducing animal-related methane emissions is being investigated.

Partitioning and Processing Alternatives

Obviously, from previous discussion of nutrient management, land application for fertilizer recovery is the primary method used to manage manure. In order to allay fears that unprocessed animal manures are a risk factor to humans, it is

Flow of DM and OM (volatile solids)

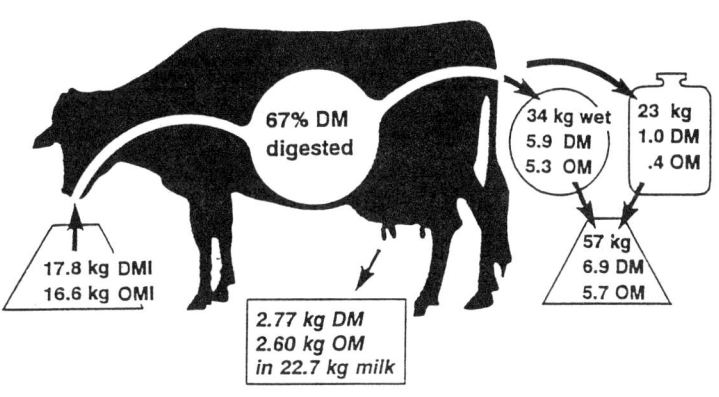

Carbon and Energy Balance

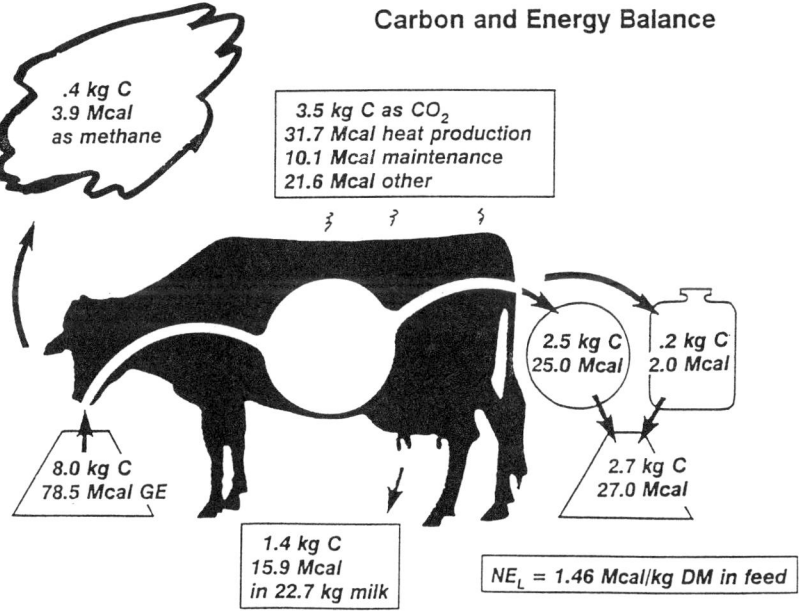

Figure 2. Estimated daily flow of DM, organic matter, energy, and carbon through typical Holstein cow (typical of year-round amounts when extrapolated to 365 d). (Reproduced with permission from ref. 4. Copyright 1994, *J. Dairy Sci.*). DM = dry matter, DMI = DM intake, OMI = organic matter intake.

important to point out the many natural biological pathways at work in pastures and in soils utilized for field crop production that effectively process manures. Anaerobic processing (degradation) of undigested fecal organic matter is already underway in the lower gut of animals before feces are voided. This process continues after defecation to the extent that anaerobic conditions are maintained, e.g., the centers of manure droppings firm enough to stack may remain anaerobic. Wherever oxygen permeates, aerobic microbes take over the degradation process, and anaerobic activity decreases. Insects, such as dung beetles and worms also contribute. Aerobic systems and soil associations effectively oxidize most odorous compounds rapidly. When cattle distribute manure on pastures, it is spread across enough area that remaining odorous compounds are effectively diluted and odors are usually not a problem.

When manure is collected from cattle housed in lots and barns, some partitioning or preprocessing before soil application may be done to facilitate collection and transport, reduce volume, and, when necessary to send nutrients to other farms for utilization, prepare suitable products to move off-farm.

Anaerobic Treatment. Anaerobic lagoons represent the most common method of anaerobically processing cattle manure or liquid components collected from lots and barns. Digestion of organic components of manure ultimately occurs, oxidatively or anaerobically, at some stage in any system selected, e.g., in the soil if not removed previously. Anaerobic lagoons are utilized primarily to facilitate handling of an already wet product (fresh cattle manures, feces plus urine, are 85 to 90% water) that can be easily moved with additional flushwater to clean pens. Anaerobic fermentation is easily established simply by permitting the anaerobic microorganisms already active in the lower gut of the animal to compete in a lagoon environment. A regional research project on manure management in warm, humid climates (*21*) found, overall, that three-pond anaerobic lagoon systems reduced chemical oxygen demand 75%, total solids 48%, organic solids (also called volatile solids) 46%, total N 69%, total P 47%, total plate count 85%, total coliform count 99%, fecal coliform count 98%, and fecal *Streptococcus* count 99%. Reduction in organic matter implied that lost carbon was transferred to the atmosphere in volatile, carbon-containing gases (mostly CO_2, CH_4, and volatile fatty acids) or retained in lagoon sludge. Reduction in P implied that P was in the sediment and sludge retained in the lagoon. In lagoon systems for cattle operations, the effluent usually is dilute and is land-spread via an irrigation system to supply fertilizer nutrients for crops and supplemental water.

Well managed anaerobic lagoons or anaerobic digestion systems also reduce odors. Powers et al. (*15*) showed with lab-scale anaerobic digesters that odor intensity of digester effluent, as judged by a human panel, was reduced linearly by increasing the average hydraulic retention time from 0 to 6, 10, 15, or 20 days. As mentioned previously, odors primarily are from released intermediary volatile compounds accumulated during anaerobic degradation of organic materials. Containment of substrates within the digester for longer periods allows more of these compounds to be digested to non-odorous end products, i.e., methane and CO_2.

Deciding whether or not to utilize the methane from anaerobic digestion as a fuel to produce heat or electricity currently is an economic decision. For example, Van Horn et al. (4) estimated that 2.005 m^3 of biogas containing 60% methane could be derived from the 5.73 kg of volatile solids (Figure 2) which could be converted to 10.71 megacalories of gross energy or to 2.15 kilowatt hr of electricity (1.0 kWh/.934 m^3 biogas); estimated value would be $.129 ($.06/kWh) to $.215/d ($.10/kWh). This value converts to $47 to $78/yr per cow. Relative returns may be even greater if the biogas can be utilized as a substitute for other fuels used to produce heat. These values have not been an economic incentive to establish anaerobic digestion for methane recovery and utilization.

Solids Separation. Removal of manure from animal pens by flushing with water is an easy and clean way to handle manure. However, this process results in a larger volume of manure slurry to be managed. Separation of the coarse solids from flushed manure is potentially important for several reasons:
1. To remove large particles and sand that could plug or damage distribution nozzles in irrigation equipment.
2. To reduce organic loading on anaerobic and aerobic lagoons.
3. To capture a fibrous by-product with some N and mineral content for uses such as bedding for free stalls, part of the feed for cattle on maintenance diets, plant-potting compost, and fertilizers.

The primary benefit of separation of solids from liquid is the production of two fractions that are inherently more manageable than the original slurry. The most popular systems used to remove a portion of the solids from manure slurries are mechanical separation and sedimentation basins.

Stationary screens, which are most common, usually remove 20 to 30% of the organic matter from liquid dairy manure but 80 to 95% of the N and P stay with the effluent (22).

Powers et al. (22) and Barrow et al. (23), in laboratory experiments, showed that sedimentation for 20 min without additives removed approximately 65% of total solids, 20% of N, 60% of P, and 40% of K. Additions of ferric chloride and calcium oxide (23) improved removals up to 93% of total solids, 51% of N, 91% of P and 60% of K. Field trials determined amounts of organic solids removed from flushwaters by screening with an Ag Pro® Manure Extractor with screen size of 1.5 mm to be approximately 30% and removal by screening plus sedimentation in flow-through traps was approximately 53%. The estimated economics of flocculation with Fe and Ca salts were not favorable based on the estimated cost of chemicals added versus the value of fertilizer nutrients sedimented. Flocculation has the potential to concentrate sedimented nutrients for food-animal producing units that must export nutrients off-farm to meet environmental regulations and choose to move manure from pens and animal holding areas with flushwaters.

Although not tested by Barrow et al. (23), it is likely that Al^{+3} compounds would function similarly to ferric iron compounds. Moore (24), for example, showed that alum-amended poultry litter decreased P runoff and decreased NH$_3$ volatilization. Additionally, weight gains of birds and feed conversion were improved, presumably because of decreased NH_3 levels in poultry houses. The reduced NH_3 may have been a function of reduced pH in the litter.

Alternatives that Permit Export of Nutrients Off-Farm. *Manure Application on Nearby Farms.* Large food-animal producing units vary greatly in land resources that are available on the same farm to produce crops that will consume the manure nutrients produced. For example, most dairy farmers have sufficient forage needs so that they have traditionally maintained a sizeable farming operation in conjunction with the dairy. Thus, most of these dairies, but not all, can recycle their manure N nutrients on-farm if they maximize crop production intensity on the land they have. Many dairies cannot, however, recycle all of their manure P. Large beef cattle feedlots, however, almost assuredly will need to export manure nutrients. Based on excretion estimates of about 50 kg N/steer-yr, a feedlot of 50,000 head with 80% occupancy will generate about 2,000,000 kg N/yr. If 50% of the N is utilized effectively as fertilizer (50% volatilized) for crops requiring 200 kg N/ha, 5,000 ha cropland is needed for utilization of the N. If the feedlot is in a dry area, irrigated cropland will be required or application rates reduced accordingly to match productivity of the dry land. One significant advantage of locating large feedlots in dry regions is that the manure can be scraped and hauled off-site very easily, as compared with feedlots located in wet regions. And earthen structures to contain runoff are very modest in size compared to high-rainfall areas.

Burning. Some regions that do not have sufficient crop production near the animal production unit have needed to find other means to utilize or transport manure nutrients off-farm. Burning manure is a possibility. The first large-scale resource recovery project in the world to burn cattle manure as fuel to generate electricity was in the Imperial Valley of southern California (Van Horn, 1990 personal communication with Western Power Group and National Energy Associates, El Centro, CA). It was designed to utilize manure from the many beef cattle feedlots in the valley. Use of animal manure for fuel has been reviewed (25). Utilization of poultry litter for fuel is expected to approach 80% of the litter produced in the United Kingdom within 5 to 10 yr, due to favorable (23 to 42%) internal rates return. When manure is burned, the ash nutrients still need to be managed accountably.

Composting. A significant amount of dried manure, composted manure, or a combination of dried and composted manure is bagged and sold as organic fertilizer. An example with dairy manure is a dairy cooperative in the Chino Valley in California which was set up to move manure off of large, intensive drylot dairies located in an urban area. The dairies pay a fee to the cooperative to pick up their manure, take it to a central location where it is processed, bagged, and marketed.

Composting is a logical way to process wetter manures (but not slurries) when livestock producers must create a product that must move off-farm and be stable enough so that suburban users or agricultural users near urban centers want to utilize it. Composting is relatively costly, labor intensive, and some of the most valuable fertilizer constituent, N, is driven off to the atmosphere during processing. Therefore, dairies and feedlots usually consider the process only if a marketable product is created that will help them remove the excess nutrients from the farm that they must remove. Several advantages include: aerobic composting reduces volume and converts biodegradable materials into stable, low-odor end products; thermophilic temperatures of 54°C (130°F) to 71°C (160°F), achieved in the process, kill most weed seeds and pathogens. If moisture content is too high, anaerobic conditions develop producing odorous compounds.

The physical form of cattle manures often does not provide optimal composting conditions. Fresh manure is too wet, and screened solids are usually too low in N content and other fertilizer nutrients. Thus, mixing materials from other sources may be helpful. Supplies of manure and bulking agents, as well as market demand for the finished compost, should be investigated before a dairy invests in composting equipment.

Kashmanian (26) estimated that a minimum of 5,640 farms were composting manure, crop residues, or poultry carcasses as compared to municipal facilities composting the following materials: yard trimmings (3,014), biosolids (201), and source-separated municipal organics or solid waste (around 17). Although manure and crop residue composting has expanded, most of the recent growth in numbers of on-farm composters has concerned the composting of poultry carcasses.

Water Cleanup. Municipal waste processing systems are built to treat water sufficiently to meet accepted water quality standards before discharge. These methods generally are not needed by production agriculture because economical biosystems are utilized for manure processing and utilization. However, many dairies exist in locations where lack of agricultural land and relations with urban neighbors demand unique processing if the dairy is to continue operating in that location. Constructed wetlands and naturally occurring wetlands are being researched and utilized for water quality improvement in several regions (e.g., 27). In some cases, wetlands have been shown to be effective. However, unharvested wetlands serve as a sink for removed nutrients, and accumulated nutrients may have to be removed at some point if the nutrient content of effluents approach levels demanded by regulatory action.

An example is the Lake Okeechobee area in Florida, where several dairy farms were unable to reduce P concentrations in surface waters exiting the farm with normal nutrient management practices. Some farms utilize commercial systems employing a combination of precipitation of nutrients with $FeSO_4$ and constructed wetlands from which produced forages can be harvested periodically. Although removal of P by this system is relatively expensive compared with the value of P from other sources, such systems have reduced the P content of regulated waters sufficiently to make it possible for dairies to continue operating at those sites.

Effluents from lagoons and anaerobic digesters, and perhaps from some wetlands, contain sufficient nutrients to require further treatment if not used in irrigation for nutrient and water recycling. Such wastewaters can be used as a growth medium for microalgae which could be harvested as a high protein by-product. Production of algal biomass (*Spirulina* spp.) that contained 60% CP (DM basis) effectively removed 73.6% of ammonia N in anaerobic lagoon effluent (28). Potential uses of harvested algal biomass include biogas generation, animal feed, and fertilizer.

Discussion: Credits and Debits Associated with Cattle Manures

One of the major challenges facing society is how to improve environmental quality. Are quality and quantity of human diets and products used a part of our assessment of environmental quality? Animal products play an important part in both. The digestibilities of animal-derived foods are high and they supply many nutrients, especially limiting amino acids, many vitamins, and calcium, that are essential and required to make efficient use of the starchy dietary staples consumed by most of the world's population (29). Cattle play an especially important role increasing the quantity of nutrients available for human digestion because, in total, they deliver more digestible nutrients for human consumption than were available without including cattle in the food chain. The multitude of animal-derived products, from food to biomedical and industrial products, can enhance people's lives.

Animal agriculture often is perceived by the public as having negative environmental effects, e.g., concern with swine units in North Carolina, Iowa, and Missouri; poultry units in Georgia, Maryland, Alabama, Arkansas, and Connecticut; cattle feedlots in Texas, Oklahoma, Kansas, and Colorado; dairies in Wisconsin, California, Florida, and Washington; etc., etc., etc. This paper has attempted to show the principles that beef and dairy producers must consider when designing environmentally accountable manure management systems. Agriculture is based on biological systems that effectively process manure nutrients and other biomass in cost-effective, environmentally acceptable ways. The public sector needs to be aware of this and to monitor agricultural systems based on real concerns and not perception so as not to impose unnecessarily costly processing methodology.

A key to the public continuing to enjoy relatively cheap, nutritious animal-derived foods and other products is for the extra costs of environmental safeguards imposed on animal producers to be recoverable in the value added to their operations by the more efficient use of marginal resources. For example, better use of manure nutrients will save the purchase of equivalent commercial fertilizer nutrients and the installation of an anaerobic digester for improved odor control in a suburban setting will be partially offset by the energy value of the methane utilized from it.

One method of estimating the resource value of manure is to assign a value to the utilized N, P, and K, the most valued fertilizer nutrients. For example, based on assumed values of $.66/kg N, $1.32/kg P, and $.33/kg K, the equivalent fertilizer value for the 725 kg N from 10.4 cows in Figure 1 and partially utilized P and K would be about $760 or $73/cow-year. If more acreage for forage production were available to fully utilize all of the P and K, the value of the fertilizer equivalent N, P, and K would be expected to be about $89/cow-year. Forages produced, often with the required fertilizer coming from recycled manure nutrients, provide 35% to 60% of the nutrients for lactating cows. These values are very important to help pay for the extra expense and management to recover those nutrients to protect the environment. In reality, most cattle producers do not recover as high a percentage of the excreted N as estimated in Figure 1, which is a best-case scenario. Recoverable values usually are much less.

Consumer and community effects also are possible. Boggess et al. (*J. Dairy Sci.*, in press) reported a study to evaluate the impact of three water quality programs designed to reduce P runoff from dairies and other agricultural lands into Lake Okeechobee in South Florida. Implementation of the programs cost just over $51 million from 1987 to 1993. Dairy operators paid approximately 40% of the total mandatory changes with state agencies providing the balance. Dairies that complied with the Dairy Rule spent on average $1.14 per cwt. of milk ($.10/gallon) with a net of approximately 30% cost-shared for both mandatory and optional components. This additional cost was more than offset by the 13% increase in average milk production experienced as a result of the Dairy Rule investments. In general, these dairies went to more confinement to capture manure to avoid losses and added shade and cooling fans (optional) to improve the physical environment for the cows which improved performance. Thus, the losses for the watershed primarily reflect the relocation of dairies to other regions rather than losses in income to remaining dairies. The exceptions would be those dairies that did not experience significant increases in milk productivity as a result of their Dairy Rule investments.

Direct economic impacts included average annual reductions in milk sales of $28 million in the watershed and annual reduction in employment in the dairy industry of 274 jobs. Total direct, indirect, and induced economic impacts were approximately 1.5 times larger than the direct effects. Many of the community's economic losses were realized as gains in other regions where dairies that left the region relocated.

Lest some conclude that agriculture is the major problem relative to water quality, the public sector should recognize that in agriculture, sufficient land and crop production suitable for recycling those nutrients exist. Waste nutrient flows from human activities are similar to waste nutrient flows from farm animals but processing differs and nutrient recycling is more difficult. Additionally, urban use of commercial fertilizer is appreciable even if small relative to commercial agriculture (*30*). Frink (*31*) also underscored that municipal wastes and urban runoff were major contributors to N and P loads in surface waters.

The urban population may benefit from an assessment of the ability of agriculture to process urban wastes. That avenue has potential to reduce costs of processing urban wastes and, at the same time, give better environmental accountability. This already is happening, with some municipalities managing agricultural land or contracting with farmers to utilize treated wastewater (reclaimed water) and sewage sludge (residuals) monitored to contain safe levels of heavy metals [e.g., (*32, 33*)]. If most consumers really understood and evaluated what is done with their own waste stream of sewage and solid wastes and the costs of processing, they probably would be much more open to discussion with agricultural planners about how to create a system that better serves all of us to achieve a more sustainable world.

The value of commercial fertilizer N sales represents a market opportunity for urban taxpayers to recover the resource value in urban wastes if they can be processed to be a safe and effective fertilizer and offered at a price attractive to fertilizer users. This value also represents an opportunity to livestock producers if they are not making full use of potential fertilizer value of manures now. If livestock producers must transport excess nutrients from their farms, they will need to produce a marketable fertilizer product. If urban dwellers and farmers can reduce their net expenses in processing wastes by marketing fertilizer products economically, all will benefit.

An important point to include in environmental considerations for food-animal producing units, particularly with regard to ruminants, is the positive contribution that animals now make in providing cost-effective disposal of many wastes from other food and fiber industries. For example, citrus pulp, cane molasses, cottonseed hulls, soybean hulls, wheat middlings, rice bran, hominy feed, corn gluten feed, grain screenings, distillers' grains from the drinking alcohol industry, distillers' grains from fuel ethanol production plants, brewers' grains, outdated bakery products, whey, culled vegetables and discarded supermarket vegetables, corn gluten feed, blood meal, fish meal, meat and bone meal, feather meal, cottonseed meal, peanut meal, canola meal, soybean meal, yellow grease, tallow, animal-vegetable mixed fats, waste candy, almond hulls, all are byproducts that some other industry at one time had to dispose of as waste products that are utilized in feeding programs for cattle.

Conservative survey estimates (Hall, M. B., University of Florida, personal communication, 1996) indicate that each year, dairy cattle in Florida consume byproducts produced in the state by people-related industries containing more than 150,000 metric tonnes of dry matter, 4,300 tonnes of N, 2,800 tonnes of P, 5,600 tonnes of K, and having a volume of over 460,000 cubic meters. Additionally, Florida dairymen import about three times this many byproducts from other states so that byproducts contribute about half of the nutrients that lactating dairy cows consume. If not consumed by livestock, these byproduct feeds would require landfilling, or some other method of disposal. Dairies and many other food-animal producing units should earn "environmental credits" for providing environmentally sound, and sometimes profitable, waste disposal for many other industries. Additionally, these units are required to recycle the manure produced. Thus, dairy farms and other food-animal production farms usually have a positive effect, environmentally, on overall nutrient cycling rather than negative as many perceive.

How important is it to create a partnership between farmers and the public sector to create a more sustainable world? It is more important to consider how agriculture can help improve society's sustainability than it is to worry specifically about a sustainable agriculture. Figure 3 attempts schematically to show how cattle and crops can be utilized to keep the wastes of people and their industries from polluting the environment while, at the same time, maintaining the environmental integrity of agriculture per se.

Food production on our remaining agricultural land must be increased. It is a challenge to do that and maintain all of the other environmental qualities that are important. Achieving those desired environmental qualities will require some regulations. However, skillful use of incentives and regulatory standards based on desired outcome rather than process will give farmers much more

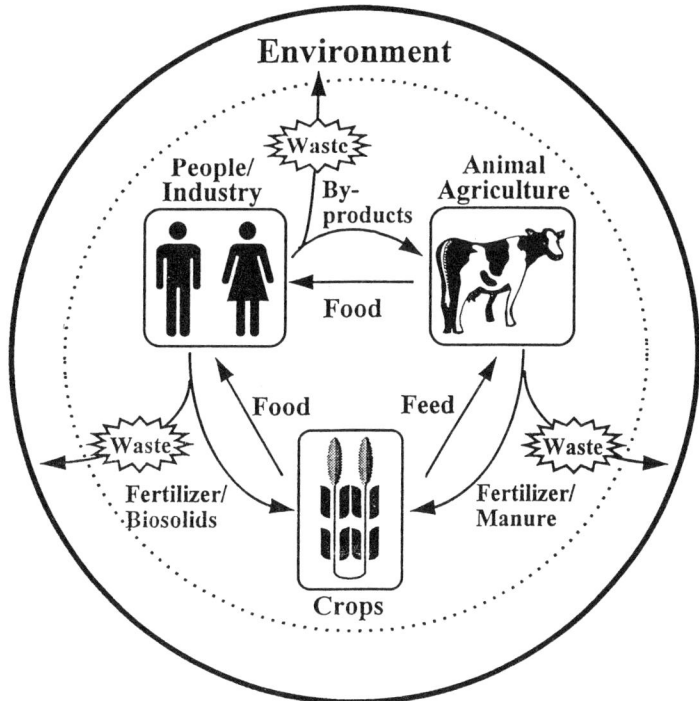

Figure 3. Example of interdependence between People/Industry and agricultural production by food animals and crops to minimize waste flows to the environment.

freedom to increase food production while at the same time enhancing the public's environmental accountability.

Literature Cited

1 Steenvoorden, J.H.A.M. In *Proc. Dairy Manure Management Symp.*, Publ. No. 31, Northeastern Reg. Agric. Engin. Serv., Cornell Univ., Ithaca, NY, 1989; pp 15.
2 Sweeten, J. M. 1988. *Environ. Health.* **1988,** *50(5)*,282.
3 *Reducing Methane Emissions from Livestock: Opportunities and Issues.* EPA 400/1-89/002, US Environ. Protection Agency, Offic. Air Radiat., Washington, DC, 1989.
4 Van Horn, H. H.; Wilkie, A. C.; Powers, W. J.; Nordstedt, R. A. *J. Dairy Sci.* **1994**, *77,*2008-2030.
5 Watts, P. J.; Gardner, E. A.; Tucker, R. W.; Casey, K. D. In *Proc. Great Plains Anim. Waste Conf. on Confined Anim. Prod. and Water Quality.*

Balancing Animal Prod. & the Enviornm. GPAC Publ. No. 151. Great Plains Agric. Council, Fort Collins, CO, 1994; pp 27-33.
6 *Nutrient Requirements of Beef Cattle (6th Ed.).* National Academy Press, Washington, DC, 1984.
7 *Nutrient Requirements of Dairy Cattle (6th Rev. Ed.).* National Academy Press, Washington, DC, 1989.
8 Morse, D.; Head, H. H.; Wilcox, C. J. *J. Dairy Sci.* **1992**, *75,1979-1986.*
9 Van Horn, H. H.; G. L. Newton, and W. E. Kunkle. *J. Animal Sci.* **1996**, *74,3085-3105.*
10 Moore, J. A.; Gamroth, M. J. In *Proc. Western Large Herd Dairy Management Conf.* The Dairyman Publ., Corona, CA, 1993, pp 102.
11 Newton, G. L.; Johnson, Jr., J. C.; Davis, J. G.; Vellidis, G.; Hubbard, R. K.; Lowrance, R. In *Proc. Florida Dairy Production Conf.,* Dairy and Poultry Sci. Dept., Univ. Florida, Gainesville, 1995, pp 113-123..
12 Johnson, J. C., Jr.; Newton, G. L.; Butler, J. L. In *Proc. 28th Annual Florida Dairy Prod. Conf.,* IFAS, Univ. FL, Gainesville, 1991, pp 41.
13 Klausner, S. D.; V. R. Kanneganti, D. R. Bouldin. *Agron. J.* **1994**, *86,897-903.*
14 Sweeten, J. M.; Miner, J. R. *Paper No. 924514.* Int. Winter Mt. Am. Soc. Agric. Eng., Am. Soc. Agric. Eng., St. Joseph, MI, 1992.
15 Powers, W. J.; Van Horn, H. H.; Wilkie, A. C.; Nordstedt, R. A. *J. Anim. Sci.* **1996**, *74(Suppl. 1),126.* (Abstract)
16 Apsimon, H. M.; Kruse-Plass, M. In *Odour and Ammonia Emissions from Livestock Farming.* V. C. Nielsen, J. H. Voorburg, and P. L'Hermite, ed. Elsevier Appl. Sci. Publ., London, England, 1991, pp 14.
17 Elliott, H. A.; Brandt, R. C.; Martin, K. S. *Atmospheric disposal of manure nitrogen.* Tech. Note No. 8, Pennsylvania Bay Educ. Office, Harrisburg, 1990.
18 Johnson, F. R., Hill, T. M.; Ward, G. M. 1992. In *Proc. Minnesota Nutr. Conf.,* Minnesota Ext. Serv., Univ. Minnesota, St. Paul, 1992. pp 81
19 Waggoner, P. E.; Baldwin, R. L.; Crosson, P. R.; Drabenstott, M. R.; Duvick, D. N.; Follett, R. F.; Jense, M. E.; Marland, G.; Peart, R. M.; Rosenberg, N. J.; Ruttan, V. W. *Preparing U. S. Agriculture for Global Climate Change.* CAST Task Force Rep. 119. Counc. Agric. Sci. Technol., Ames, IA, 1992.
20 Crutzen, P. J.; Aselmann, I.; Seiler, W. *Tellus* **1986**, *38B:271.*
21 Hill, D. T., Newton, G. L.; Nordstedt, R. A.; Payne, V.W.E.; Ramsey, D. S.; Safley, L. M.; Sutton, A. L.; Westerman, P. W. 1990. In *Proc. 6th Int. Symp. Agric. Food Proc. Wastes.* Am. Soc. Agric. Eng., St. Joseph, MI, pp 515.
22 Powers, W. J., Montoya, R. E.; Van Horn, H. H.; Nordstedt, R. A.; Bucklin. R. A. 1995. *Appl. Engin. in Agric.* **1995**, *11(3),431-436.*
23 Barrow, J. T.; Van Horn, H. H.; Nordstedt, R. A.; Anderson, D. L. 1995. *J. Dairy Sci.* **1994**, *78(Supplement 1),182.* (Abstract)
24 Moore, P. A., Jr. In *Proc. Conf. on Nuisance Concerns in Animal Manure Management: Odors and Flies,* Florida and Georgia Coop. Ext., Dairy and

Poultry Sci. Dept. Univ. Florida, Gainesville, 1995; pp 126-134.
25   Badger, P. C.; Lindsey, J. K.; Veitch, J. D. In *Animal Waste and the Land-Water Interface;* Steele, K. Ed.;Lewis Publishers, Boca Raton, Fl, 1995; pp 475-484.
26   Kashmanian, R. M. In *Farm Scale Composting,* JG Press, Inc., Emmaus, PA, 1995; pp 20-22.
27   Cathcart, T.; Pote, L.; Strong, L.; Ulmer, R. In *Proc. Environmentally Sound Agric. Conf.,* Florida Coop. Ext. Serv., Univ. Florida, Gainesville, 1991; pp 341.
28   Crawford, J.J.W.; Wilkie, A. C.; Lincoln, E. P. In *Am. so. Microbiol. Abstr.,* Am Soc. Microbiol. Washington, DC, 1993; pp 425.
29   Van Horn, H. H., Cunha, T. J.; Harms, R. H. *BioScience 1972, 22(12),710-714.*
30   Van Horn, H. H.; Sloan, D. R.; Kunkle, W. E.; Nordstedt, R. A.; Ott, E. A.; Chambliss, C. G.; Kidder, G.; Jackson, L. K.; Hochmuth, G. J. In *Proc. 31th Ann. Florida Dairy Production Conf.,* Florida Coop. Ext. Serv., IFAS Dairy Science Dept., Univ. Florida, Gainesville, 32611, 1994; pp 109-113.
31   Frink, C. R. *CSRS/USDA Agric. Sci. Rev. 1971, 9(2),11-31.*
32   Jackson, J. L.; Cross, P. In *Proc. Conf. Environmentally Sound Agric.,* Florida Coop. Ext. Serv., Univ. Florida, Gainesville, 1991; pp 470.
33   Vidak, W.; Roberts, A. In *Proc. Conf. Environmentally Sound Agric.,* Florida Coop. Ext. Serv., Univ. Florida, Gainesville, 1991; pp 490.

Chapter 8

Agricultural and Environmental Issues in the Management of Swine Waste

Robert L. Mikkelsen

Department of Soil Science, North Carolina State University, 3205 Williams Hall, Raleigh, NC 27695-7619

The concentrated and regionalized nature of the U.S. swine industry results in large quantities of animal waste being produced in a relatively small area. Although animal wastes can be used for a variety of purposes (e.g. animal feed, biogas generation, and composting) their primary use has been as a source of plant nutrients for crop production. Since the concentration of essential plant nutrients is generally low in manures compared with inorganic fertilizers, there is often little incentive to recover the full agricultural value from these materials. Waste is commonly stored in pits or in lagoons prior to application, depending on the type of production system. Manure management practices may have significant impacts on surface water, groundwater, and atmospheric emissions. When swine manure is applied to crop land for extended periods of time, an accumulation of nutrients can occur in excess of the plants nutritional requirement.

Swine production is an important agricultural enterprise throughout the United States. During 1995, over 60 million swine were produced, with the majority of the production occurring in the Midwestern U.S. (Table 1). Traditionally, swine production was characterized by relatively small numbers of animals located on farms in proximity to corn production, a primary feed source. However modern swine production has become more intensive and centralized, typically characterized by large farms that are dependent on purchases of animal feed from off of the farm. Increasing numbers of animals are raised in specialized housing facilities that allow more control over the production environment. Additionally, integration of the swine industry has become common, where the supply of feed and the sale of the animals is handled by a contractor. However, swine production remains a major agricultural activity world-wide (Table 2) which is difficult to generalize due to the extreme diversity in production and management practices. This discussion will focus primarily on swine-production issues in the United States.

Table 1. Inventory of hogs produced in the top ten swine-producing states in the U.S during 1996 (*1*)

State	Millions of hogs	State	Millions of hogs
Iowa	13.0	Nebraska	3.7
North Carolina	9.3	Missouri	3.6
Illinois	4.7	Ohio	1.9
Minnesota	4.9	South Dakota	1.5
Indiana	4.0	Kansas	1.3

As swine production continues to become geographically concentrated and the number of animals on an individual farm increases, problems associated with both air and water quality can become a serious challenge. Odor control has become a major concern on large swine production facilities in relation to nearby neighbors. Ammonia loss from animal waste has been a major environmental issue in Northern Europe during the past decade and will likely be the future subject of regulation in nutrient-sensitive areas. The production of other gases associated with animal production (such as methane) also remain important environmental concerns.

Table 2. Inventory of hogs produced in the top ten swine-producing countries in 1995 (*2*)

Country	Millions of hogs	Country	Millions of hogs
P.R. China	415	Poland	19
United States of America	60	Spain	18
Brazil	32	France	15
Russian Federation	25	Ukraine	14
Germany	25	Netherlands	14

Modern nutrition of swine generally consists of concentrated, high-energy grain-based feeds with only a small proportion of roughage. Proper nutrition is particularly essential for swine because they grow much faster in proportion to their body weight than larger farm animals and also produce young at an earlier age (*3*). However in comparison with other commonly raised animals, swine excrete larger amounts of manure that must be dealt with (Table 3). The majority of the nutrients initially present in the feed pass through the swine digestive system and are ultimately excreted in the manure and urine (more than 80% of the ingested N and P, and more than 90% of the K is typically excreted). This relatively inefficient nutrient recovery by swine results in large amounts of manure-derived nutrients that are available as a valuable resource if managed properly. Nutrients arrive on the farm as feed and mineral additives, then subsequently converted into swine body mass and waste. The nutrients that are not removed when the animal is sold will remain on the farm and must be incorporated into a nutrient management plan. Unless these wastes are properly managed, the nutrients may accumulate to concentrations harmful to growing crops and become detrimental to surface and ground water. The most common method of managing the nutrients is to apply the wastes to cropland and attempt to remove the nutrients in the harvested portion of the plant for sale off of the farm. This practice ideally allows the redistribution of nutrients back over a large area when the crops are sold and dispersed.

Table 3. Annual manure production by various animals (equalized on 1000 pounds of animal mass)

Animal Species	Annual Production (Tons manure/year)
Broiler	4.5
Sheep	6
Horse	8
Beef	8.5
Dairy	12
Swine	16

Manure management practices vary tremendously depending on the objectives of the swine producer (Figure 1). For example, the primary objective in manure management on a specific farm may be to optimize nutrient recovery and utilization, to minimize land or labor inputs, to control odor, or perhaps some combination of these objectives. By combining the management objectives with factors such as the amount of available land, labor, climate, management skills, and local regulations, an appropriate waste utilization plan can be developed. A successful waste management plan will include consideration of the form and source of nutrients, the amount of nutrient that must be managed, the appropriate placement of the nutrients, and the timing of nutrient applications.

Swine manure is most commonly handled as either a slurry (4 to 15% solids) or a liquid (<1% solids). For slurried swine manure, under-floor storage pits are used to collect manure that drops through a slotted floor where the pigs are housed. The storage pit is scraped at least daily to an outside storage facility where it is collected prior to field application. Storage capacity of these pits typically ranges from 120 to 180 days of manure collection. This amount of storage necessitates manure application onto suitable cropland two or three times a year. This semi-solid manure is generally applied to the surface of crop land using a manure spreader or may be injected beneath the soil surface to improve nutrient recovery and reduce odor concerns.

Liquid-based manure management systems involve flushing the manure through slotted floors several times each day with a sufficient quantity of water to rinse the floors and transport the manure to a storage pit or to an outside lagoon. This management system requires a greater capacity for manure storage due to the large amounts of water added for flushing. However, recycling the lagoon water back through the swine house for flushing the floor will minimize the use of fresh water and the manure storage requirement. Due to the large volumes of water involved and the relatively low nutrient content, the dilute effluent is typically applied through an irrigation system onto adjacent fields. Anaerobic lagoons are especially popular for treatment of swine waste in areas where suitable land for application is limited, because this management system favors high N losses from the storage lagoon. Since manure application rates are typically determined by the N requirement of the growing crop (Table 4), an anaerobic lagoon requires significantly less land for manure management than other management systems.

Table 4. Typical annual nutrient removal in the harvested portion of selected field crops

Crop	Average Yield	Nutrient Removal		
		Nitrogen	P_2O_5	K_2O
		------ Pounds/acre ------		
Bermudagrass	4 tons/acre	184	48	200
Soybean	40 bu/acre	160	32	56
Corn	100 bu/acre	75	44	30
Cotton	1.5 bale/acre	47	18	21
Wheat	50 bu/acre	58	28	17

Anaerobic treatment lagoons are becoming the most common management system for swine manure. However, care needs to be taken during initial siting and construction to prevent leakage into groundwater. Lagoon treatment offers the advantages of low labor requirements, convenient waste treatment, flexibility in storage and application, and a reduced amount of land required for disposal. Anaerobic bacteria present in the lagoon are quite efficient at decomposition of swine manure. However, anaerobic treatment results in incomplete decomposition of organic matter and the production of potentially offensive by-products such as hydrogen sulfide, NH_3, and volatile amines. Because the bacterial action required for waste treatment is partially regulated by temperature, lagoon treatment of swine waste is more common in the warmer southern states than in cooler parts of the country.

Anaerobic lagoons are especially popular for treatment of swine manure in areas where suitable land for effluent application is limited. Anaerobic lagoons typically lose between 60 and 80% of the entering N to the atmosphere via NH_3 volatilization, thereby reducing the amount of manure-derived N that must be applied to land (Table

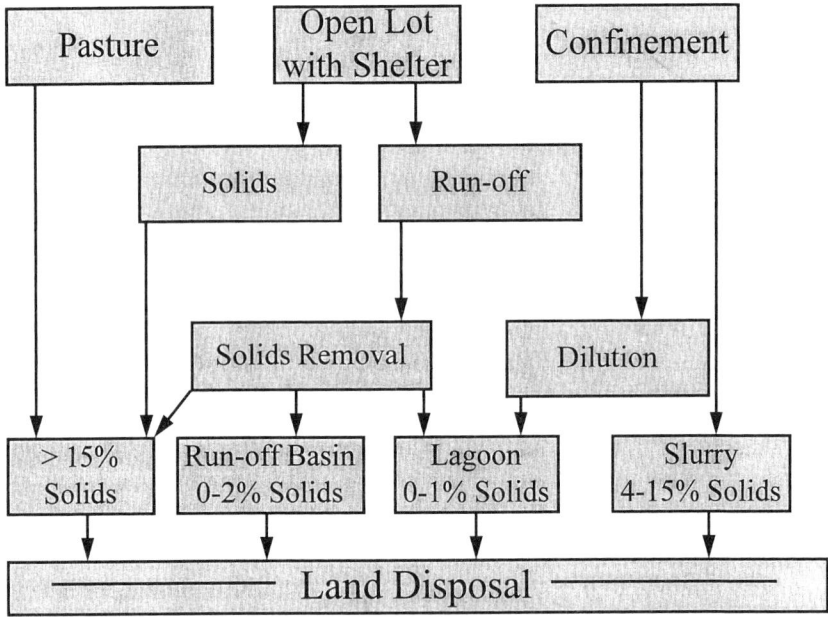

Figure 1. Common swine manure management systems that result in land application of waste (*adapted from 4*).

5). Since manure application rates are typically determined by the N requirement of the growing crop, an anaerobic lagoon requires significantly less land for manure management than for other common manure handling systems. While the majority of N is lost via volatilization from the lagoon, the P and K remain in the lagoon, therefore the ratio of plant nutrients is frequently not balanced with crop needs. In swine lagoons, a high proportion of the P remains in the sludge and is removed during periodic clean-out. Potassium in effluent is generally soluble and is not lost from the lagoon liquid. If the lagoon is properly sized for the number of animals on the farm, sludge clean-out should not be required at less than five to ten-year intervals.

Table 5. Typical nitrogen loss during storage and application of swine waste with different manure management systems (*adapted from 5*)

Management System	Nitrogen Loss (% of total)
Scrape and Haul	15-35
Bedded Manure	20-40
Above-ground storage	10-30
Earth Storage	20-40
Anaerobic Lagoon	60-80

The ratio of essential plant nutrients contained in manure is generally not well balanced with the requirements of plants. Since volatile losses account for a major reduction of the N concentration, there are proportionally greater amounts of P and K remaining than the plant requires. This inherent over application of P and K will result in an accumulation of these elements, along with trace elements present in the waste (such as Cu and Zn). The long-term application of swine manure to soil results in nutrient imbalance and accumulation within the root zone. For example, solid swine manure contains a $N:P_2O_5$ ratio of approximately 1.3:1, while anaerobic swine lagoon effluent typically contains a ratio of 2.5:1. However, a crop such as bermudagrass hay accumulates N and P_2O_5 in a ratio in excess of 5:1. Long-term application of swine waste as a primary N source for plants can quickly lead to excessive P accumulation in the soil with many crops.. The buildup of P in the soil may eventually have adverse impacts on water quality if the P can be transported to surface water. Additionally, extensive downward movement of P has been found in sandy soils that have received long-term additions of swine manure. The long-term implications of this soil enrichment beyond the needs of the plant are not well known, but P loss from these manured soils can potentially degrade water quality and excessively high metal concentrations can impair plant growth.

The form and concentration of N in swine waste is strongly influenced by the manure management system used. For example, slurry manures typically have at least 50% of the total N in the ammonium form. Ammonium generally constitutes at least 90%

of the total N present in anaerobic lagoon effluent. The remaining N is present in organic compounds that require mineralization prior to plant uptake.

The technique used for applying swine manure to the field will have a large impact on N availability to plants. For example, swine manure broadcast directly onto the soil surface may have NH_3 volatilization losses of 75% or more. However, injection of manure beneath the soil surface can reduce the NH_3 loss to less than 5% of the total N applied. When anaerobic lagoon is applied through an irrigation system, NH_3 volatilization losses can exceed 50% of the total N applied. The spatial placement of the applied manure relative to plant roots may also influence nutrient availability.

Loss of N via NH_3 volatilization is clearly a major pathway of nutrient removal from swine farms. The long-term environmental impact of NH_3 emissions is unclear. Ammonia emissions to the atmosphere are important because it is the most important alkaline constituent in the atmospheric boundary layer and can have a significant neutralizing effect on acid rain. However, NH_3 volatilization and subsequent redeposition may provide a significant N input into terrestrial and aquatic environments. Recent European surveys estimate that animal wastes and fertilizers are responsible for 90% or more of the anthropogenic NH_3 emissions in parts of that continent. The encouragement of NH_3 volatilization as a mechanism for reducing the nutrient load on land will require close examination in the future to determine the overall environmental impacts.

Swine manure can be an excellent source of nutrients for crop production systems in agriculture, horticulture, or forestry, but each potential use must be examined individually. There are potential limitations to using swine manure, or any organic material as a source of plant nutrients compared with manufactured fertilizers. For example, swine manure contains a relatively low concentration of essential plant nutrients (Table 6). This is especially true for liquid from anaerobic lagoons which typically contain less than 600 mg N/L (ppm N). The low concentration of nutrients in the manure presents many practical limitations in handling and transporting the large quantity of nutrients required for production of high-yielding crops. It may also

Table 6. Average nutrient composition of swine manure as influenced by handling and storage methods (*adapted from 6*)

Manure Source	Total Solids	Total N	Total NH_3	Total P_2O_5	Total K_2O	$N:P_2O_5$
	---------- ppm ----------					
Scraped lot	180,000	6,500	2,800	6,600	4,500	1:1
Liquid slurry	51,000	3,200	2,000	2,300	1,800	1.4:1
Lagoon liquid	3,200	560	460	220	580	2.5:1
Lagoon sludge	100,000	2,900	710	6,300	780	0.5:1

be difficult to achieve uniform application of the manure across the field, resulting in areas that may have excess nutrients and other areas that are deficient in nutrients. Such practical problems with efficient and convenient utilization of manure-derived nutrients have presented substantial barriers to efficient use of swine and other animal wastes.

The rate of nutrient release from the manure and its subsequent availability for plant uptake is governed by many soil and environmental factors that are often difficult to accurately predict prior to application. The application rate of swine waste applied to soil is generally determined by the N requirement of the target crop. However, the exact amount of N that will be subsequently supplied by the manure to the crop is not easily determined. The fraction of N that is initially present in the organic fraction of the swine waste will only become available for plant uptake after mineralization by soil microorganisms. The extent of N mineralization will depend on the specific properties of the waste and the soil environment. For example, the manure application method, soil temperature, moisture, and pH, will all influence the rate of N mineralization. Some studies suggest that depending on the application method, approximately 30 to 90% of the total N in swine manure will be available for plant uptake during the season of application, with a smaller fraction released in subsequent years (Table 7). However, such estimates tend to over-simplify the complex interactions associated with predicting nutrient availability from organic materials. Development of a rapid, accurate assay for predicting potential nutrient release would promote better waste utilization and nutrient management.

Table 7. Fraction of nitrogen available for plant uptake from swine manure during the first year following application (first-year availability coefficient) *(adapted from 6)*

Manure Type	Method of Application			
	Injection	Incorporation	Broadcast	Irrigation
		N availability		
Scraped paved surface	-	0.6	0.4	-
Liquid slurry	0.8	0.7	0.4	0.3
Anaerobic lagoon liquid	0.9	0.8	0.5	0.5
Anaerobic lagoon sludge	0.6	0.6	0.4	0.4

Many factors combine to determine the amount of nutrients in manure that will ultimately be available for plant uptake. For example, the concentration of nutrients contained in swine manure will vary over the year and an accurate nutrient analysis is essential for proper nutrient utilization. Since each swine operation is unique, the specific characteristics of the swine waste will also be unique. Summaries of average nutrient composition of swine manure are available based on common manure handling practices (e.g. Table 6) that may be helpful for planning purposes, however it is essential that manure be analyzed to determine the specific nutrient concentrations prior to application.

Swine manure contains significant quantities of P, since 70 to 90% of P ingested is excreted in the manure and urine. Phosphorus in swine manure is initially present in both organic (up to half of the total P) and inorganic compounds (primarily phosphate). The availability of the manure P for plant uptake depends on the mineralization of organic P compounds and on the specific adsorptive capacity of the soil. Although many experiments have been conducted to compare the use of manure P with inorganic fertilizer P for crop production, the results have not been consistent. However, over several years the nutrient value of both P sources for crops should be similar. It must also be remembered that application of animal manures may also change the physical and chemical condition of the soil. Therefore, factors such as the soil atmosphere, pH, microbial population, bulk density, organic matter content, and water holding capacity will also change following manure application, all of which may influence nutrient availability for plants.

Most of the P in swine feed is present as readily soluble dicalcium phosphate. However, in the feed grains, 40 to 70% of the P is present in the form of phytin, the calcium and magnesium salt of the hexaphosphate ester of inositol In order for the phytate P to be assimilated by swine, it must be enzymatically hydrolyzed to inorganic phosphate by the phytase enzyme. Since the production of phytase is limited in the swine digestive tract, most of the phytin P passes through the animal without hydrolysis. Considerable effort has been directed at the development of a supplemental source of phytase enzyme that could be added to the feed to improve the digestibility of plant P. Similarly, efforts to breed grain crops that contain lower concentrations of phytin have met with recent success.

Alternative methods of managing swine waste are attracting attention as substitutes for land application. For example, major efforts are underway to make composted swine manure an acceptable option for waste disposal. The use of worms for the production of vermicompost from swine manure is also receiving attention. The use of constructed wetlands may provide an attractive option for treatment of liquid swine wastes. The chemical treatment and concentration of nutrients from the waste into another useable form (such as struvite ($MgNH_4PO_4$) also is an attractive option. The lack of economic incentives to manage the swine waste efficiently has been the major constraint to the development of these new technologies. Clearly, land application is a simple and relatively inexpensive method of disposing of swine wastes and will remain the primary method of disposal for the foreseeable future. However, as environmental concerns and regulations related to swine production issues increase, new management options will be needed to provide additional flexibility in dealing with these issues. As innovative waste management technologies are developed in the future, it may become easier to redistribute the manure-derived nutrients away from the swine farm and utilize the manure as a valuable nutrient resource.

Literature Cited

1. U.S. Department of Agriculture. Agricultural statistics 1996. U.S. Government Printing Office. Washington D.C. 1996.
2. U.S. Department of Agriculture. Agricultural statistics 1995. U.S. Government Printing Office. Washington D.C. 1995.
3. Whittemore, C. The science and practice of pig production. Longman Sci. Technical. Essex, UK. 1993.
4. Melvin, S.W., F.J. Humenik, and R.K. White. Swine waste management alternatives. PIH-67. Pork Industry Handbook. North Carolina Agricultural Extension Service. Raleigh, NC. 1979.
5. CAST. Waste management and utilization in food production and processing. Council Agric. Sci. Technol. Report 124. Ames, IA. 1995.
6. Zublena, J.P., J.C. Barker, J.W. Parker, and C.M. Stanislaw. Swine manure as a fertilizer source. AG-439-4. North Carolina Agric. Exten. Serv. Raleigh, NC. 1993.

Chapter 9

Composition and Uses of Organic Fertilizers

E. E. Huntley[1,2], Allen V. Barker[3], and M. L. Stratton[4]

[1]Department of Soil and Water Sciences, University of Florida, 2169 McCarty Hall, Gainesville, FL 32611-0290
[2]Florida Certified Organic Growers & Consumers, Inc., P.O. Box 12311, Gainesville, FL 32604
[3]Department of Plant and Soil Sciences, Bowditch Hall, University of Massachusetts, Amherst, MA 01003
[4]Range Cattle Research and Education Center, Institute of Food and Agricultural Sciences, University of Florida, 3401 Experiment Station, Ona, FL 33865-9706

> The objective of this chapter is to present facts that lead to a scientific understanding of organic fertilizers. Organic fertilizers may generally be defined as "plant-nutrient-bearing, naturally occurring materials with low solubility in water, with low nutrient concentrations, or with both of these traits". Grouped as materials derived from animal products, plant products, and rocks and minerals, the organic fertilizers are evaluated for their nutrient content and availability.

Organic agriculture has no widely accepted definition because there are multiple conceptions of the basic nature of the term organic among biologists, chemists, and practitioners. Organic farming recognizes the views of biologists, chemists, and practitioners and, in some cases, regulators. Practitioners use broad terms to define organic agriculture as a system that utilizes natural materials and practices that develop, nurture, and protect biological systems for optimum soil health and crop ecology (*1,2*). General agreement among practitioners and scientists is that organic agriculture is a production system that avoids or largely excludes the use of synthetically compounded fertilizers, pesticides, growth regulators, and feed additives (*1,2,3,4*).

The concept that organic gardening involves only natural materials is based on ancient practices dating back to the beginning of crop cultivation (*5*). However, studies of soil fertility and plant nutrition are relatively new sciences (*6*). Early users of fertilizers mixed various organic wastes. Symbiotic fixation of nitrogen by legumes and bacteria was the principal means of returning nitrogen to the soil, when fertilizer nitrogen was very expensive (*7*). Sir Albert Howard, an English agricultural adviser in India, was a prominent, early (1930s and 1940s) advocate of natural systems of farming to maintain organic matter and fertility in soil (*8,9*). In the impoverished region where he was working, farmers could not afford synthetic fertilizers or deal with bulk wastes. Composting permitted the farmers to break down bulk materials and create a valuable soil conditioner and fertilizer to replace humus and nutrients removed by crops. J.I. Rodale was influenced by Howard's work and first used the word organic to describe natural methods of growing (*10*). Rodale's publishing and experimentation popularized and developed organic farming in the United States (*4,10,11*).

In the past twenty years, the market for organically grown produce has expanded from a niche to include mainstream citizens (*12,13,14*). This demand has heightened needs for focusing on a definition for organic farming to protect the public against fraud and to provide fairness in marketing and production (*4,14*). The absence of regulations in most states and differences in regulations among states has led to the development of private certifying organizations for definitions of organic farming (*2,16,17,18*). Only a few of these certifying organizations have state charters or legislative mandates (*19*). Organic food Production Act (OFPA), federal legislation from the 1990 Farm Bill, exists to define and unify organic farming standards among states and certifying organizations (*20,21*). Not yet implemented, OFPA has encountered many procedural, legal, and cost barriers (*21,22*). Not strictly defining organic farming, certifying organizations provide guidelines, listing practices and materials that growers can use for production of organic food. The guidelines are inconsistent, at least in details, among organizations, and some include materials, such as hydrogen peroxide, sulfur, copper salts, and soaps, that may be toxic and not naturally occurring. Most guidelines use terms such as *allowed*, *permitted*, and *prohibited* materials and practices (*1,23,24*). Certification can be expensive and demanding and must be repeated annually. Although a certain amount of reciprocity exists among organizations, certifications by various organizations have different levels of prestige and costs.

Organic farming requires definition of a number of agricultural materials, including pesticides or pest-control devices, growth regulators, adjuvants, and fertilizers. This chapter will deal only with organic fertilizers, which have a working definition of being "plant-nutrient-bearing, naturally occurring materials with low solubility in water, with low nutrient concentrations, or with both of these traits". "Naturally occurring" means that human intervention has not occurred in the development of the fertilizers other than physical processing of the materials or physically extracting them essentially unmodified from a stream of wastes or by-products. In contrast to fertilizers, organic amendments refer to natural materials that impart improvements in soil fertility through physical or chemical modifications. Soil amendments may carry one or more plant nutrients and will be discussed in this chapter as appropriate with fertilizers.

Fertilizing organically does not involve efforts to find a fertilizer for each of the fourteen soil-derived plant nutrients (N, P, K, Ca, Mg, S, Fe, Cu, Zn, Mn, Mo, Ni, B, and Cl). Nitrogen, phosphorus, and potassium are the primary plant nutrients sought in organic fertilizers. Calcium, magnesium, and sulfur are added with soil amendments or as co-constituents with plant nutrients in fertilizers. Concentrated chemical fertilizers today are essentially void of sulfur with these fertilizers being composed of sulfur-free compounds, such as urea, ammonium nitrate, potassium chloride, and concentrated superphosphate (25). Micronutrients generally are not added intentionally as organic fertilizers, although in some cases additions of specific micronutrient fertilizers or sprays of metallic chelates may be permitted in cases of documented deficiencies (2,17,18,26). The organic farmer relies on micronutrients being present in organic fertilizers or amendments in sufficient quantities to replenish those removed by crops, whereas chemical fertilizers may have micronutrient fertilizers that are available for addition separate from mixed fertilizers (25,27).

Crops on organic farms generally are well fertilized with nitrogen. Organic management with use of cover crops, green manures, composts, and farm manures and an abundance of organic fertilizers for nitrogen have a surplus of nitrogen inputs over nitrogen removal by crops (28,29,30,31,32,33,34,35,36). Organic farms tend to have deficits of phosphorus and potassium inputs with phosphorus removal exceeding additions by over 10% and potassium removal exceeding additions by as much as 40% (36,37). Depletions of sulfur, calcium, magnesium, and minor elements in organic systems are not considered to be problems or are lesser problems than in conventionally managed systems.

Release of plant nutrients from fertilizers or soils into a form that plants can absorb is called nutrient availability. Nutrient availability from organic fertilizers is slow and nutrient contents of organic fertilizers are highly variable (38). Considerations of nutrient concentration are important in selecting an organic fertilizer, for concentration governs nutrient availability, the bulk of material that must be applied, and any added benefits that the fertilizer may have as a soil amendment.

Within the following discussion, organic fertilizers are divided into three main groups: (a) Fertilizers derived from animal products; (b) Fertilizers derived from plant products; (c) Fertilizers derived from rocks and minerals. Within each group, fertilizers are evaluated for their nutrient content and availability, especially with respect to nitrogen, phosphorus, and potassium. As will become apparent in this review, many of the organic fertilizers are not well balanced with respect to macronutrients, and micronutrients are present in only low concentrations. Blended fertilizers are mixtures of plant-, animal-, or rock-and-mineral-derived materials with relatively high nutrient contents and are designed to mimic synthetic fertilizers with rapid nutrient availability.

All of the fertilizer analyses reported are in terms of elemental analyses of actual N, P, K, Ca, Mg, or S, although in the trade analyses may be reported or guaranteed in units of N, P_2O_5, K_2O, $CaCO_3$ or CaO, and MgO.

Organic Fertilizers

Fertilizers Derived from Animal Products. Fertilizers derived from animal products are valued for nitrogen, phosphorus, calcium, and magnesium contents (39). Animal products are low in potassium with the exception of farm manures, which are essentially plant-derived products since most livestock and poultry are fed vegetative diets. Many of the animal-derived organic fertilizers have long histories of use, which have continued today as organic gardeners search for quality materials of natural origins (Table I).

Table I. Organic fertilizers derived from animal products and concentrations of primary plant nutrients on an elemental basis

Fertilizer	Primary nutrient concentration, % dry mass		
	Nitrogen	Phosphorus	Potassium
Farm manure			
Livestock	1 to 3	0.4 to 2	1 to 2.5
Poultry	3 to 5	1 to 3	1 to 2
Guano			
High N	10 to 12	5 to 6	2 to 3
High P	1	6 to 7	nil
Sewage biosolids	1 to 4	0.5 to 2	nil
Steamed bone meal	1 to 2	9 to 13	nil
Dried blood	6 to 12	0.5 to 1.5	0.5
Hoof and horn meal	10 to 16	2 to 3	nil
Feather meal	10 to 16	1 to 2	nil
Hair, wool, and silk	8 to 16	1 to 2	nil
Tankage, dried meat, fish scraps	4 to 12	3 to 4	nil

Farm manures. Manures, consisting of feces and urine from livestock and feces from poultry, are common organic materials used as fertilizers, as soil amendments, or in composts (*38*). The main types of manures are those produced by large animals (dairy and beef cattle, hogs, horses, sheep, goats) and by poultry (broilers and other chickens, turkeys, ducks) (*38*). In most of the U.S., manure production, and consequently, total nutrient value, by large animals exceeds that from poultry (*38*). On small farms, manure is usually handled as litter, a mixture of animal manure, bedding material (which may reduce nutrient concentration because of dilution), and waste feed and by-products (*40*). In large production systems, livestock and poultry likely are grown and handled without litter, and the manure is placed in lagoons (cesspools) or tanks before land application. Adding manure or litter regularly to arable land increases soil organic matter or slows its loss (*41,42*). Both products are valued for their potential for improving soil fertility, through better soil structure, aeration, water-holding capacity, for example, and enhancements in crop yields (*34,43*). Most commonly in commercial agriculture, farmers are likely to apply manures from lagoons or tanks to land. This manure will be 90% or higher in water, and the nutrient concentrations in Table I will be diluted accordingly (*44*). Manures from barns, stables, or piles are over 80% moisture, except for chicken manures which are about 35% moisture, and will be diluted with bedding. Large animal manure handled in this manner has nutrient concentrations of about 10 lb. N, 2 lb. P, and 8 lb. K per ton wet mass. Poultry manure will have about 30 lb. N, 7 lb. P, and 8 lb. K per ton wet mass. The high nitrogen concentration with a high percentage of ammonium universally present or formed by mineralization must be considered with caution, for it is the factor attributed to burning noted with liberal use of poultry manure. A caution to users of manures from large farm animals is to avoid or sparingly use manures from livestock, hogs particularly, that have been given dietary supplements of copper or zinc salts (*39,45,46*).

The potential for fertilization of farm land with manures is large. The total farm manure production in the United States currently is about 1.4 billion wet tons (272 millions dry tons) annually (*32*). With good recovery of nutrients, manures may average, on a dry basis, 2.0% N, 0.5% P, and 1.5% K, giving total primary nutrient contents in farm manures of 5.44 million tons of N, 1.36 million tons of P, and 4.98 million tons of K. A typical application of wet farm manure to farm land is 20 tons per acre annually to meet the nitrogen requirement, assuming that half of the nitrogen is available in the year of application. The total manure production could potentially fertilize 70 million acres of farm land.

Manures are most valued for their nitrogen contents. Phosphorus concentrations in manures are considered too low for short term fertilization, although in the long term, phosphorus from manure applications will accumulate in soil (*47*). If farm manures are applied to meet the nitrogen requirement of crops a large fraction of the other element requirements will be met. Farm manures have 1 to 3% Ca, 0.4 to 1% Mg, and about 0.4% S on a dry mass basis.

Only a fraction, perhaps 2.5%, of the total manure production in the U.S. is utilized fully on farm land (*33*). Dairy, livestock, and poultry farmers do not have adequate areas of land on which manures can be applied. Often, livestock production does not occur in regions with large acreages of crop land, and shipping costs discourage distribution of the manures. The imbalances between distribution of livestock and crop land and ineffective means of collection, transportation, and application contribute to losses of nutrients through volatilization and leaching (*38*).

Organic certification standards sometimes limit applications of uncomposted or raw manures to food crops (*17,18*). Generally, green manures and crops not for human consumption can receive unrestricted applications of manure, but food crops often have the restriction that manures be applied 60 to 120 days before harvest.

Guano. The somewhat fossilized deposits of excreta and carcasses of sea fowls, marine animals, and bats has recorded use dating back to the 1100's and with extensive use developing in the 1800's (*7*). Guano is considered a nitrogen fertilizer (8% N, 6% P, 2% K) if from arid regions (South America west of the Andes, mainly Peru and some caves of southwestern USA). Some deposits from humid regions (South Pacific) are high in phosphorus (1% N, 6.5% P, 0%K), with nitrogen and potassium being leached from the deposits (*48*). Guano has much available nitrogen in soluble forms, likely as nitrate. Ammonium may be high in damp guano. Expensive today, guano is scarce material sold in small bags to gardeners. Release of nutrients from guano is rapid with much of the nutrients being water soluble, giving the material quick acting properties similar to those of chemical fertilizers.

Sewage biosolids. Sewage biosolids are organic and inorganic compounds removed during treatment of wastewater from domestic and industrial discharges. The production of sewage biosolids is about 7 million dry tons per year (*49*). Biosolids vary widely in nutrient composition being on average about the same as farm manures, except for potassium, which is partitioned into the aqueous fraction or effluent at the wastewater treatment plant. Nitrogen that remains is largely in organic compounds, but phosphorus is largely in inorganic phosphates of iron, aluminum, and calcium, which may be added during treatment of wastewater. Availabilities of nitrogen correlate with concentration of nitrogen present and range from 10 to 50%. Phosphorus availability would be high in biosolids, but fixation in soil would make biosolid phosphorus little or no different from that of other fertilizers. Trace metals including essential elements and nonessential elements (As, Cd, Cr, Hg, Pb, Se) are present in sewage biosolids. Concentrations of trace elements are among the deciding factors as to whether or not biosolids can be utilized on land (*50*). Federal regulations controlling what is called clean sludge or exceptional quality biosolids give protection against contamination of soil with trace elements. Generators of clean sludge are required to have an ongoing monitoring program to ensure that their biosolids meet standards of clean sludge (*51*). Biosolids that meet the federal standards for clean sludge can be applied without restrictions to land. Use of sludge is not allowed in all organic certification standards. If sludge is allowed, organic growers must follow guidelines established by certifying organizations (*18*).

Bone meal. Bones, commercially called steamed bone meal or bone phosphate, are valued as fertilizer for their phosphorus content. The principal source of bones is slaughterhouses. In processing, raw or green bones are steamed at high pressures to remove fats and proteins, which go into gelatins and glues (*52*). Organic solvents may be used in extraction along with the steam, which are used to drive off the solvent. The term raw bone refers principally to bones of newly slaughtered animals but also to bones which have been buried or weathered for a long time. Nonetheless, because of the value of the extracted fats, gelatins, and glues and because of advantage of killing of pathogens by steaming, raw bone meal is seldom available in the market and should be avoided. The grease in raw bones makes them difficult to grind and resistant to weathering. Although raw bone meal will have about twice the nitrogen as steamed meal, steamed bone meal is richer in phosphorus, because of the concentration of phosphorus with removal of moisture and organic matter in steaming. Bone meal is chemically similar to tricalcium phosphate, which is the major mineral constituent of bone. Steamed bone meal is about 1 or 2% nitrogen, about 10 to 13% actual phosphorus, and is virtually void of potassium (*24,52*). Although bone has slightly lower phosphate availability than basic slag or superphospates, the calcium phosphate of bone is much more available than fluoridated compounds in rock phosphate.

Bone meal has slight value as a lime (4%, in raw bone, to 8% calcium and magnesium carbonates). Liming of soil with the application of bone meal generally depresses the solubility of phosphorus in the meal. Bone meal is valued by organic gardeners because of its high phosphorus concentration and availability without having burning or desiccating effects on crop foliage or roots. It can be used rather indiscriminately and can be mixed closely with seeds, bulbs, or transplants without salt damage. Bone meal represents only a small fraction of the total phosphorus fertilizers used today. Competing industries, such as production of animal feeds, bone black, knives, cutlery, and buttons, combined with rather low supplies of bones contribute to high prices for bone meal as a fertilizer.

Dried blood. Commonly called blood meal and collected from slaughterhouses, dried blood is valued for its relatively high nitrogen concentration for an organic fertilizer (~12%). Some dried bloods, called black blood meals because of aging, may have 6% nitrogen. Because of low phosphorus and potassium concentrations, analyses on bags of market dried blood usually do not guarantee phosphorus and potassium concentrations. Almost all (>70%) of the nitrogen in dried blood is available in the first year of application; therefore, dried blood is considered as a rapid-release organic fertilizer and not as one to build soil by retention of residual nitrogen (*7*). It gives nitrogen availability faster than any other organic fertilizer. In practice of use, dried blood should be treated as an ammoniacal chemical fertilizer, applied to soil one to two weeks before planting. Use of dried blood requires that soil conditions (moisture, temperature) be favorable for its mineralization and nitrification.

Other animal tissues. Leather, felt, wool, hair, silk, feathers, hooves and horns, tankage, dried meat, and fish scraps, emulsions (>90% water), or gurries (>80% water) are examples of packing-house refuse products suitable as organic fertilizers. These materials are valued for their nitrogen and phosphorus contents. Supplemental potassium fertilization is required with use of these fertilizers unless the soil is inherently rich in potassium. Usually less active as nitrogen sources than dried blood, these materials can be used for increasing nitrogen storage in depleted soils. The nitrogen in hoof and horn meal is about 70% available in the first season. Leather and felt scraps or meal, wool, hair, silk, and similar wastes are tough and have poor mechanical condition as fertilizers (*7*). Without processing, the availability of nitrogen in these materials is less than 30%. These materials may be treated with sulfuric acid to form rough ammoniates, which are not certifiable as organic because of their chemical treatment. Users should be alert to the possibilities of chromium being present in tannery by-products. Some fish emulsions may be so dilute in plant nutrients that they have little value as a fertilizer. Also, some of the emulsions or slurries may be fortified with urea, phosphoric acid, and potassium chloride to raise their analyses and to give the fertilizers balance with primary macronutrients. These fortified materials should not be considered as organic. Fish emulsions without fortification are organic materials and are used for foliar fertilization or on the soil (*3,17,18,26,43,53*). Sprays with these materials are only supplemental to soil management and are not substitutes for soil-applied fertilizers. Foliar sprays of fish emulsions are used also by organic gardeners for suppression of plant diseases (26,54).

Fertilizers Derived from Plant Products. These products include many commercially available and farm grown materials (Table II). The commercial materials include by-products of agricultural manufacturing processes and composts. By-products include seed meals (cottonseed meal, soybean meal, castor pomace, linseed meal) from vegetable oil processing industries. Farm-grown materials include crops residues and green manures. Making up about 50% of organic waste produced in the United States, roughly 70% of crop residues are recycled back to the land in nutrient-conserving practices, returning to the soil an estimated 4.4 million tons of nitrogen, other nutrients, and organic matter each year (55,56,57,31).

Table II. Organic fertilizers derived from plant products and concentrations of primary plant nutrients on an elemental basis

Fertilizer	Primary nutrient concentration, % dry mass		
	Nitrogen	Phosphorus	Potassium
Seed meals			
Cottonseed	5 to 7	1	2
Soybean	6 to 7	1	2
Castor bean	4 to 7	1	2
Linseed	5 to 7	1	2
Tobacco stems	2	0.2 to 0.5	5 to 9
Compost	0.3 to 3	0.1 to 1.7	0.1 to 2
Hay			
Grass	1 to 3	0.1 to 0.5	1.5 to 4
Legume	2 to 5	0.1 to 0.5	1.5 to 4
Kelp	1	0.1 to 0.5	1.5 to 13
Garbage tankage(food wastes)	1 to 3	0.2 to 1	1 to 3
Wood ashes	0	1 to 2	1.5 to 10

Seed meals. Seed meals are made from seeds of crops valued for their oil. Seeds after they are freed of hulls, pods, lint, and the oil are ground into meal. Without extraction of oil, seeds decompose rather slowly in soil. Relatively little of these products, such as protein rich cottonseed meal, soybean meal, and linseed meal, reach the fertilizer trade because of their value as livestock feed. Castor pomace is poisonous to mammals because of its content of the toxic alkaloid ricinine (58). Castor pomace is cheaper because it is not a livestock feed. It is highly poisonous to humans and should be handled accordingly, assuring that none of the dust is inhaled. Seed meals are valued for their nitrogen contents (4% to 7%) but are not valued for phosphorus and potassium concentrations. The availability of nitrogen from seeds is expected to be about 65% to 70% in the growing season of application.

Tobacco stems. Lightly fertilized with nitrogen and phosphorus fertilizers, tobacco is commonly fertilized with cottonseed meal and linseed meal for nitrogen. Tobacco is heavily fertilized with potassium for leaf quality (*24*). Potassium accumulation in stems makes them an excellent source of potassium, since all potassium in the stems is water soluble (*59*). Tobacco stems are a rare product today.

Green manures and cover crops. Green manures are crops grown specifically for the addition of organic matter to soil. Nitrogen is added if the green manure crop is a legume (*58*). The amounts of nitrogen added by legumes vary from below 50 to in excess of 150 lb. per acre per year (Table III). Grasses and other nonlegumes do not add nitrogen to the soil. If the primary goal of green manuring is to build up organic matter (*59*) (humus) in soil, nonlegumes or grasses and legumes are best. The nonlegumes or mixtures yield more biomass than legumes alone.

Table III. Nitrogen fixation by common legumes

Legume	Fixation, lb. N.acre-1.yr-1
Alfalfa (*Medicago sativa* L.)	>150
Hairy vetch (*Vicia villosa* Roth.)	>150
Cowpea (*Vigna sinensis* Salvi)	>150
Sweet clover (*Melilotus officinalis* Lam. or *M. alba* Desr.)	100 to 150
Red clover (*Trifolium pratense* L.)	50 to 150
Soybean (*Glycine max* Merr.)	50 to 150
White clover (*Trifolium repens* L.)	50 to 150
Crimson clover (*Trifolium incarnatum* L.)	50 to 150
Garden bean (*Phaseolus vulgaris* L.)	<50
Garden pea (*Pisum sativum* L.)	<50

If nitrate in soil is high, nitrogen fixation will be low, as the legumes will use the soil-available nitrogen rather than fixing it (*60*). Under conditions of low soil nitrate, from 66% to 80% of the nitrogen in legumes comes from biological fixation; the remainder is supplied by the soil. It is essential that the legume be plowed under for the nitrogen to be added. The nitrogen does not reside in the nodules of the roots as some practitioners believe. About 20% to 50% of the nitrogen in green manure will be available to the next crop (*61*). The composition of the shoots of legumes resembles that of legume hay (Table II). Harvesting of legume hay adds no net nitrogen to soil, whereas incorporation of legume or grass hay produced off-site would enrich the soil with nitrogen and organic matter.

Green manure crops are turned under at an immature state into the ground in contrast to crop residues, which are the materials left after harvest. Green manures are turned under at initiation of bloom or at about half height of the mature crop, whichever state is easiest to recognize and manage. Cover crops are grown to protect soil against erosion (*62,63*), although the term cover crop often is used interchangeably with the term green manure (*64*). Green manures and cover crops can help to enrich soils by reducing the loss of soil nutrients (*65*). The fibrous roots of rapidly establishing grasses and cereal grains are much more effective in restricting leaching than the roots of legumes, which may be slow growing (*65,66,67*).

Composts. Compost is a mixture of decomposed organic matter, similar to humus in soil (*35*). Made in on-farm activities (*71*) or manufactured off-farm (*69,70*), compost can be made of essentially any decomposable organic waste. Plastics will not decompose in composting. The formation of compost is related to the composition of the feed stock into the compost (*1,18,29,35*). Nitrogen is the principal nutrient in compost. Composts should have a minimum of 1% total N; otherwise, the compost is suitable only as a mulch, or supplemental fertilization must be applied with compost. Phosphorus and potassium are low in compost due to their low concentrations in compostable feedstocks. Potassium also can be lost during composting. Other nutrients vary with the feedstock. In addition to its nutrient content, compost is valued for its organic matter content. Compost has wide acceptance in the nursery, landscape, and turf industries for land applications and for substitutions for peatmoss. The process of composting is important to organic farmers because it reduces bulk matter into manageable substrates, kills weed seeds and pathogens, and destroys pesticide residues (*57,71,72,73,74,75,76,77*). Composts are suggested to be strongly suppressive of soil borne pathogens, such as several fungal diseases and nematodes (*35,,75,78*).

Wood ashes. Wood ashes are one of the best organic sources of potassium. The potassium in ashes ranges widely but typically is about 3% for softwood ash and 8% for hardwood ash. Present largely as potassium oxide or carbonate, potassium in wood ashes is water soluble and is immediately available for plant nutrition. Wood ashes also have value as lime, their value being about 20% calcium (50% $CaCO_3$) or about half as effective as agricultural limestone (*7,48*). Long-term applications of wood ashes can lead to substantial increases in soil pH. Coal ashes are poor sources of nutrients being less than 1% K.

Food wastes. Food wastes result from the processing and trimming of foods and from leftovers (*56,79*). Most (89%) of the wastes are from fruits and vegetables, and production is estimated at 3.4 millions tons per year (*56,79,80*). Similar to the case with animal manure production, availability of land for application is limited (*56*). Much of the by-product is wasted, or land near by is overloaded (*56,81*). Most of collected waste (84%) is used for animal feed, and about 16% is available for land application (*56*). The wastes can be liquid or solid from food processing industries to include pomace (residue from extraction of fruit juices), pulp, skins, seeds, and cull fruits. Liquid material may be used in irrigation (*82*). Nutrient content is low, but organic matter may be considerable in solid matter (*56,82*). Vegetative parts may be valued for nitrogen and potassium. Hulls and skins are valued for potassium, which may be 20 to 40% K in the ash of banana and citrus skins and seed hulls (7). Sometimes composting is recommended to remove sugars and destroy pesticide residues (18,29).

Kelp. Seaweed or kelp are of marine origin. The biomass of the seaweed is dried and ground and marketed as meals for land application. Supposedly, if the grower is close enough to coastal areas for harvesting seaweed, it could be applied fresh and in bulk to land. Aqueous extracts of seaweed are marketed as liquid products or as dried meal from the extracts for foliar sprays. Potassium concentration varies with species, but marketed products are about 4.5% K. Kelp does not contain appreciable sodium as sodium, being external on the kelp, is washed off easily (83). Kelp extracts refer to aqueous extracts or dried powders of aqueous extracts of kelp which contain dilute concentrations of nutrients and are very expensive. Kelp extracts are marketed for their trace element, growth hormones, vitamins, amino acid, and mineral contents, and disease-suppressing powers (26,54). Users should take caution when vendors advertise products as containing over 70, or even as much as 93 minerals or trace elements, knowing that only 14 elements are essential for plants.

Fertilizers Derived from Mineral Sources. Natural mineral powders are staples in organic soil management. Rocks and minerals are specifically sources of calcium, magnesium, potassium, phosphorus, and sulfur and also carry micronutrients. Physical processing to finely divide minerals is accepted as an organic practice, although chemical processing, heating, or burning may cause loss of organic classification. These materials are generally free of nitrogen. Some mineral-derived materials are listed in Table IV.

Table IV. Organic fertilizers derived from mineral or rock products and concentrations of primary plant nutrients on an elemental basis

Fertilizer	Primary nutrient concentration, % dry mass				
	P	K	Ca	Mg	S
Rock phosphate	13	nil	30	tr	tr
Colloidal rock phosphate	9	nil	15	nil	nil
Granite dust	nil	5	tr	tr	nil
Basalt dust	tr	0.7	2	0.5	tr
Glacial dust	tr	1	1-3	1	tr
Greensand	nil	7	nil	nil	nil
Gypsum	nil	nil	23	nil	13
Calcitic limestone	nil	nil	40	4.9	nil
Dolomite	nil	nil	22	12	nil
Langbeinite	nil	18	nil	11	23

Mean values are reported in this table are for marketed grades of fertilizers although the actual concentrations vary around the reported means. tr: trace or less than 0.5%

Rock phosphate. Rock phosphate is a mined material apparently originating from weathered precipitates from waters of ancient seas (*25,52*). Excavation occurs from shallow strip mines. The deposits of high-grade rock phosphate in the world are enormous, and the low-grade deposits are inestimable. Rock phosphates among these sources vary greatly in phosphorus concentrations, but marketed grades have 13% to 15% actual P. Rock phosphate is predominately a mineral called francolite, which is calcium carbonate fluorapatite [$Ca_{10}F_2(PO_4)_{6-x}(CO_3)_x$]. It is a very hard mineral similar to enamel of teeth. Two products are available for commercial markets in the United States, the brown rock phosphate from Tennessee and waste-pond phosphates from Florida. The latter is called colloidal rock phosphate. Rock phosphate must be ground to silt-sized particles to increase phosphorus availability. In rock phosphate, available P is only about 1% actual P (0.9 to 1.3, depending on total P), whereas in the colloidal rock phosphate, available P would be proportionally less than 1%.

Incorrect use of rock phosphate, such as using it in well-limed soils, not applying it at several times the recommended rate for commercial fertilizers, failure to mix it well in soil, or to add organic matter, can give results not better than applying no phosphorus fertilizer. Finely pulverized rock phosphate or colloidal rock phosphate must be used on acidic soils (not much above or much below pH 5.5) and mixed well into the soil. The rock phosphates must be applied at two to four times the amount of application recommended by soil tests, which are calibrated for use of superphosphates (*7,25*). Organic matter should be used liberally with applications of rock phosphate. The acids of decaying organic matter assist in dissolution of rock, and the organic matter chelates (forms complexes with) soluble iron and aluminum, restricting phosphorus fixation in soil (*84*). A common practice in use of rock phosphates is to mix the powder with farm manures or with composts or to add it during composting so that the availability of phosphorus is increased. About 50 lb. of rock phosphate should be mixed with each wet ton of manure or compost. Use of rock phosphate with a green manure crop, such as sweet clover, improves effectiveness of the rock. Fibrous-rooted crops and ones that absorb a lot of calcium, an action that might increase dissolution of rock, are reported to make efficient use of rock phosphate (*7*). Due to the low solubility of rock phosphate, its residual value in the soil after the first year is no better, perhaps worse, than that from any other phosphorus fertilizer (*25*). Superphosphates, ammonium phosphates, and pyrophosphates manufactured by acidulation, heating, or ammonification are not considered organic products (*18,25,52*).

Granite dust. Tested with varying outcomes, granite dusts (sand-sized particles of feldspars and some micas) have substantial amounts of potassium (5% K). Yet, the availability is nil so that bags of marketed granite dust have no guaranteed analysis of available potash. This means the material is not a fertilizer and has no short term value. The principal form of potassium is primary minerals, resembling those of granite dust. The total potassium in the primary mineral form would be little affected by the addition of granite dusts at moderate applications of tons per acre. Other potassium bearing rocks are alunite [$K_2Al_3(OH)_6(SO_4)_3$], an ore for alum], leucite [$KAl(SiO_3)_2$], and potassium-bearing shales. The ores have potential for extraction of potassium by the fertilizer industry if current sources should be exhausted (7).

Greensand. Large quantities of easily accessible greensand are found off the Eastern Seaboard in New Jersey, Delaware, Maryland, and Virginia. As the name implies, the texture of greensand is that of mixed sands. The principal mineral in greensand is glauconite [$K_2(MgFe)_2Al_6(Si_4O_{10})_3(OH)_{12}$]. Greensand has slightly higher total potassium (7% K) and available potassium than granite dust. Greensand has value in its potential for long-term enrichment of soil with potassium.

Basalt rock dusts. Basalt is rock with minerals of plagioclase (Ca and Na feldspar), augite (Mg, Fe, Al pyroxene), magnetite (iron oxide), and others, including phosphorus-, potassium-, and micronutrient bearing minerals. It is a finely ground (diameters of very fine sand or silt) by-product of the aggregate industry. It is a slightly alkaline material giving it some liming potential. Trials by the University of Massachusetts (Barker, A.V.) and Remineralize the Earth (Campe, J., Soil remineralization research packet; Agriculture; Northampton, MA, personal communication) demonstrated soil fertility with respect to K, Ca, Mg, P, Mn and Fe. Typical applications are 4 to 20 tons rock dust per acre, with the small amounts used for repeated annual applications.

Glacial dust. Glacial dust is a mixture of quartz (silica), feldspars, micas, and other minerals from washing of glacial till in the aggregate industry. It varies in quality with respect to size of particles and composition, and its value comes from the K, Ca, Mg, and micronutrients in the minerals. It is essential that the material be fine, and since in many cases the material is coarser than rock dusts from the rock-crushing industries, the glacial dusts are slower to release nutrients than the rock dusts.

Gypsum. Gypsum is a soft mineral ($CaSO_4 \cdot 2H_2O$) occurring in rock form or in sands. Its use was introduced by German farmers who noted that crops grew better in the paths of mortar mixers working with plaster (7). The nutritive value of gypsum comes from its calcium content with secondary effects arising from its sulfur content. Use of gypsum is questioned in some publications by organic gardeners (85) but not by others (86). The problem seen with gypsum was its sulfate content. Some organic gardeners attributed the sometimes suppressive effects of ordinary superphosphate to its sulfur content. Ordinary superphosphate is 60% gypsum by mass. However, sulfate has low phytotoxicity, especially in the presence of calcium, which keeps the concentration of sulfate low (87). The toxic effects of ammonium sulfate on earthworm populations are attributed to the toxicity and acidifying effects of ammonium, not to the sulfate (88).

Gypsum has fallen into disuse as a fertilizer, because the same effects can be produced in most cases more cheaply by lime and ordinary superphosphate. Modern usage would be in cases where calcium nutrition is required without the need for raising the pH of soil or where calcium is needed for its ameliorative effects on soils of poor physical structure (25).

Limestone and other limes. Limes are calcium- and magnesium-containing materials that are capable of neutralizing soil acidity. The activity of limes comes from their contents of carbonates, hydroxides, or oxides of calcium and magnesium, not from the actual calcium and magnesium present. Agricultural limestone, marble, quicklime, hydrated lime, chalk, bones (for P), marl, marine shells, wood ashes (for K), and slag (for P) are limes. Agricultural limestones are the most commonly used limes and are mixtures of calcite ($CaCO_3$) and dolomite ($CaCO_3 \cdot MgCO_3$) (89). The nutrient concentrations vary from 40% Ca and 0% Mg in pure calcite to 22% Ca and 13% Mg in pure dolomite. Pure calcite and dolimite deposits are seldom found. Agricultural limestones are mixtures (intergrades) of calcite and dolomite and have varying concentrations of calcium and magnesium. Users should check the analysis on labels and make purchases according to whether or not magnesium fertilization is needed in addition to the calcium and neutralizing capacity of the limestone. The liming potential of calcite and dolomite are virtually equal, particularly if the limestone is ground to pass a 60 to 100 mesh screen (<0.4-mm diameter particles). No distinction is made between agricultural limestones used by conventional and organic farmers.

Quicklime (CaO) is burnt lime, produced by heating or burning agricultural limestones. Quicklime is amorphous and with less mass per unit of neutralizing capacity than agricultural limestones. Quicklime is more reactive than agricultural limestone, with a unit of quicklime being equivalent to 1.8 units of limestone. Use of quicklime is not permitted in most organic practices (85). The reasoning for this prohibition is arbitrary with no scientific basis or consistency. Wood ashes, basically of the same composition and reactivity as quicklime, are permitted. Hydrated lime [$Ca(OH)_2$], formed by slaking quicklime, is usually prohibited as an organic fertilizer.

Marble, marls, bones, shells, wood ashes, slag, and other products are less effective in neutralizing acidity than agricultural limestone because of their hardness or low calcium carbonate equivalency. However, these materials would be expected to be adequate calcium and magnesium fertilizers.

Langbeinite and other potassium-containing ores. Mineral beds of potassium-containing ore are found across the world. Although these beds are deep into the earth (800 to >5,000 ft), ample supplies of potassium fertilizers are available. Organic gardeners utilize very few of the materials from these sources, which are extracted from the earth by solid ore recovery or by solution mining (25). Carnallite ($KCl \cdot MgCl_2 \cdot 6H_2O$), sylvinite (mix of KCl and NaCl) and langbeinite ($K_2SO_4 \cdot 2MgSO_4$) are the principal potassium-bearing minerals from the mines. Potassium chloride (~50% K) and potassium sulfate (~39% K) are produced by dissolution, recrystallization, and washing of the salts in the ores. Langbeinite is purified by washing of the ores to remove chloride salts, which are more water soluble than the double salt (sulfate of potassium-magnesia). The purified sulfate of potassium-magnesia (18% K, 11% Mg, 23% S) is sold under several commercial registered trade names and is accepted by many certifying organizations as organic (*23*). Potassium chloride and potassium sulfate are not certified, generally, because of their high solubilities in water and high concentrations of potassium or chloride (*16*). The lack of certification of potassium chloride, a natural product, because of its chloride, the rejection of ordinary superphosphate, a manufactured product, because of its sulfate, the questioning of gypsum because of its sulfate, and the acceptance of langbeinite, a water-soluble sulfate, are matters that give scientific puzzlement to certification standards. These inconsistencies are particularly severe, considering that most organic growers are extracting more potassium from the soil than they are returning (*37*). Manure salts are unpurified potassium ores, about 20% K, 0.25% Mg, 22% Na, 50%, Cl, and 0.2% S among other elements (*7*). The high sodium concentration might present problems in agriculture.

Blended Fertilizers. These fertilizers are mixtures of nutrient-rich (N, P, K) organic materials derived from plant, animal, or rock and mineral sources. All growers recognize the need for additions of all nutrients to replenish those removed by crops from soils. Few of the organic products, with notable exceptions of composts and farm manures, provide much balance in supply of the primary macronutrients. The fish product failed as a fertilizer, ultimately, because it provided no potassium and was used in soils that had inherently low capacity to supply potassium. The commercial potassium fertilizer was slow to develop because scientists believed that the total potassium in soil was high, failing to recognize that 90% to 98% of the soil potassium was in primary minerals and virtually unavailable (*90*). Blended organic fertilizers are used to give a balanced analysis of primary nutrients, similar to synthetic fertilizers (*1,91*). Dried blood, bone meal, dry climate guano, and kelp might be compatible materials for blending. Wood ashes, because of their alkalinity, would not be acceptable for blending with nitrogen- or phosphorus-rich organic materials. Many commercial blends are available in the retail store and catalog markets. Often the materials blended in the fertilizers are not identified. Users should strongly question organic blends that guarantee minimum available analyses of more than 16% N, 1% P, and 5% K, being suspicious that synthetic materials have been used in the manufacture of the blends. These cautions apply to dry or liquid materials and also to nonblended materials.

Opting for Organic Agriculture

Only the practices for fertilizing crops organically are considered in this chapter. However, fertilizers alone are only part of organic farming. Organic farmers are expected to develop ecologically sound farm plans that include testing of soil (*2,17,18,28*), integrating crop rotations, returning crop residues, potentially using farm manures, composts, off-farm wastes, and accepted organic fertilizers, and controlling pests without the use of manufactured pesticides (*1,4,11,62,95,97,98*). A uniting or meeting of organic and conventional practices is the basis of integrated pest management and sustainable agriculture (*4,12,13*).

The strict interpretation of organic gardening was viewed as being inconsistent, nonscientific, and even incorrect. A critique of publications such as The Encyclopedia of Organic Gardening (85) reveals reasons for that view. Organic certifying agencies were formed to bring consistency and fairness to interpretations of organic farming (17,18). Criteria are developed for organic certification through multiple-party input from producers, processors, consumers, government agencies, researchers, professors, and organic trade associations. As support for organic agriculture grows among the public and as more and stronger organizations are formed, continued research in organic farming and communication among the involved parties are required to meet the increasing demand for organic produce (4,14).

Literature Cited.
1. *The Real Dirt*; Smith, M., Ed.; Northeast Region Sustainable Agriculture Research and Education Program: Burlington, VT, 1994, 294 p.
2. U.S. Department of Agriculture. *Report and Recommendations on Organic Gardening*; U.S. Department of Agriculture: Washington, DC, 1980; 94 p.
3. Organic Foods Production Association of North America. *Allowed and Prohibited Materials List. Guidelines for the Organic Foods Industry*, 2nd ed.; OFPANA-Organic Trade Association: Greenfield, MA, 1992; pp 32-38.
4. Harwood, R. *Am. J. Altern. Agric.* **1993**, 8, pp 150-154.
5. Doran, J.W.; Sarrantonio, M.; Liebig, M. *Adv. Agron.* **1996**, 56, pp 1-54.
6. Reed, H.D. *A Short History of the Plant Sciences*; Chronica Botanica: Waltham, MA, 1942; 320 p.
7. Bear, F. *Theory and Practices in the Use of Fertilizers*; Wiley: London, 1938; 360 p.
8. Howard, A. *An Agricultural Testament*; Oxford: London, 1940; 253 p.
9. Golueke, C.G. *Composting, a Study of the Process and its Principles*; Rodale: Emmaus, PA, 1972; 110 p.
10. Rodale, R., Ed.; *The Basic Book of Organic Gardening*; Ballantine: New York, 1971, 377 p.
11. Rodale, J.I. *Pay Dirt, Farming and Gardening with Composts*; Rodale: Emmaus, PA, 1945; 245 p.
12. Hess, C.E. In *Sustainable Agriculture Research and Education in the Field*; Board on Agriculture, National Research Council, Ed.; National Academy: Washington, DC, 1991; pp 13-31.
13. OTA. *The Organic Report*; Organic Trade Association: Greenfield, MA, 1995; March, 12 p.
14. OTA. *The Organic Report*; Organic Trade Association: Greenfield, MA, 1995; July, 12 p.
15. U.S. Department of Agriculture, Agric. Marketing Serv. *Federal Register.* 1995, Monday, March 27, 60, 58, pp 15744-15745.
16. International Federation of Organic Agriculture Movements. *Basic Standards of Organic Agriculture and Food Processing*; IFOAM General Assembly, Christchurch, New Zealand; Tholey-Theley, Germany, 1994; 32 p.

17. Florida Certified Organic Growers and Consumers. *Policy and Procedures Manual*; Gainesville, FL, 1996, 53 p.
18. *Guidelines for the Development of National Standards for Organic Food Production*; Prolman, G. Ed.; Made in Nature: San Rafael, CA, 1995, pp 1-11.
19. Landay, J.S. Organic farmers to Washington: regulate us. *The Christian Science Monitor*; Boston, MA, 1996; Monday, December 30, p 3.
20. Merrigan, K. *Am. J. Altern. Agric.* **1993**, 8, 4, pp 158-160.
21. Sullivan, T. *Minnesota Family Farm Law Update.* Farmers' Legal Action Group, Inc.: St. Paul Minnesota, 9, 3, pp 3-9.
22. Sullivan, T. *Minnesota Family Farm Law Update.* Farmers' Legal Action Group, Inc.: St. Paul Minnesota, 9, 4, pp 1-9.
23. Simmonds, B. and Brosten, D. Defining "organic". *Agrichemical Age*, **1990**, 34, 3 pp 12-14.
24. Garrett, J.H. *The Dirt Doctor's Guide to Organic Gardening*; University of Texas Press: Austin, TX, 1995; 216 p.
25. Jones U.S. *Fertilizers and Soil Fertility*; Reston Publ. Co.: Reston, VA, 1979; 368 p.
26. Senn, T.L. *Seaweed and Plant Growth*; Clemson University: Clemson, S.C., 1987; 166 p.
27. Murphy, L.S.; Walsh, L.M. In *Micronutrients in Agriculture*; Mortvedt, J.J., P.M. Giordano, W.L. Lindway, Eds.; Soil Science Society of America: Madison, WI, 1972; pp 347-387.
28. Gershuny, G.; Smillie, J. *The Soul of Soil: a guide to ecological soil management*; agAccess: Davis, CA, 1992, 174 p.
29. Stevenson, F.J. *Humus Chemistry: genesis, composition, reactions*, 2nd ed.; John Wiley & Sons: New York, NY, 1994; 496 p.
30. Cox, C. *J. of Soil and Water Cons.* **1995**, 50:3, p 223.
31. Power, J.F.; Legg, J.O. In *Crop Residue Management Systems*; Oschwald, W.R., Ed.; Am. Soc. Agron. Spec. Publ. 31: Madison, WI, 1978; pp 85-100.
32. Eck, H.V. and B.A. Stewart. In: *Soil Amendments and Environmental Quality*; J.E. Rechcigl, Ed.; Lewis Publishers, Boca Raton, FL, 1995; pp 169-198.
33. Safley, L.H, Nelson, D.W. and Westermann, P.W. *Transaction of Am. Soc. Agri. Eng.*, **1983**, 26, pp 1166-1170.
34. Rausmussen, P.E.; Allmaras, R.R.; Rhode, C.R.; Roager, N.C. *Soil Sci.Soc. Am. J.* **1980**, *44*, pp 596-600.
35. Stratton, M.L.; Barker, A.V.; Rechcigl, J.E. In *Soil Amendments and Environmental Quality*, J.E. Rechcigl, Ed.; Lewis Publishing: Boca Raton, FL, 1995; pp 249-309.
36. Lockeretz, W.; Klepper, R.; Commoner, B.; Gertler, M.; Fast, S.; O'Leary, D.; Blobaum, R. In *Agriculture and energy*; Lockeretz, W., Ed.; Academic Press: New York, NY, 1977; pp 85-101.
37. Lengnick, L.L.; King, L.D. *Am. J. Altern. Agric.* **1986**, 1, 3, pp 108-114.
38. U.S. Department of Agriculture. *Improving soils with organic wastes*, Report to the Congress in response to Section 1461 of the Food and Agriculture Act of 1977 (PL 95-113); U.S. Government Printing Office: Washington, D.C. 1978, 157p.

39. Elliot, L.F.; Swanson, N.P.; *Land application of waste materials*; Soil and Water Conservation Society national conference proceedings; Soil and Water Conservation Society: Ankeny, IA, 1976, pp 80-90.
40. Bosch, D.J.; Napit, K.P. *J. Soil Water Conserv.* **1992**, 47, pp 342-346.
41. Russell, E.W. *Soil conditions and plant growth*, 9th ed.; Jarrold and sons: Norwich, 1961, 480 p.
42. Karlen, D.L.; Cambardella, C.A. In *Structure and Organic Matter Storage in Agricultural Soils*; Lewis Publishers: New York, NY, 1996; pp 395-420.
43. Swisher, M.E.; Monaghan, P.F. *Quart. J. Fl. Acad. Sci.* **1995**, 58, 1, pp 1-9.
44. R.H. Follett; Murphy, L.S.; Donahye, R.L. *Fertilizers and soil amendments*; Prentice-Hall: Englewood Cliffs, NJ, 1981, 557 p.
45. Moore, J.A.; Gamroth, M.J. *Calculating the fertilizer value of manure from livestock operations*; Oregon State University Extension Service: Corvallis, OR, 1989; 8 p.
46. Allaway, W.H. In *Soils for management of organic wastes and waste waters*; L.F. Elliot; Stevenson. F.J., Eds.; Soil Sci. Soc. Am.: Madison, WI, 1977; pp 282-298.
47. Steele, K. *Animal wastes and the land-water interface*; Lewis Publishers: Boca Raton, FL, 1995, 589 p.
48. Lyon, T.L.; Fippin, E.O.; Buckman, H.O. *Soils*. Macmillan: New York, NY, 1916, 764 p.
49. Hue, N.V. In J.E. Rechcigl Ed.; Soil amendments and environmental quality; Lewis Publishers: Boca Raton, FL, 1995, pp 199-247.
50. U.S. Environmental Protection Agency. *Federal Register.* **1993**, Friday, February 19, 58, 32, pp 9387-9404.
51. U.S. Environmental Protection Agency. *Biosolids recycling: beneficial technology for a better environment*; EPA 832-R-94-009; U.S. Environmental Protection Agency: Washington, D.C., 1994; 32 p.
52. Waggaman, W.H.; *Phosphoric acid, phosphates and phosphatic fertilizers*; Reinhold Publ. Corp: New York, NY, 1952; 683 p.
53. Simpson, K. *Fertilizers and manures*. Longman Group Limited: New York, NY, 1986; 254 p.
54. Biddington, N.L.; Thomas, T.H. *Botanica Marina.* **1976**, 18, pp 163-165.
55. Parr, J.F.; Miller R.H.; Colacicco, D. In *Organic farming: current technologies and its role in a sustainable agriculture*; Bezdicek, D.F.; J.F. Power; D.R. Keeney; M.J. Wright, Eds.; Am. Soc. of Agron. Spec. Publ. no. 46; ASA: Madison, WI, 1984; pp 83-95.
56. U.S. Department of Agriculture. *Improving soils with organic wastes*; Report to the Congress in response to Section 1461 of the Food and Agriculture Act of 1977 (PL 95-113); U.S. Government Printing Office: Washington, D.C., 1978; 157 p.
57. Andow, D.A.; Davis, D.P. In *Food and Natural Resources*; Pimentel, E.; Hall, C.W., Eds.; Academic Press, Inc.: San Diego, CA, 1989; pp 191-234.
58. Robinson, T.; *The organic constituents of higher plants*, 2nd ed.; Burgess Publ. Co.: Minneapolis, MN, 1967; 352 p.

59. Wyn Jones, R.G.; Brady, C.J.; Spiers, J. In *Recent Advances in Biochemistry of Cereals*; Laidman, D.L.; Wyn Jones, R.G., Eds.; Academic Press: London, 1979; pp 63-103.
60. Havelka, U.D.; Boyle, M.G.; Hardy; R.W.F. In *Nitrogen in agricultural soils*; Stevenson F.J. Ed.; American Society of Agronomy: Madison, WI, 1982; pp 365-422.
61. Heichel, G.H.; Barnes, D.K. In *Organic farming: current technologies and its role in a sustainable agriculture*; Bezdicek, D.F.; Power, J.F.; Keeney, D.R.; Wright, M.J., Ed; Am. Soc. Agron. Spec. Publ. No. 46; ASA: Madison, WI, 1984; pp 49-59.
62. Zhu, J.C.; Gantzer, C.J.; Anderson, S.H.; Alberts, E.E.; Beuselinck P.R. *Soil Sci. Soc. Am. J.* **1989**, 53, p 1210.
63. Papendick, R.I.; Elliott, L.F. In *Organic farming: current technologies and its role in a sustainable agriculture*; Bezdicek, D.F.; Power, J.F.; Keeney, D.R.; Wright, M.J., Ed; ASA Special Publication No. 46; ASA: Madison, WI, 1984; pp 69-81.
64. Sarrantonio, M. *Northeast cover crop handbook*; Rodale Institute: Emmaus, PA, 1994; 118 p.
65. Reeves, D.W. Cover crops and rotations; In *Crop Residue Management*; Hatfield, J.L.; Stewart, B.A., Eds.; Advances in Soil Science, 7; Lewis Publishers: Boca Raton, FL, 1994, pp 125-172.
66. Meisinger, J.J.; Hargrove, W.L.; Mikkelsen, R.L.; Williams, J.R.; Benson, V.W. *Cover Crops for Clean Water Proceedings of an International Conference*, Jackson, TN; Soil and Water Conservation Society: Ankeny, IA, 1991; pp 57-68.
67. Shipley, P.R.; Meisinger, J.J.; Decker, A.M. *Agron. J.* **1992**, 84, pp 869-876.
68. Kashmanian, R.M.; Rynk, R. J. Soil Water Conserv. **1996**, 51,3, pp 194-201.
69. Haug, R.Y. *The practical book of compost engineering*; Lewis Publ.: Boca Raton, FL, 1993; 717 p.
70. Tyler, R.W. *The organics game*; ASHS Press: Alexandria, VA, 19 : 269 p.
71. Chaney, D.E.; Drinkwater, L.E.; and Pettygrove, G.S. *Organic Soil Amendments and Fertilizers*; Division of Agriculture and Natural Resources; University of California: Oakland, CA, 1992; 36 p.
72. Mandelbaum, R.; Hadar, Y.; Chen, Y. *Biol. Wastes*. **1988**, 26, p 261.
73. Nelson, E.B.; Hoitink, H.A.J. *Phytopathology*. **1982**, 72, p 275.
74. Spring, D.E.; Ellis, M.A.; Spotts, R.A.; Hoitink, J.A.J.; Schmitthenner, A.F. *Phytopathology*. **1980**, 70, p 1209.
75. Hoitink, H.A.J.; Fahy, P.C. *Annu. Rev. Phytopath.* **1986**, 24, p 93.
76. Lemmon, C.R.; Pylypiw, H.M. *Bull. Environ. Contam. Toxicol.* **1992**, 48, p 409.
77. Hobson, P.N.; Wheatley, A.D. *Anaerobic Digestion Modern Theory and Practice*; El Sevier Applied Science: New York, NY, 1993; 269 p.
78. Rich, J.R.; Hodge, C.H. *Nematropica*. **1984**, 14, p 10.
79. Katsuyana, A. M. et al; *Solid Waste Management in the Food Processsing Industry*; NTIS Publ. No. PB-219 019; National Canners Association: Berkeley, CA, 1973; 306 p.

80. Cooper, J.L. *The Potential of Food Processing Solid Waste as A Source of cellulose for Enzymatic Conversion*; Biotechnical and Bioeng. Symp. No. 6; John Wiley and Sons, Inc.: New York, NY, 1976; pp 251-271.
81. U.S. Environmental Protection Agency. *Pollution abatement in the fruit and vegetable industry: wastewater treatment*. Environmental Research Information Center, Technology Transfer; U.S. Environmental Protection Agency: Washington, D.C., 1975; Vol 3, 131 p.
82. Hunt, P.G.; Glide, L.C.; Grancingues, N.R. Land treatment and disposal of food processing wastes. In *Land Application of Waste Material*; Soil Conservation Society of America: Ankeny, IA, 1976; pp 112-135.
83. Epstein, E. *Mineral nutrition of plants*; John Wiley and Sons: New York, 1972; 412 p.
84. Bohn, H.; McNeal; B; O'Connor, G. Soil Chemistry; John Wiley and Sons: New York, NY, 1979; 329 p.
85. *The encyclopedia of organic gardening*; Rodale, J.I., Ed.;. Rodale Books: Emmaus, PA, 1969; 1145 p.
86. Gershuny, G. *Start with the soil*; Rodale Press: Emmaus, PA, 1993; 274 p.
87. Gaugh, H.G. *Inorganic plant nutrition;* Dowden, Hutchinson & Ross: Stroudsburg, PA, 1972; 488 p.
88. Edwards, C.A.; Bohlen, P.J. *Biology and ecology of earthworms*, 3rd ed.; Chapman & Hall: London, 1996; 426 p.
89. Barber, S.A. In *Soil acidity and liming*, 2nd ed.; Adams, F. Ed.; Agronomy 12; Am. Soc. Agron.: Madison, WI, 1984; pp. 171-209.
90. Sparks, D.L. Chemistry of soil potassium in Atlantic Coastal Plain soils: a review; *Commun. Soil Sci. Plant Anal.* **1980,** 11, p 435.
91. Lloyd, S.L. *Mining and manufacture of fertilizing materials and their relation to soils*; D. Van Nostrand Co.: New York, NY, 1918, 153 p.
92. Carson, R. *Silent Spring*; Houghton Mifflin: Boston, MA., 1962; 368 p.
93. Brklacich, M.; Bryant, C.R.; Smit, M. *Environmental Management.* **1991,** 15, 1, pp 1-14.
94. Faeth, P. *J. Soil Water Conserv.* 1993, 48, 2, pp 94-99.

Chapter 10

Composition and Uses of Compost

Allen V. Barker

Department of Plant and Soil Sciences, Bowditch Hall,
University of Massachusetts, Amherst, MA 01003

Composts compostions vary with raw materials and processes used to make composts. Uses, primarily agronomic or horticultural, are based on chemical composition. Applications include uses as fertilizers, mulches, potting media, or soil amendments. Nitrogen is the main component governing uses. Composts vary from about 0.5% to 3% N on a dry mass basis. Composts with over 1% N may be used as fertilizers; those under 1% N should go to other uses. Composts of farm manures, sewage biosolids, and food wastes generally have over 1% N, and composts of dead leaves, yard wastes, and municipal solid wastes usually have less than 1% N. Nitrogenous components (total N, ammonium) and C:N ratios are critical for assessing compost maturity, which affects most composts uses. Other plant nutrients, P, K, Ca, Mg, S, and minor elements, are considered in compost applications relative to concentrations for sufficiency, toxicity, and total loading onto land. Some elements (Pb, Cd, As, Cr, Hg) are considered as pollants, and their presence may restrict composts to nonagricultural uses. Salinity and alkalinity of composts are also factors in uses of composts. Before land applications are made, compost users should have analyses presenting compostions of nutrients and pollutants.

In general terms, composting is an engineered practice to reduce organic wastes microbiologically into humic substances (*1, 2*). Composting is an ancient practice with origins in agriculture (*3*); however, today, composting is adapted for treatment of many solid wastes, including crop residues, farm manures, sewage biosolids, industrial sludges, septage, garbage, paper, yard wastes, petroleum sludges, explosives, and other diverse mixed materials (Table I) (*4, 5, 6*). Composting reduces the volume of materials, kills plant and animal pathogens, reduces the carbon concentration, increases concentrations of plant nutrients, and destroys organic compounds considered as environmental hazards. Composted materials usually have lesser unfavorable impacts in the environment than raw compostable substrates or feedstocks. Composting of these diverse materials creates products with beneficial uses in agriculture or land remediation (*2, 3, 7-12*).

Table I. Estimates of potentially compostable organic materials in the United States

Waste	Components	Production, Mt/yr	
		Generated	Collected
Agricultural	Farm manures animal carcasses, crop residues	590	50
Logging and wood manufacturing	Bark, chips, scraps, sawdust	55	5
Municipal	Paper, cloth, yard refuse, leaves, garbage, landscape refuse, wood	125	70
Biosolids	Municipal sewage sludges	9	9
Septage	Domestic septic tank sludges	3	NE
Industrial	Petroleum, paper, food processing wastes, textile, pharmaceutical	45	6
Special & miscellaneous	Hydrocarbon-contaminated soil, TNT, petroleum, pesticides, pesticide-contaminated wastes	50	6

Adapted from Haug 1993 (4); Hyatt 1996 (9); Slivka et al., 1992 (16)
NE, no estimate.
Mt, million common tons

Composting essentially converts materials that are unfit for land application into materials that are safe for land application. Land application of composted organic wastes is considered an efficient method for saving energy and costs of solid waste disposal. Hence, what was once a small-scale practice by farmers or more recently by organic gardeners has become a major industry involving municipalities, corporations, and farms (Table II) (*13-15*). The largest potential users of composts from any of these major operations are the crop-producing agricultural industries (*12, 16, 17*). Potential national use of compost in agricultural industries far exceeds, perhaps by a factor of ten, the potential for compost production in the United States (Tables II & III) (*9, 16*).

Enhanced yields usually result from applications of high-quality, mature composts to crop land. Numerous investigations have shown that improved soil fertility or quality, as measured by physical and chemical properties, result from land applications of compost (*2, 17*). Compost functions as an agent to condition soil and to suppress soil erosion (*2, 17, 18*). Bulk density, water-holding capacity, porosity, and aggregate stability are among physical factors improved by compost additions to land (*17, 18*).

Table II. Current and potential production of composts

Compost	Production, Mt/yr	
	Current	Potential
Municipal solid waste	1	30
Biosolids	2	3
Horticultural	5	15
Agricultural	0.3	3
Total	8.3	51

From Slivka et al., 1992 (16)
Mt, million common tons

Often the effects due to improvement of physical properties result from use of large or prolonged applications, but once these effects are imparted, they may be long-lasting (*18-20*). Compost carries plant nutrients in various concentrations and functions as a fertilizer (*2, 12, 15, 16, 21, 22*). The principal nutritional benefit given to crops from composts usually is the nitrogen (N) carried in slow-release form (*18, 23, 24*). Major attention has been given to micronutrient and trace metal concentrations in composts (*15, 21, 25*). Regulations governing land applications of sewage biosolids are applied to biosolids-based composts and are based on concentrations of trace metals as well as on pathogenicity of the products (*26*).

Table III. Current and potential demand for composts

Activity	Consumption, Mt/yr	
	Current	Potential
Agriculture	5	500
Forestry	0.5	50
Landscaping & Nursery	3.7	19
Reclamation & Landfills	0.1	1
Total	9.3	570

Adapted from Slivka et al., 1992 (16)
Mt, million common tons

Carbon (C) is a key requirement for composting (27). Carbon is an energy source for microorganisms that do the composting. The C concentration in compost gives an indication of the amount of organic matter and the effects that compost will have on soil fertility. Ratios of C to other plant nutrients indicate the bioavailability of nutrients and govern the value of compost as a slow-release fertilizer (Table IV). Composts with wide ratios of C:N may not be suitable for incorporation into soil. These composts might be considered immature, and their agricultural applications might be limited to uses as mulches.

Compost is a form of humus, which is ill-defined, but which is a dark-colored, complex of amorphous, organic substances derived from biological decomposition of plant and animal residues, and which usually is associated with soils (28). The humus of soils is stable organic matter, relative to freshly incorporated plant and animal residues. In soils, humus-forming processes are called humification. The release of plant nutrients from any organic residues, including humus, is termed mineralization. Composting is a managed process in which humification and mineralization occur. The feedstocks entering into composting and their processing and stage of decomposition govern the composition and utility of composts. The following text addresses the composition of organic matter, plant nutrients, and trace metals as factors governing the use of composts in agricultural operations.

Organic Matter

Humus has important chemical and physical properties and is an indicator of soil fertility (28, 29). Humus is a storehouse of plant nutrients, which are released slowly by mineralization. Humus has high cation exchange capacity, which buffers against changes in soil acidity, holds bases for plant nutrition, and complexes toxic materials injurious to plants. Humus changes soil structure allowing for improved tilth, increased fine pores to hold water, and increased large pores for good aeration. The humus content of soils depends on soil management (*30-32*). Tillage of soils increases aeration and microbial activity and diminishes the humus content in soils. Continuous monoculture or production of crops, such as vegetables, that leave small amounts of residues, leads to depletion of humus. Replenishing humus in soils takes years of additions of raw plant materials, green manures, or farm manures, for decomposition of these materials leaves little in the way of stable residues in a year or two after their addition (*28, 31, 32*).

Compost, although not a biologically static material, is a humus, and its addition to soils improves soil fertility by adding materials that have been formed by the general processes of humus formation in soils. Hence, composts quickly impart to soils, physical, chemical, and biological factors of fertility and give such improvements more reliably than additions of many uncomposted organic wastes (*33, 34*). Composts vary greatly in composition and physical properties depending on the feedstocks and procedures used in their manufacture. Finished products vary in quality, with differences occurring due to variations in maturity and residues of uncomposted material resistant to decay (*17*). Screening helps to improve quality by removal of uncomposted material; however, recalcitrant materials become more brittle after composting than they were in the original state (*4, 35*).

Nitrogen

Carbon:nitrogen ratios. Temperature, moisture, air, and C, N, and other nutrients are key parameters in substrate biodegradation in compost piles or in soils (*4, 27, 36*). Carbon and N are the two most important elements in composting. Carbon-containing substrates are sources of energy and contribute to the biomass of microbial populations. Nitrogen, a constituent of proteins and genetic matter, is critical for microbial growth. Nitrogen may be considered a more critical factor in composting than C. If N is limiting, microbial populations will remain small, and composting will proceed slowly. The weight percentage ratio of C to N (C:N ratio) in compostable material is a guideline for estimating rates of composting (Table IV). Materials with wide C:N ratios, e.g., greater than 100, will compost slowly. For optimum composting, the C:N ratio in the starting material should be about 25 (*1, 27*). The C:N ratio of finished compost is also about 25, although this ratio can vary substantially.

Generally, materials to be composted do not start with a C:N ratio as narrow as 25. Solid organic waste is typically about 50 % C with N being the variable component of the ratio. Nitrogen can vary from nil to several percent on a dry weight basis in organic matter, giving a wide range of C:N ratios (*35*). A purpose of composting is to narrow the C:N ratio, for conditions in soils or media amended with materials with ratios exceeding 35 may result in N immobilization, leading to deprivation of plants for nitrogen (*36*). Mixed solid waste organic substrate typically has a C:N ratio of 50 to 100. In composting, C is lost faster as carbon dioxide than N is lost as ammonia; hence, the ratio narrows (Figure 1).

Table IV. Approximate carbon:nitrogen ratios of compostable organic substrates

Material	C:N	Material	C:N
Biosolids, raw	10	Seed meals	7-8
Animal tankage	4	Grass clippings	15-18
Dried blood	4	Vegetable wastes	12-20
Fish scrap	4-5	Food wastes, mixed	18-20
Poultry carcasses	5-6	Hulls, various	120-150
Cow manure	18	Dead leaves	100-180
Horse manure	25	Straw, cereal grain	120-180
Hog manure	18	Sawdust	300-500
Stable manures	40-50	Paper	500-5000
Poultry manure	12		
Poultry litter	15		
Digested sewage	6-8		
Urine, dry	1		

When N concentrations are low in organic materials, N will not be released into the medium (soil or compost) by ammonification (mineralization). Inorganic N in the medium will be incorporated into microbial bodies, hence, immobilized. Generally if the C:N ratio is wide, the ratios of C to other plant nutrients also are wide, and plants will be deprived by immobilization of more nutrients than just N. After incorporation into soil, a compost, depending on its stability or maturity, will release from less than 10% to more than 30% of its N in a growing season (17, 37). Mineralization rates of composts will be slower than rates of some fresh nitrogenous materials but will exceed rates of carbonaceous materials (38).

A lot of attention has been given to defining maturity of composts (39-43). Carbon:N ratio of compost is used as an assessment of maturity, and a wide ratio is considered as an indication of immaturity. Low N concentrations are the most common factors causing wide C:N ratios. Hence, a material with a wide ratio has a low concentration of N, and a lot of material will have to be delivered to carry sufficient N to fertilize a crop. Additionally, incorporating compost with a wide C:N ratio will immobilize N. Compost users should consider that composts with wide C:N ratios or low N concentrations are suitable only as mulches, which are left on top of the ground. Utilization of composts in crop production often requires

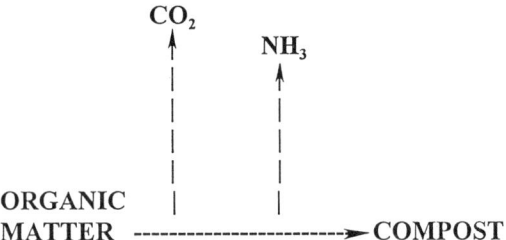

Figure 1. Diagram of losses of carbon as carbon dioxide and nitrogen as ammonia. Carbon losses exceed those of nitrogen so that C:N ratio is narrower in compost than in the original organic matter.

ENERGY:NUTRIENT RATIO

WIDE: Cellulose, starch, sugar

NARROW: Proteins, lignin, polyethylene

Figure 2. Examples of organic substrates with wide or narrow energy:nutrient ratios.

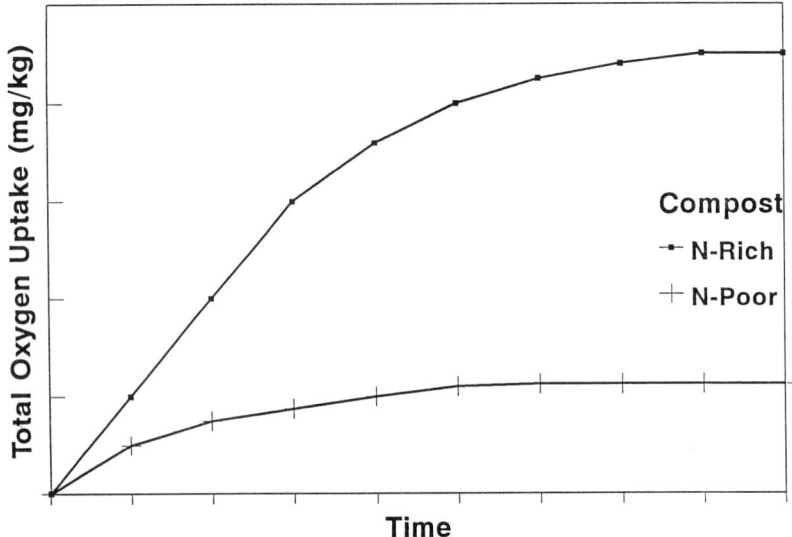

Figure 3. Model of oxygen consumption during composting of nitrogen-rich (narrow energy:nutrient ratio) or nitrogen-poor substates (wide energy:nutrient ratio).

supplemental applications of N to enrich the compost or to accelerate the rate of decomposition of compost and its release of nutrients (*12, 44, 45*).

Energy-nutrient relationships must be considered to evaluate whether the product will contribute to plant nutrition by mineralization or deplete available nutrients by immobilization (Figure 2). Jansson and Persson (*36*) suggest use of an energy:N ratio to assess potential for mineralization-immobilization transformations of N. Although this ratio is difficult to measure, it is based on the proportions of the amounts of N-free or low-N decomposable organic matter to the amounts of N-rich decomposable organic matter in soil, compost, or other media. Oxygen consumption during composting is an indication of energy:nutrient ratios (Figure 3) (*43*). Easily decomposable carbonaceous N-free materials (sugars, starches, cellulose, lipids) impart a wide energy:N, as well as a wide C:N, ratio into feedstocks. Feedstocks with a wide C:N ratio attributed to N-free decomposable materials derived from wastes such as dead leaves, sawdust, or mixed municipal solid wastes start at C:N ratios exceeding 100 and will be slow in reaching a C:N ratio of 25. Even at that level, composts may not be stable and fit to incorporate into crop land, if a large proportion of easily degraded materials remains relative to the proportion of slowly degraded materials (*4, 36, 42*). The N-poor lignins of raw plant materials and peats are poor energy sources and undergo decomposition slowly due to their low energy:nitrogen ratios.

The degree of division affects bioavailability of C. Large chips of wood are degraded slowly because of small surface areas for attack by microorganisms. Increasing the surface area by grinding or shredding will accelerate decompostion if N is available (*4, 27*). In composting of municipal solid waste, shredding and additions of N-rich feedstock, such as biosolids or farm manures or even urea fertilizer, accelerate decompostion. On the other hand, wood chips, leaves, fruit pomace, or other carbon-rich feedstocks are utilized as bulking agents in composting of wet or N-rich feedstocks. Each of these practices affect the C:N and energy:nutrient ratios, and usually gives a higher quality of compost for agricultural application than composting of the components separately. Large bulking agents, such as wood chips, can be screened from the final product, thus narrowing the C:N ratio of the final product, and can be returned to composting. The recycled wood chips will have different energy:N ratios and will have different decay rates than fresh chips (*4*).

Ligneous materials with low energy:N ratios are used commonly in making of soilless peat-lite mixes (peat, vermiculite, perlite) as media for container-grown plants (*46, 47*). These materials decompose slowly, and N immobilization is not a problem in peat-based media. Attempts to use composts as components of potting media, substituting for the peats, have been successful (*2, 44, 48*) but perhaps do not give the quality of media as the low energy:N materials. Composts even in fully finished states are still at rather high energy:N ratios. Composts in potting media continue to decompose and do not impart physical stability to the media as peats do. Composts alone likely do not give adequate structure for aeration, and peats, perlite, or other structural agents must be added with compost. Media generally on a volumetric basis can be about half to two-thirds compost with the remainder being peat, perlite, or other structural mixes. The precise proportions of composts to other ingredients will depend on the nature of the finished compost.

Table V. Approximate concentrations of plant nutrients in composts

Element	Concentration, % dry wt
Nitrogen	1.0 to 3.0
Phosphorus	0.4 to 1.0
Potassium	0.5 to 1.5
Calcium	1.0 to 2.0
Magnesium	0.5 to 1.0
Sulfur	0.5 to 1.0
Minor elements	Present, nontoxic

An advantage of composts over peats, however, is that as the composts decompose, nutrients are released, and plant nutrition is improved (Table V) (*45, 49*). Enough nutrients may be released from composts to nourish a crop (*48, 50-52*). Since N is likely to be the limiting factor in media and since N-poor media are likely to be low in other nutrients, composts selected for formulation of media for container production of crops should be over 1%, preferably 1.5%, total N. Media made from leaf composts are generally below this threshold and would require supplemental N fertilization for crop production. Media with high concentrations of N, *e.g.*, 3% total N, are likely to be immature and ammoniacal. Composts with over 1,000 mg NH_4-N per kg dry weight probably are phytotoxic (*51-55*). These composts can be used successfully only after further composting or providing conditions in which ammonia is oxidized or volitalized. At least, one week of time in the containers is required for dissipation of ammonia to below phytotoxic levels (Figure 4) (*55*, O'Brien and Barker, University of Massachusetts, unpublished data). Peats are essentially void of plant nutrients, as the elements are leached under the wet conditions in which peats are formed. Bark, wood chips, and sawdust have wide energy:N, so their use in media would have nutrient-immobilizing effects relative to compost and would require supplemental fertilization similar to or above that given to crops in peat-based media (*35, 56*).

Nitrogenous constituents. When N is relatively unavailable in compostable materials, due to wide C:N or energy:N ratios, release of N from organic combination is low. Carbon is lost from the medium by microbial respiration of the substrates, and the C:N ratio narrows with composting (Figure 1). Composting of

N-rich substrates (sewage biosolids, farm manures, grass clippings) often is problematic because N is in excess of the needs of microorganisms. The excess N may be lost by ammonia volatilization, being wasted and possibly polluting to the atmosphere (57, 58). Since N and sulfur (S) are largely in proteins of organic substrates, a high N content indicates a high S content. Composting of N-rich materials can lead to release of odorous volatile nitrogenous and sulfurous gases (2, 5, 57). Ammonia losses from N-rich materials can be so high that composts made from N-rich feedstock may end up with a lower total N concentration than composts made from materials with a lesser N concentration (48). A bulking agent needs to be used with N-rich feedstock to lessen ammonia losses and to conserve N (4).

The total N concentration of composts indicates not only the N that will be delivered by an application of compost, but it also allows for a quite good estimate of C:N ratio. Determination of total C or organic C in organic materials is relatively difficult, employing semiquantitative procedures or expensive instrumentation for analysis (59). Some C analyses are based on determining ash and N concentrations in materials (13, 35). For uncomposted organic materials, C:N ratios can be estimated from the knowledge that organic matter is about 50% C and N is variable. The ratio is estimated by dividing 50% N by the percent concentration of total N, which is easily determined by Kjeldahl analyses (60, 61). Organic matter with 1% total N would have a C:N ratio of about 50. Composts vary widely in total N, but generally if the composting process proceeds long enough, perhaps years, total N concentrations stabilize at about 1%. The C:N ratios of composts cannot be estimated by the rule-of-thumb applied to uncomposted organic matter. Composts with 1% total N will vary from 10 to 50 in C:N ratios, depending on the feedstocks, maturity, and inorganic material (dirt, soil, sand, etc.) added to composts.

The absolute concentration of total N must be considered in using composts in agriculture. Generally, if the compost is less than 1% total N on a dry basis, supplemental N fertilization will be required for soil incorporation or for compost-based potting media even with well-matured composts (Figure 5). If total N is about 0.6%, as is the case with many leaf composts, potential for N immobilization is high, and these composts may be best used as mulches. With mulch-applied compost, contact with the soil is at the soil-mulch interface, and little immobilization occurs. Nitrogen deprivation can result if roots grow into thick mulches of low-N composts, rather than into the underlying soil. A mulched application of composts with more than 1% total N will provide enough N for crop nutrition. Mineralization and nitrification release mobile forms of N that will leach into the soil.

Composts with 1.5% or more total N should be incorporated into soil or in potting media to conserve N. Mulching with these high-N composts may not impart plant injury, but may permit losses of N by ammonia volatilization (31, 32, 62, 63). In some uses, surface application of high-N composts may be desirable. Composts may be laid on plastic for production of sods for harvest (51, 64, 65). Plantings can be made directly into thick mulches of N-rich composts, which control weeds and provide plant nutrition (66). Thick, surface applications of fertile composts will provide nutrition to plants growing directly in the compost and in the soil underlying the mulch, in contrast to N-poor composts, sawdust, bark, woodchips, or other plant

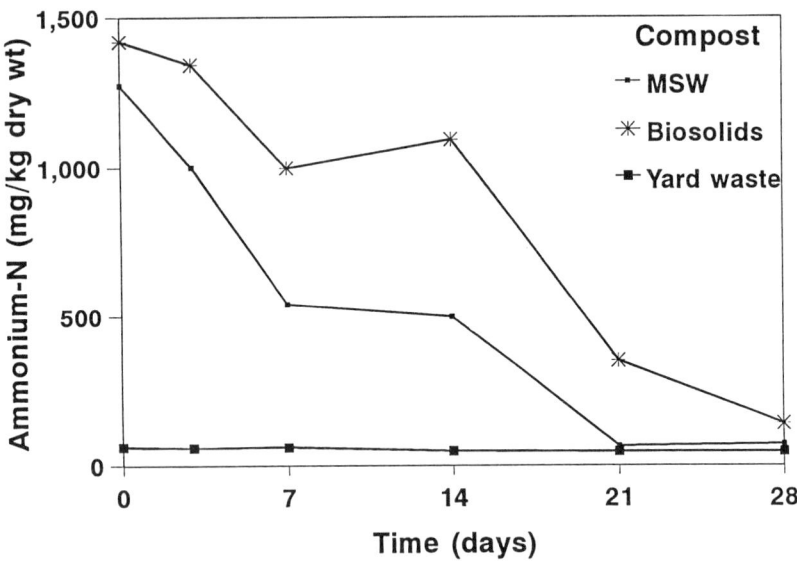

Figure 4. Concentrations of ammonium-N in composts of municipal solid wastes, biosolids, and yard wastes as a function of time after application to surface of crop land.

MINERALIZATION-IMMOBILIZATION TURNOVER

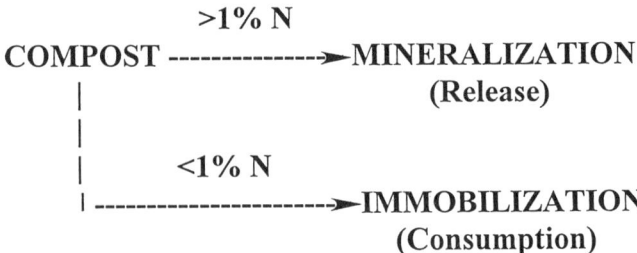

Figure 5. Mineralization-immobilization turnover of composts in relation to nitrogen concentration in composts.

residues in which rooting plants will starve and which will release few nutrients into the underlying soils. With N-rich immature composts, phytotoxicity may occur if sufficient time is not allowed for dissipation of ammonia (*51, 52*).

In a finished, mature compost, ammonia (or ammonium, depending on the acidity of the medium) concentrations should be low. Ammonia produced by mineralization of organic substrates will be lost by volatilization, consumed by microbial utilization, or oxidized to nitrate (*67*). Composts produced in vessels may be high in ammonia, following removal of the composts from the vessels. A curing time of several weeks, during which aerobic composting occurs, is needed for dissipation of ammonia and other reduced products to produce a material suitable for agricultural use (*2*). Even finished composts from vessels or aerobic piles may contain more ammonia and less nitrate than is optimum (*44*).

Composting of N-rich substrates in piles leads quickly to anaerobiosis in piles unless aeration is vigorous. Ammonia in these composts can remain at phytotoxic levels even after long periods of curing. High ammonium in compost hinders seed germination and injures seedlings and soil fauna (*51-54, 68, 69*). Phytotoxicity from high ammonia in composts is confused often with saline injury. Composts delivered from composting plants in spring seasons may not be adequately matured and often have high concentrations of ammonia. Growers err in trying to leach soluble salts from these composts in containerized media and make matters worse by filling pores with water and giving inadequate oxygen for transformation of ammonia to nitrate. Dissipation of perceived saline injury with time with compost-based media may be based on ammonia transformation or volatilization (*65, 68, 69*). A waiting period of several days or weeks between application of composts to land or containerized media may be needed for dissipation of ammonia (*51, 55*).

Nitrate is low in composts almost regardless of maturity (Figure 6). Materials entering into composts may contain more nitrate than mature composts. Microorganisms utilize nitrate in their growth during composting and will soon exhaust any free nitrate supply. The high C content of composts and the anaerobic environment of static piles create an ideal reducing environment for denitrification and gaseous losses of relatively inert nitrogen gases (N_2, N_2O, NO), which would not be retained (*70*). Nitrate may be leached from unprotected piles, since nitrates in composts are water-soluble and are not held to exchange sites in organic matter. One would expect that leaching losses would be low compared to losses by denitrification. High ammonium relative to nitrate, especially in immature composts in which absolute ammonium concentrations are high, can give problems of phytotoxicity (*44, 55, 71*).

Phosphorus and Sulfur

Phosphorus (P) concentration in adequately nourished plants ranges from 0.2% to 0.4% (*72,73*). In farm manures, P concentrations average about the same as in plant materials (*32*). Phosphorus is higher in biosolids, about 1% or 2% of dry mass (*23, 24*). In composting, C is lost, and P is not leached or volatilized and is conserved (even more so than N) so that composts are richer in P than the original materials.

The relative concentration of P may be up to twice that in the original material, considering that about half of the original weight will be lost during composting (4). Most of the P in biological materials is held in organic complexes (75). Characteristic C:P ratios of composts are 80 to 100. Ratios wider than these values might indicate that P immobilization would occur. The maximum C:P ratios for optimum composting was deemed to be 120 (76). Energy:P ratios, as with energy:N, should be considered, with easily decomposable materials potentially immobilizing P and with recalcitrant materials not immobilizing P. Supplemental P fertilization with compost use, as a rule, does not increase plant growth (45, 49).

Sulfur concentrations in plants and in composts are about equal to those of P (15, 19, 21, 77, 78). Some composts of paper sludges and municipal solid wastes may have lower concentrations of S than those from plant residues, manures, or biosolids. In residues of biological origin, most of the S is in organic combinations and must be released by mineralization. Carbon:S or energy:S ratios must be considered in S mineralization, and composting will narrow these ratios, allowing for release of S for plant nutrition once the compost is matured. Supplemental S fertilization with compost application is not necessary, and an application of compost to meet the N requirement of crops would be expected to meet the S. The chemistry of S in composts, soils, or media is more like that of N than that of P. Most sulfate salts are relatively water-soluble compared to phosphate salts, and sulfate is not subject to fixation in soils. Lost of volatile S-containing compounds can lead to problems with odors in composting.

Potassium, Calcium, and Magnesium

Potassium (K) is one of the most abundant nutrients in plants. Potassium in biological materials is not metabolized into covalent bonds in organic molecules and remains exchangeable (75). In dead organic materials, K can be leached with water or with dilute salt solutions. In compost, K does not need to be mineralized before it is available for plant nutrition, since it remains in water-soluble forms. Composts are not necessarily K-rich substrates, for much of the K may have been leached from the feedstocks or during composting. With biosolids, K would be lost in effluent at the wastewater treatment plants. With farm manures, failure to recover urine results in fairly low K concentrations. Generally, one should not expect composts to be good sources of K for plant nutrition. Potassium concentrations in finished composts are usually less than those of N or P (15, 21, 23, 44, 48, 79). In uncomposted carbonaceous materials, although K is water-soluble, K immobilization could occur if the C:K ratio (energy:K) is wide, but in mature composts, immobilization should not occur. Plants grown in containerized media should respond to K fertilization, because of the low K of the media. In soils, the response would be related to the supply of available (soluble, exchangeable, or nonexchangeable) K in the soil.

Calcium is in various states in biological materials. Part of the Ca is bound firmly in organic structures, some is exchangeable, and some is held in sparingly soluble complexes (75). Calcium concentrations in composts vary from about 1% to 4% of dry mass (*15, 48, 79*), which is likely higher than that in the original materials, for Ca is conserved and concentrated as carbon dioxide is lost. These concentrations exceed those in most soils. Also, lime often is added to composts or feedstocks (biosolids) for vector control, pathogen control, or acidity control thereby increasing the concentration of Ca (*26, 35, 74, 80*). Composts made from lime-stabilized biosolids can be very high in Ca, perhaps 10% of dry mass, depending on the amounts of other feedstocks added for composting. Calcium is not an ion that causes much of a specific toxicity in soil salinity (*81*); so a high concentration of Ca in compost is generally considered as a benefit. Peat-based media supply inadequate Ca for container-production of crops because of the lack of Ca in the media and because of the effects of fertilization with ammoniacal fertilizers on Ca nutrition of crops (*82-84*). Use of compost as a substitute for peat in containerized media would have an advantage of supplying Ca.

Magnesium (Mg) is held in biological materials in organic complexes, in exchange, and in sparingly soluble salts (*75*). Magnesium concentrations in composts average about 0.2% to 0.4% of dry mass, about the same the original materials (*15, 48, 79*). These concentrations may be lower than those in soils, but with the quantities of materials used in soilless media or on land, Mg in compost should improve plant nutrition. Composts of lime-stabilized biosolids may be fortified with Mg if Mg-containing limes are used.

Micronutrients and Trace Elements

In the context here, micronutrients refer to chemical elements essential for plant growth (Cu, Zn, Mo, Ni, Mn, Fe, B, Cl) and sometimes to additional ones required for mammalian nutrition (Cr, Se, Na). Trace elements is a general term that includes most of these and other elements that are nonessential (As, Cd, Hg, Pb) for plant or animal nutrition. These elements are ubiquitous in the environment, being found in soils, fertilizers, and composts. The elements enter composts from the feedstocks in which the elements are accumulated in various ways. Soils and fertilizers contain various levels of micronutrients and trace elements (*85-87*). Although somewhat selective in absorption of ions, plants take up almost anything in solution and will accumulate these elements from soil or other media. Some fungicides contain copper, zinc, or manganese, and their residues may remain on organic matter (*88, 89*). Farm manures will have these elements from feeds and forages, and the elements may be more concentrated in manures than in the feeds. Copper is added to diets of hogs (>200 mg Cu/kg feed) to stimulate growth and for antibiotic effects (*90-93*). Trace elements end up in municipal solid wastes from yard wastes, pigments, plastics, batteries, metals, inks, and many other sources (*2, 15, 50, 85*). Trace elements enter sewage biosolids through industrial and domestic output (*94*). Much has been written about the potential hazards of using organic wastes with concentrations of trace elements above background levels in the environment (*95*). Trace elements known as heavy metals (Cd, Cu, Hg, Pb, Zn, Ni, Mo) have represented the primary concern in the uses of composts in agriculture (*9, 96*). Regulations have been written to govern concentrations of trace elements in

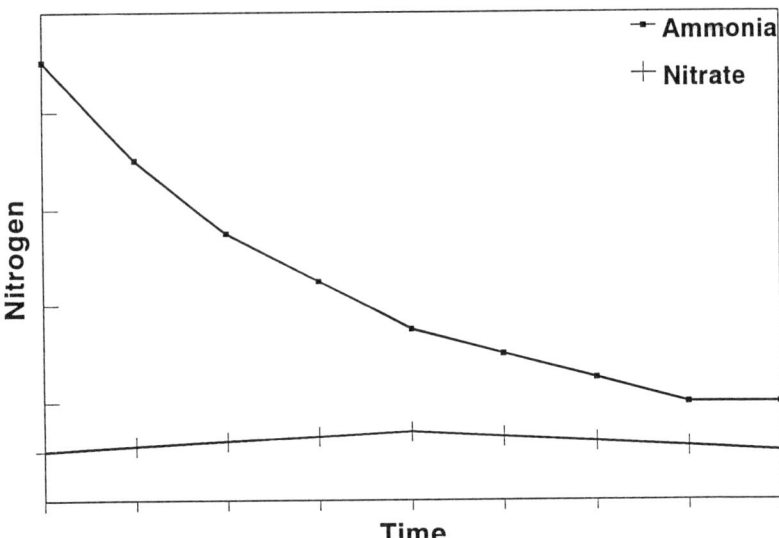

Figure 6. Model of trends for nitrate and ammonia concentrations in composts over the duration of the process of composting from addition of raw materials to maturity. Time and nitrogen concentrations are in arbitrary units.

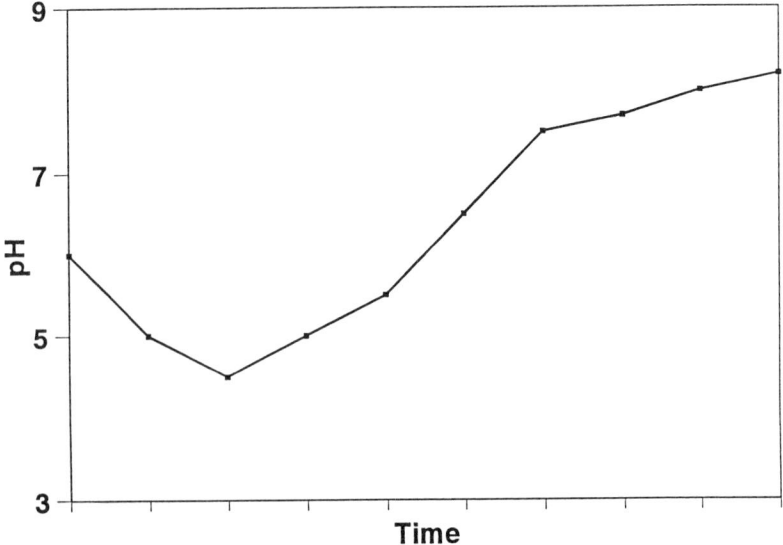

Figure 7. Model of trends of pH in composts during the process of composting from addition of raw materials to maturity. Time is in arbitrary units.

sewage biosolids, and by inference, in composts applied to crop land (Table VI) (26). Adherence to these regulations and source separation of materials entering into feedstocks have helped to make composts safe for land application (97, 98).

During composting, the concentrations of trace elements will be increased because of loss of C relative to losses of trace elements, although the water-soluble trace elements may be decreased (48, 99). Applications of organic waste might be considered as crude methods of applying micronutrients because of the bulk of materials that would be needed (90). Although their concentrations of micronutrients are minimal, composts are reasonable sources of micronutrients, and applications to land or incorporation in soilless media would be expected to increase crop growth and yield, in part due to compost-borne micronutrients without deleterious effects on plant or composition or on soils (15, 49, 100). Most risk analyses of composts indicate that health risks from metals in composts are low. Metals added to soils or media as parts of composts of municipal solid wastes or biosolids have low bioavailability (8, 25, 85). Composts reduce the bioavailability of metals in media or in diets (25).

Table VI. Concentrations of trace element pollutants allowed in biosolids for land applications

Pollutant	Conc., mg/kg	Pollutant	Conc., mg/kg
As	41	Hg	17
Cd	39	Mo	18
Cr	1,200	Ni	420
Cu	1,500	Se	36
Pb	300	Zn	2,800

US EPA 503 Regulations, 1993 (26)

Alkalinity

Most mature composts are slightly alkaline materials of about pH 7.5 (Figure 7). Initially, production of organic acids from degradation of easily decomposable materials increases acidity. Addition of limes may accelerate rates of decompostion of organic wastes, but the costs of lime and enhanced loss of ammonia outweigh the benefits of increasing alkalinity in composts (1, 3). The alkalinity of composts would not be expected to increase alkalinity of soils or other media by more than one pH unit. Sandy, unbuffered soils may increase in pH more than finer soils. Ordinarily, compost is not a static humus and is weakly buffered. Its further decomposition in soils and media could be acidifying. On the other hand, composts with elevated lime levels can increase soil pH substantially (*12, 22*). Composts based on lime-stabilized biosolids are strongly buffered with alkalinity. Lime-stabilization typically involves massive adding of limes of calcium hydroxide (about 1:4 w:w lime:dewatered biosolids) or kiln dust (1:1 lime:dewatered biosolids) to raise alkalinity of biosolids to pH > 12 for a few days before stabilizing at about pH 11 (*75, 80*). These materials have calcium carbonate equivalency 10 to 20 times that of unlimed biosolids. The uncomposted biosolids or their composts have potential to raise soil pH and could be used to remediate acid soils. The liming potential of these materials might be a limiting factor in the amounts of lime-stabilized materials that can be used in container media or applied to land, particularly sandy soils (*12, 22, 95*).

Soluble Salts

Composts may have considerable salinity which is attributed to extensive mineralization of organic substrates. Suppressions in crop yields have been attributed to soluble salts in composts, if the composts constitute a high proportion of container mixes or are added in high amounts to land (*39, 51, 52, 101-103*). The growth-limiting effects of salinity may be due to osmotic factors or to specific ion effects (*e.g.*, B, Na, NH_4^+). In container media, the suppressive effects of salinity are difficult to delineate from those due to poor aeration, compost immaturity, ammonia toxicity, and other conditions imparted by a high proportion of compost in the media. Salinity in composts can vary with the sources of compost and can reach electrical conductivities of about 10 dS/m (saturated paste extracts) (*48, 51, 52, 101*). Green yard wastes and biosolids can deliver compostables that result in build up of salinity (*12, 22, 51, 52*). Composts of food residuals and leaf composts may be low in soluble salts (48, 51, 52). Generally, electrical conductivity should be measured in composts for land application or for container media, and composts with values above 4 dS/m in either 1:2 compost:water (w:v) or saturated paste extracts should be evaluated for phytotoxicity before heavy applications are made to soils or container media.

Organic Compounds

Contents of persistent organic chemicals have important roles in quality determinations of composts. These chemicals may be introduced into the composts as xenobiotic constituents of the original materials, or they may be formed during composting (Table VII). Vogtmann and Fricke (*104*) reviewed the subject of xenobiotic organic substances in composts. Their general conclusion was that mature or immature composts are low in xenobiotic organic compounds and that setting of maximum tolerance levels for these substances was unneeded.

Chlorinated pesticides, polychlorinated biphenyls (PCBs), and polycyclic aromatic hydrocarbons were extremely low in composts and did not increase in composts even if materials containing these compounds were added. These materials were degraded or volatilized during composting in varying but in substantial amounts so that no concentrating of these xenobiotics occurred (*104*). Dioxins, which are ubiquitous because of their persistence, were suggested as entering composts through dust particles and were not enriched in compost to the extent to consider regulations. Chaney et al. (*105*) noted that in countries in which manufacture of PCBs is prohibited, these xenobiotics are very low in biosolids and represent no threat to highly exposed individuals or to the general public. With composting of hazardous materials (*e.g.*, trinitrotoluene), the objective is to convert these materials into innocuous end-products which are likely returned to their sites of origin and not to be applied to crop land (*106*).

Table VII. Examples of xenobiotic and synthesized organic compounds in composts

Xenobiotic	Synthesized
Chlorinated pesticides	Amines
Polychlorinated biphenyls	Sulfides, disulfides, thiols
Polycyclic aromatic hydrocarbons	Short chain organic acids
Polychlorinated dibenzidioxins	Volatile fatty acids
Dibenzofuranes	Aromatic hydrocarbons
Plastics	Phenols
	Aldehydes, ketones

Biological oxidation of organic material generally is less complete in anaerobic than in aerobic environments. A number of intermediate metabolites, such as methane, organic acids, ammonia, amines, and reduced sulfurous compounds, may accumulate or be evolved in gaseous forms. Evolution of these compounds gives odors that have presented problems with some composting operations (*2,57*). If these substances remain in anaerobic composts, phytotoxicity can occur in containerized media or soils. In the reducing conditions of anaerobiosis, soluble levels of Mn and Fe may increase and might pose problems of specific ion phytotoxicity. In well-aerated composts, destruction of volatile organic carbon compounds, ammonia, amines, and reduced inorganic substances will occur or these materials may be sorbed to surfaces of solid organic matter (*57*). Aeration of composts is important in odor control and in the production of nonphytotoxic materials for agricultural use. A curing time in piles or on land before planting may be needed to avoid toxicity from inadequately aerated composts.

Literature Cited

1. De Bertoldi, M. In *Composting and Compost Quality Assurance Criteria*; Jackson, D. V.; Merillot, J.-M.; L'Hermite, P., Eds.; Commission European Communities: Luxembourg, 1992, pp. 85-93.
2. Stratton, M. L.; Barker, A. V; Rechcigl, J.E. In *Soil Amendments and Environmental Quality*; Rechcigl, J. E., Ed.; Lewis Publ.: Boca Raton, FL, 1995, pp 249-309.
3. Poincelot, R. P. *The Biochemistry and Methodology of Composting*; Connecticut Agric. Expt. Stn Bull. 754; 1975.
4. Haug, R. T. *The Practical Book of Compost Engineering*. Lewis Publ.: Boca Raton, FL; 1993.
5. Hoitink. H.A.J.; Keener, H. M. *Science and Engineering of Composting*; Renaissance Publ.: Worthington, OH, 1993.
6. Goldstein, J. *Compost Sci.*, 1980, 21(4), 28-32.
7. Bergman, J.; Steen, E. *Environ. Pollul*,. 1989, 56, 127-144.
8. Chaney, R. L. *BioCycle*, 1990, 31(9), 54-59.
9. Hyatt, G. W. In *Agricultural Utilization of Urban and Industrial By-Products*, ASA Special Publ. 58, American Society of Agronomy: Madison, WI, 1995, pp 19-72.
10. Poincelot, R. P. *Compost Sci.*, 1974, 15, 24-31.
11. Scanlon, D. H.; Duggan, C.; Bean, S. D. *Compost Sci.*, 1973, 14(3), 4-8.
12. Shiralipour, A; McConnell, D. B.; W.H. Smith, W. H. *Biomass Bioenergy*, 1992, 3, 267-279.
13. Brake, J. D. 1992. *A Practical Guide for Composting Poultry Litter*; Mississippi Agric. Forest Expt. Stn. Bull. 981, 1992.
14. Goldstein, N. *BioCycle*, 1980, 30(4), 34-38
15. He, X. T.; Traina, S. J.; Logan, T. J. *J. Environ. Qual.*, 1992, 21, 318-329
16. Slivka, D. C.; McClure, T. A.; Buhr, R.; Albrect, R. *Biomass Bioenergy*, 1992, 3, 281-299.
17. Tyler, R. *Winning the Organics Game*; ASHS Press: Alexandria, VA, 1996.
18. Dick, W. A.; McCoy, E .L. In *Science and Engineering of Composting*; Hoitink, H. A. J.; Keener, H. M., Eds. Renaissance Publ.: Worthington, OH, 1993, pp 622-644.
19. Mays, D. A.; Terman, G. L.; Duggan, J. C. *J. Environ. Qual.*, 1973, 2, 89-92. 20. Khaleel, R.; Reddy, K. R.; Overcash, M. R. *J. Environ. Qual.*, 1981, 21, 318-329.
21. He, S. T.; Logan, T. J.; Traina, S. J. *J. Environ. Qual.*, 1995, 24, 543-552.
22. Shiralipour, A.; McConnell, D. B.; Smith, W. H. *Biomass Bioenergy*, 1992, 3, 261-266.
23. Ozores-Hampton, M.; Schaffer, B.; Bryan, H. H. *HortScience*, 1994, 29, 785-788.
24. Wen, G.; Bates, T. E.; Voroney, R. P. *J. Environ. Qual.*, 1995, 24, 527-534.

25. Chaney, R. L.; Ryan, J. A. In *Science and Engineering of Composting*; Hoitink, H. A. J.; Keener, H. M., Eds. Renaissance Publ.: Worthington, OH, 1993, pp 451-506.
26. U. S. Environmental Protection Agency. *Federal Register*, 1993, 58, 9387-9404.
27. Richard, T. L. *Biomass Bioenergy*, 1992, 3, 163-180.
28. Waksman, S. A. *Humus*; Williams & Wilkins, Baltimore, MD, 1936.
29. Allison, F. E. *Soil Organic Matter and its Role in Crop Production*; Elsevier: New York, NY, 1973.
30. Cooperative Extension, University of Illinois. *The Morrow Plots*. University of Illinois, College of Agriculture Circ. 777, 1968.
31. Salter, R. M.; Schollenberger, C. J. In *Soils and Men*; Yearbook of Agriculture; U.S. Government Printing Office: Washington, DC, 1938, pp 445-461.
32. Salter, R. M.; Saltenberger, C. J. *Farm Manure*. Ohio Agr. Expt. Stn. Bull. 605, 1939.
33. Hill, R. L.; James, B. R. In *Soil Amendments and Environmental Quality*; Rechcigl, J. E. Ed.; Lewis Publ.: Boca Raton, FL, 1995, pp 311-325
34. Parr, J. F.; Willson, G. B. *HortScience*, 1980, 15, 162-166.
35. Diaz, L. F.; Savage, G. M.; Eggerth, L. L.; Golueke, C. G. *Composting and Recycling Municipal Solid Waste*; Lewis Publ.: Boca Raton, FL, 1993.
36. Jansson, S.L.; Persson, J. In *Nitrogen in Agricultural Soils*. Stevenson, F.J., Ed.; Agronomy 22; American Society of Agronomy: Madison, WI, 1982; pp 229-252.
37. Hadas, A.; Portnoy, R. *J. Environ. Qual.*, 1994, 23, 1184-1189.
38. Castellanos, J. Z,; Pratt, P. F. *Soil Sci. Soc. Amer. J.*, 1981, 45, 354-357.
39. Chanyasak, V.; Katayama, A.; Hirai, M. F.; Mori,S.; Kubota, H. *Soil Sci. Plant Nutr.*, 1983, 29, 239-259.
40. Chen, Y.; Inbar, Y. In *Science and Engineering of Composting*; Hoitink, H. A. J.; Keener, H. M., Eds. Renaissance Publ.: Worthington, OH, 1993, pp 551-600.
41. Cheneby, D.; Nicolardot, B. In *Composting and Compost Quality Assurance Criteria*; Jackson, D. V.; Merillot, J.-M.; L'Hermite, P., Eds. Commission European Communities: Luxembourg, 1992, pp 188-196
42. Cole, M. A. *Technical Symposium, The Composting Council's Third National Conference*;The Composting Council: Washington, DC, 1992, pp 31-36.
43. Iannotti, D. A.; Grebus, W. E.; Toth, B. L.; Madden, L. V.; Hoitink, H.A. J. *J. Environ. Qual.*, 1994, 23, 1177-1183.
44. Bugbee, G. J. and C. R. Frink. *HortScience*, 1989, 24, 625-727.
45. Terman, G. L.; Soileau, J. M.; S.E. Allen, S. W. *J. Environ. Qual.*, 1973, 2, 84-89.
46. Baker, K. F. *The U. C. System for Producing Healthy Container-grown Plants*; California Agric. Expt. Stn. Manual 23, 1957.
47. Boodley, J. W.; Sheldrake, R. S. *Cornell Peatlite-mixes for Commercial Plant Growing*. Cornell Univ. Plant Sci. Info. Bull. 43, 1972.

48. *Composting Source Separated Organics*; Staff of BioCycle, Ed. The JG Press: Emmaus, PA, 1994.
49. Chaney, R. L.; Munns, J. B.; Cathey, H. M. *J. Amer. Soc. Hort. Sci.*, 1980, 105, 485-492.
50. Heckman, J. R.; Kluchinski, D. *J. Environ. Qual.*, 1996, 25, 355-362.
51. O'Brien, T. A.; Barker, A. V. *Compost Sci. Util.*, 1995, 3(3), 53-65.
52. O'Brien, T. A.; Barker, A. V. *Compost Sci. Util.*, 1995, 3(4), 69-77.
53. Barker, A. V. *Agronomy Abstr.*, American Society of Agronomy, Madison, WI, 1993, p. 26.
54. Barker, A. V. *HortScience*, 1993, 28, 256.
55. O'Brien, T. A.; Barker, A. V. *Commun. Soil. Sci. Plant Anal.*, 1996, 27, 57-85.
56. Pokorny, F. A.; Henny, B. K. *J. Amer. Soc. Hort. Sci.*, 1984, 109, 770-776.
57. Kissel, J. C.; Henry, C. L.; Harrison, R. B. *Biomass Bioenergy*, 1992, 3, 181-194.
58. Ladd, J. N.; Jackson, R. B. In *Nitrogen in Agricultural Soils*;Stevenson, F.J., Ed.;Agronomy 22;American Society of Agronomy:Madison, WI,1982,pp173-228.
59. Black, C. A. Ed. *Methods of Soil Analysis. Chemical and Microbiological Properties*; Agronomy 9, Part 2; American Society of Agronomy, Madison, WI, 1965.
60. Bradstreet, R. B. *The Kjeldahl Method for Organic Nitrogen*; Academic Press: New York, NY, 1965.
61. Jones, J. B., Jr. *Kjeldahl Method for Nitrogen (N) Determination*; MicroMacro Publishing: Athens, GA, 1991.
62. Klausner, D. D.; Guest. R. W. *Agron. J.*, 1981, 73, 720-723.
63. Smith, J. H.; Peterson, J. R. In *Nitrogen in Agricultural Soils*; Stevenson, F. J., Ed.; Agronomy 22; American Society of Agronomy: Madison, WI, 1982; pp 791-831.
64. Cisar, J. L.; Snyder, G. H. *HortScience*, 1992, 27, 219-222.
65. Mitchell, W. H.; Molnar, C. J.; Barton, S. S. 1994. *BioCycle*, 1994, 35, 62-63.
66. Barker, A. V.; O'Brien, T. A. *Proc. Northeast. Weed Sci. Soc.*, 1995, 49, 56-60.
67. Schmidt, E. L. In *Nitrogen in Agricultural Soils*; Stevenson, F.J., Ed.; Agronomy 22; American Society of Agronomy: Madison, WI, 1982; pp 253-288.
68. Zucconi, F.; de Bertoldi, M. In *Compost Specifications for the Production and Characterization of Compost from Municipal Solid Waste: Production, Quality and Use*; de Bertoldi, M.; Ferranti, M. P.; L'Hermite, P.; Zucconi, F., Eds.; Elsevier: London, 1987, pp 30-50.
69. Edwards, C. A. *Agric. Ecosyst. Environ.*, 1988, 24, 21-31.
70. Firestone, M.K. In *Nitrogen in Agricultural Soils*; Stevenson, F. J., Ed.; Agronomy 22; American Society of Agronomy, Madison, WI, 1982, pp 289-326.

71. Barker, A. V.; Mills, H. A. *Hort. Rev.*, 1980, 3, 395-423.
72. Jones, J. B., Jr.; Wolf, B.; Mills, H. A. *Plant Anlysis Handbook*; Micro-Macro Publishing, Athens, GA., 1991.
73. Bingham, F. T. In *Diagnostic Criteria for Plants & Soils*. H.D. Chapman, H. D., Ed.; Chapman: Riverside, CA, 1965, pp 324-361.
74. Logan, T. J.; Harrison, B. J. *J. Environ. Qual.*, 1995, 24, 153-164.
75. Marschner, H. *Mineral Nutrition of Higher Plants*; Academic Press: New York, NY, 1995.
76. Brown, K. H.; Bouwkamp, J.; Gouin, F. R. *The Composting Council's Fifth National Conference*; The Composting Council: Washington, D.C., 1994, p 25.
77. Bellamy, K. L. *J. Environ. Qual.*, 1995, 24, 1074-1082.
78. Eaton, F.M. In *Diagnostic Criteria for Plants and Soils*. Chapman, H. D., Ed.; Chapman: Riverside, CA, 1965, pp 444-475.
79. O'Brien, T. A. *Evaluation of Composts for Production of Sod and Groundcover Crops*. Master's thesis, University of Massachusetts, Amherst, 1995.
80. Sloan, J. J.; Basta, N. T. *J. Environ. Qual.*, 1995, 24, 1097-1103.
81. Gaugh, H. G. *Inorganic Plant Nutrition*; Dowden, Hutchinson & Ross: Stroudsburg, PA, 1982.
82. Barker, A. V.; Volk, R. J.; Jackson, W. A. *Soil Sci. Soc. Amer. Proc.*, 1966, 30, 228-322.
83. Maynard, D.N.; Barham, W. S.; McCombs, C. L. *Proc. Amer. Soc. Hort. Sci.*, 1957, 69, 318-322.
84. Pill, W. G.; Lambeth, V. N.; Hinckley, T. M. *J. Amer. Soc. Hort. Sci.*, 1978, 103, 265-268.
85. Epstein, E.; Chaney, R.L.; Henry, C.; Logan, T. J. *Biomass Bioenergy*, 1992, 3, 227-238.
86. Lee, K. W.; Keeney, D. R. *Water, Air, Soil Pollut.*, 5, 109-112.
87. Williams, C. H.; David, D. *J. Soil Sci.*, 1976, 121, 86-93.
88. Graham, R. D.; Webb, M. J. In *Micronutrients in Agriculture, Second Edition*; Mortvedt, J. J., Ed.; Soil Science Society of America: Madison, WI, 1991, pp 329-370.
89. Thomson, W. T. *Agricultural Chemicals. Book IV--Fungicides*. Thomson Publ.: Fresno, CA, 1993.
90. Allaway, W. H. In *Soils for Management of Organic Wastes and Waste Waters*; L.F. Elliot, L. F.; Stevenson, F. J., Ed.; Soil Science Society of America: Madison, WI. 1977, pp 282-298.
91. Cheung, Y. H.; Wong, M. H. *Agric. Wastes*, 1983, 5, 63-81.
92. Tam, N. F. Y.; Tiquia, S. Resourc. Conserv. *Recycling*, 1994, 11, 261-274.
93. Underwood, E. J. *Trace Elements in Human and Animal Nutrition*; Academic Press: New York, NY, 1971.
94. Hue, N. V. In *Soil Amendments and Environmental Quality*; Rechcigl, J. E., Ed.; Lewis Publ.: Boca Raton, FL, 1995, pp 199-247.

95. Chaney, R. L.; Ryan, J. A. *Proc. Internat. Symp. Land Application of Organics*; Council for Promotion of Untilization of Organic Materials: Tokyo, 1995.
96. Sanderson, K.C. *HortScience* 15:173-178. 1980.
97. Richard, T. L; Woodbury, P. B. *Biomass Bioenergy*, 1992, 3, 195-211.
98. Woodbury, P.B.; Breslin, V. T. *Biomass Bioenergy*, 1992, 3, 213-225.
99. Leita, L.; De Nobili, M. *J. Environ. Qual.*, 1991, 20, 73-78.
100. Sims, J. T; Kline, J. S. *J. Environ. Qual.*, 1991, 20, 387-395.
101. Gouin, F. R. *HortTechnology*, 1993, 3, 161-163.
102. Manios, V. I.; Syminis, H. I. *BioCycle*, 1988, 29(7), 44-47.
103. Rosen, C. J.; Halbach, T. R.; Swanson, B. T. *HortTechnology*, 1993, 3, 167-173.
104. Vogtmann, H.; Fricke, K. In *Composting and Compost Quality Assurance Criteria*; Jackson, D. V.; Merillot, J. M.; L'Hermite, P., Ed.; Commission of the European Communities: Luxembourg, 1992, pp 227-236.
105. Chaney, R.L.; Ryan, J.A.; O'Connor, G. A. *Proc. Intern. Symp. Organic Contaminants in Sewage Sludge*, Sci. Total Environ: Lancaster, UK, 1995.
106. Williams, R. T.; Keehan, K. R. In *Science and Engineering of Composting*; Hoitink, H. A. J.; Keener, H. M., Eds. Renaissance Publ.: Worthington, OH, 1993, pp 363-382.

Chapter 11

Composition and Uses of Uncomposted Wastepaper and Other Organics

J. H. Edwards

Department of Agronomy and Soils, 201 Funchess Hall, Agricultural Research Service, U.S. Department of Agriculture, Auburn University, AL 36849-5412

The application of uncomposted waste paper and other organics requires *in situ* composting of 4 to 6 weeks in which microorganisms incorporate C into the soil organic matter (SOM) fraction and nutrients are returned to the soil. Three annual applications resulted in an increase in soil-available P, K, Ca and Mg by a factor of 3 to 4, while SOM content was increased by a factor of 1.5 to 2. Repeated applications of uncomposted organic waste along with poultry litter (PL) as a N source resulted in buildup of soil P. Nitrogen mineralization, C mineralization and N turnover were increased more by applying N as PL when compared to soil amended with NH_4NO_3. Uses of organic wastes on agricultural land include erosion control agent, biodegradable mulch, non-chemical weed control, and biological control agent. Potential environmental problems from repeated application of organic wastes include groundwater contamination from excessive NO_3-N, increased heavy metal concentrations, and induced soil nutrient imbalances. Pelletized waste paper applied as a surface residue reduced sediment loss by wind and water, increased water infiltration, and increased total available soil water. Developing soil-crop production systems that use uncomposted organic wastes along with manures can serve as valuable sources of organic C to soils with low organic matter, improving soil quality and agricultural production.

Since the early 1960's, food and fiber production has become ever more dependant on synthetic fertilizers, and petroleum-based herbicides and pesticides for supplying plant nutrients, weed control, and crop protection. Along with the increased use of these technologies, soil cultivation has been intensified to improve weed control and seedbed preparation. These processes have contributed to the high production levels we have become accustomed to today. As a result, soil management practices that contribute to increased soil productivity and long-term sustainable agricultural production have been neglected. Failure to maintain effective soil conservation practices has resulted in a decline in soil productivity and a decrease in SOM content, as well as accelerated soil erosion and nutrient runoff losses (*1,2*).

The Waste Dilemma. Problems with landfilling municipal solid waste (MSW) have escalated because of the increasing quantity of wastes generated and decreasing availability of landfill space *(3,4)*. On average, 67% of MSW in the USA goes into landfills, 23% is recycled, and 10% is incinerated *(5)*. Even though we may be willing to pay the increased cost of landfill disposal, the U.S. Environmental Protection Agency (EPA) has established a goal of reducing the nation's dependence on landfill disposal by source reduction or recycling *(6)*. By adopting this goal, the EPA has emphasized the need for a change in the nation's approach to producing, packaging and disposing of consumer goods. To comply with EPA's national goal, many states enacted laws requiring a 30-60% reduction in the volume of MSW entering landfills by 1995 *(7)*. Currently, 30 to 40% of the MSW stream is post-consumer fiber and 15 to 20% is yard waste. These two organic wastes are targeted to help achieve mandated recycling goals.

Composting Organic Wastes. Composting the organic fraction of MSW is a management technique that is desirable because it reduces the volume and weight by 50% *(8)*. Composting organic wastes is becoming increasingly recognized as a viable method for waste management in Europe and the USA *(9)*. In the USA, a larger percentage of farmers are composting than are municipalities, commercial, or private groups combined *(10,11)*, and the number is expected to increase because of environmental restriction, loss of available crop land, urban encroachment, and the increased availability of suitable non-agricultural organic wastes to be combined with animal manures *(12,13)*. While composting improves the value of organic waste and improves crop yield *(14-16)*, it also involves additional handling, moving, and processing.

One of the main problems with the use of MSW as compost is the lack of reliable quality criteria *(17)*. The C:N ratio in solid and water phases *(18-20)*, cation exchange capacity *(21)*, humification indices *(20)*, oxygen and CO_2 respirometry *(22)*, and spectroscopic analyses of OM transformation *(23,17)* have been proposed as maturity and stability indices of MSW compost. However, compost maturity is impossible to define by only one parameter since usually several are linked for best results *(17)*. If composting organic wastes is to be successful as a means of reducing landfill disposal, consistent compost must be produced and local markets must be developed to use them as soil amendments, erosion control agents, growth media for the nursery industry, or as crop and landscape mulches.

Land Application. An alternative disposal method for post-consumer fiber and other organics such as yard waste and wood chips may be their application to agricultural land as uncomposted soil amendments and/or surface mulches *(24)*. Potential benefits may be improved plant growth by increased SOM content *(25, 26)*, reduced soil bulk density *(27)*, increased soil porosity *(28)*, increased soil pH *(29)*, increased water stable aggregates *(30,31)*, increased rooting depth *(32)*, and increased water infiltration *(33)*, all of which may help control soil erosion *(34)*. The utilization of composted and uncomposted organic wastes in agriculture has the potential of improving soil structure, increasing cation exchange capacity, enhancing plant growth *(35)*, suppressing soil-borne plant pathogens *(36)*, and increasing soil enzymatic activity *(37)*.

Benefits from applying organic wastes to land are dependant on geographical location. In the southeastern U.S., climatic conditions favor the rapid incorporation of organics into SOM where they can improve the physical and chemical properties of soil. In drier regions, applications of organic wastes may help conserve soil water resources and reduce sediment loss caused by wind erosion. What is needed in areas of limited rainfall is a soil cover that does not compete with the cash crop for water or nutrients.

Although beneficial effects of organic wastes on crop production have been demonstrated repeatedly (38-40), land application can promote degradation of water quality (41,42) and buildup of metals (43-47). Environmental problems may be created with surface-applied uncomposted organic amendments, including contamination of surface and groundwater from N and P, and the buildup of toxic levels of Cu, Zn, Mn, and Al in the soil. Using these higher C:N ratio organic wastes in combination with poultry litter (PL) or animal manures may be a means to speed decomposition, or binding into unusable forms of both, without increasing the potential for leaching of NO_3-N and loss of both N and P in surface water runoff.

Characterization of Waste Paper and Other Organics

Physical Nature and Size. The waste paper fraction of MSW, as collected by municipalities, is not suited for direct land application because it is very resistant to decomposition by soil microorganisms and can create barriers that emerging plant seedlings may not be able to penetrate. Post-consumer fiber (newsprint and telephone books) can be obtained from a local recycling plant. Different lots may contain both regular newsprint and other grades of colored paper and advertisement inserts. A processing step may include grinding, chopping, or shredding the paper; however, then it becomes very light and can be easily blown about, and storage and transportation may become problems. Some properties of uncomposted ground waste paper, i.e., low bulk density, make it very difficult to use on a large scale because specialized equipment is required for land application (37,48,35). Also, careful management is required to balance the nutrient supply to both plants and microorganisms.

Pelletized Paper. Pelletizing involves the use of a patented, pressurized extrusion process and adds an additional step and therefore cost to the processing of waste paper. This does not change the chemical content of the paper, but it reduces the active surface area of the pellets when compared to ground paper. When the organic waste is compressed by pelletization to increase its density, the number of active sites available for enzymatic reaction is reduced, as a result the release of metals contained in the organic waste is reduced, and the conversion rate into stable SOM is reduced.

Other Organics. Land application of organic wastes is an inexpensive alternative to the rising costs of landfill disposal. Organic wastes such as municipal biosolids, composted municipal solid waste, feedlot manures, poultry litter, agricultural waste by-products, and crop residues have been applied to agricultural land for decades; in many countries of the world today they are the only available fertilizers for agricultural production. Land application recycles valuable nutrients and effectively disposes of the wastes.

Yard waste, as collected by municipalities, is a mixture of grass clippings, leaves, tree stumps and limbs with a C:N ratio of about 40:1. Waste paper (newsprint, telephone books) and wood byproducts have a very high C:N ratio, 100:1 or higher. Poultry litter from broiler houses is a mixture of manure, wasted feed, bedding material and water with a C:N ratio of 9:1 to 10:1. Manure from egg-laying operations is composed of poultry excreta, wasted feed and moisture and also has a C:N ratio of 9:1 to 10:1. The combination of a high C:N ratio waste, i.e. paper or yard waste, with a low C:N ratio waste, i.e. manure, to adjust the C:N ratio to approximately 30:1 at the time of application will help to balance the nutrient supply for both plants and soil microorganisms. Since the N and P are present in inorganic and organic fractions in manures there are no simple, rapid and reproducible methods of predicting how much N and P will be mineralized.

The rate and extent that mineralization occurs will determine the usefulness of manures as a source of N for C:N adjustment of high C:N ratio cellulose organic wastes. Litter collected from 147 poultry houses in Alabama ranged from 0.61 to 3.9% P, averaging 1.6% (*34*). The P fraction is about 75% as effective as commercial fertilizer as a source of P during the year of application. The P as well as N must be taken into account in determining the application rate needed to adjust the C:N ratio of other organic wastes. If this is not done, the result may be excessive levels of P in soil and contamination of surface water.

Organic Components. Cellulose is the most abundant chemical constituent of organic wastes and accounts for 15 to 60% of the dry weight; hemicellulose 10 to 30%; lignin 5 to 50%; water-soluble fraction (simple sugars, amino acids, and aliphatic acids) 5 to 30%; alcohol-soluble fraction (fats, oils, waxes, resins, and pigments) < than 10%; N and S-containing compounds (proteins, RNA, and DNA) < 5% (*49*) (Table I). All of these compounds may not be present in all wastes. Paper is mainly cellulose and hemicellulose, while animal bedding may be composed of cellulose, hemicellulose, and lignin.

Table I. Organic Fractions in Cellulose Wastes Collected from MSW Streams and Confined Animal Industry

Organic wastes	Ash	H_2O-soluble polysaccharide	Hemi-cellulose + cellulose	Crude protein	Lignin	Total recovery
			% by dry weight of waste			
Poultry litter[1]	11.13	17.69	35.37	17.25	13.54	94.98
Newsprint #1[3]	3.31	1.35	47.73	0.75	41.22	94.36
Newsprint #2[3]	0.72	0.99	44.67	0.75	45.88	93.01
Insert adv.[3]	6.71	1.45	23.22	0.75	64.48	96.64
Cardboard[2]	9.53	1.64	23.32	0.75	53.74	88.98
Junk mail[4]	3.43	3.16	34.60	0.75	52.44	94.38
C paper[5]	8.50	2.85	34.16	0.75	50.96	97.22
Phone books[6]	6.27	1.80	25.41	0.75	58.44	92.67
C gin waste[1]	4.98	4.69	47.23	9.13	30.92	96.95
Wood chips[1]	9.86	3.26	25.92	2.82	52.40	94.26

[1] Poultry litter, cotton gin waste and wood chips collected from industries in Alabama.
[2] Cardboard collected from a local recycling firm.
[3] Local newspapers were collected for a one week period and separated into: newsprint #1 = newspaper with all advertisement inserts; Insert adv. = advertisement inserts were separated; newsprint #2 = reams of paper used by the printing company for the daily newspaper.
[4] Junk mail = bulk rate mail collected by a local homeowner for a one week period.
[5] Waste paper collected from Auburn University Computer Center.
[6] Phone books = collected during local recycling by BellSouth or GTE.

Nutrient Contents of Organic Wastes. The concentrations of macronutrients Ca, K, Mg, and P in the different sources of paper are not a problem with repeated annual applications (Table II) (*50*). The concentration of soil Cu and Zn may become a problem from repeated application when the waste contains a higher percentage of advertisement inserts, and PL is the source of N used to adjust the C:N ratio of the combination of organic amendments. Also, the concentrations of nutrients observed in PL will build up with repeated annual application, particularly soil P levels. One metal of concern is Al; its concentration is related to the manufacturing of paper (Table III).

Table II. Total Elemental Analyses of Manure from Confined Animal Industry, Newsprint and other Cellulose Organic Wastes from the MSW Stream

Organic wastes	Macronutrients				Micronutrients					
	Ca	K	Mg	P	Cu	Fe	Mn	Zn	B	Mo
	g kg^{-1} of waste				mg kg^{-1} of waste					
Poultry litter[1]	27.1	28.5	5.7	20.2	550	2144	632	533	69	5.7
Wood chips[2]	5.4	5.0	1.9	1.9	49	354	200	52	11.7	0.95
Newsprint #1[3]	0.8	0.1	0.1	0.1	22.5	136	31	51	1.8	4.2
C paper[4]	28.6	0.04	0.4	0.1	4.4	219	5	6	1.8	3.1
F container[5]	4.3	0.08	0.3	0.1	16.7	298	29	57	6.6	8.2
Insert adv.[3]	5.1	0.01	0.2	0.1	32.3	223	28	151	3.1	10.5
Junk mail[6]	10.0	0.06	0.3	0.1	12.2	159	6	10	15.7	5.6
Phone books[7]	0.9	0.09	0.2	0.04	5.7	57	45	5	1.0	1.8
Newsprint #2[3]	0.9	0.1	0.2	0.01	0.9	21	44	7	1.3	0.5

[1] Poultry litter collected from a broiler house in Alabama.
[2] Collected from a local planing mill in Alabama.
[3] Local newspapers were collected for a one week period and separated into: Newsprint #1 = newspaper with all advertisement inserts; Insert adv. = advertisement inserts were separated; newsprint #2 = reams of paper used by the printing company for the daily newspaper.
[4] Waste paper collected from Auburn University Computer Center.
[5] F container = cereal boxes and microwave dinner boxes.
[6] Junk mail = bulk rate mail collected by a local homeowner for a one week period.
[7] Phone books = collected during local recycling by BellSouth or GTE.

Metal Contents of Organic Wastes. A potential environmental concern is the concentration of micronutrients Cu, Mn, and Zn contained in newsprint. They could become a problem with repeated application of newsprint to land (*51-53*). The metals Ba, Cd, Cr, and Pb, which are contained in newspaper printed with colored inks, are also of some concern (*47,48,54*). Red inks primarily contain Ba, Cr, and Cd compounds; blue or green inks may contain Cu compounds. In colored ad inserts with a metallic finish, the heavy metals Co, Mn, and Pb could be contained in ink (*55*). The current trend in the printing industry is to use inks with a soybean oil base; these are less harmful to the environment than inks made with petroleum-based organic solvents. The concentrations of non-essential nutrients and heavy metals found in some organic wastes are given in Table III.

Method and Time of Application

Surface Application. Surface application of ground newsprint, with and without several N sources (inorganic and organic) to adjust the C:N ratio of the waste to ≤ 30:1, as well as provide N needed by crops have been evaluated by the author. Cotton (*Gossypium hirsutum* L.) yields after two annual spring applications were increased when PL, rather than NH_4NO_3, was used to adjust C:N ratios of ground newsprint (56). Phytotoxicity of newsprint to plants was ameliorated by the

Table III. Total Elemental Analyses of Manure from Confined Animal Industry, Newsprint, and other Cellulose Organic Wastes from the MSW Stream

Organic wastes	Non-essential nutrients				Heavy metals				
	Na	Al	Ba	Si	Co	Cr	Pb	Ni	Cd
	mg kg^{-1} of waste								
Poultry litter[1]	6977	2573	31.6	2075	2.0	8.5	14.6	7.6	2.4
Wood chips[2]	1369	444	15.3	734	0.4	1.1	0.4	0.9	2.4
Newsprint #1[3]	900	4525	17.2	638	0.2	1.5	8.4	0.7	0.2
C paper[4]	1012	2076	2.9	1894	0.4	3.6	7.5	0.9	0.1
F container[5]	555	10120	20.1	916	2.5	5.1	22.4	0.9	0.2
Insert adv.[3]	619	5871	28.1	648	0.3	2.2	13.9	0.4	0.2
Junk mail[6]	822	6730	20.1	1125	1.0	4.0	15.2	0.5	0.1
Phone books[7]	230	2309	18.8	417	0.1	0.8	6.3	0.3	0.3
Newsprint #2[3]	648	426	9.4	336	0.0	0.0	0.0	0.3	0.1

[1] Poultry litter and wood chips collected from the confined animal industry in Alabama.
[2] Collected from a local planing mill in Alabama.
[3] Local newspapers were collected for a one week period and separated into: Newsprint #1 = newspaper with all advertisement insert; Insert adv. = advertisement inserts were separated; newsprint #2 = reams of paper used by the printing company for the daily newspaper.
[4] Waste paper collected from Auburn University Computer Center.
[5] F container = cereal boxes and microwave dinner boxes.
[6] Junk mail = bulk rate mail collected by a local homeowner for a one week period.
[7] Phone books = collected during local recycling by BellSouth or GTE.

addition of PL as the N source. Surface application of organic waste requires more management to provide for optimum growth of the crop; the chemical composition and nutrient requirements of the microorganisms as well as the crop must be determined and used to balance the soil nutrient supply.

Vertical Trenches. Ground newsprint, with and without PL, mixed with soil in narrow vertical trenches between rows of cotton, as well as on the soil surface improved cotton rooting depth (*57*). Cotton roots grew to the bottom of the 122 cm deep trenches. Under the row (undisturbed soil), rooting depth was confined to the top 15 cm of soil. The primary reason for the proliferation of roots in the trenches was the difference in soil bulk density due to trenching between the disturbed (1.3 Mg m^{-3}) and undisturbed (1.55 Mg m^{-3}) soil.

Yields ranged from a low of 29 kg ha^{-1} of lint with 4.9 kg m^{-2} surface-applied newsprint without PL, to a high of 920 kg ha^{-1} of lint for cotton adjacent to the 61 cm deep trench backfilled with soil, ground newsprint, and PL. Total yield of the surface-applied ground newsprint was 470 kg ha^{-1} of lint, and the highest cotton yields occurred when PL was mixed with soil, newsprint and PL and was independent of trench depth (*50*).

Other experiments determined the effects of ground newsprint in vertical trenches dug to a depth of 61-cm between rows of grain sorghum [*Sorghum bicolor* (L.) Moench] and soybean [*Glycine max* (L.) Merr.] (*58*). An implement was designed and built to disrupt the soil hardpan and incorporate ground newsprint into the trench in one pass. At the end of the growing season, measurements showed reduced reconsolidation occurred in trenches formed with newsprint as compared to trenches formed without newsprint. Vertical trenching of organic waste provides a more positive rooting environment, but at a higher energy expenditure.

Placement of organic wastes in vertical trenches, or in the excavated channel of a subsoiler shank, uses specialized equipment with high energy requirements. Trenching restricts the amount of waste that can be applied in each pass. However, the adjustment of the C:N ratio of the organic waste is not as important as it is in surface application because only a small portion of the total plant root system comes in contact with the waste, and there is very little competition between microorganisms and crops for available nutrients.

Vertical trenching of waste paper could help to restore soil productivity in eroded fragipan soils (*59*). Loss of SOM content contribute to: a) increased surface sealing and runoff, b) crusting formation, c) reduced seedbed for germination of seeds. Application of organic wastes to improve the SOM content may be a method of maintaining productivity in eroded fragipan soils (*34*).

Time of Application. When ground newsprint less than 6 mm in diameter was applied to the surface and incorporated into the top 15 cm of soil at a rate of 4.9 kg m^{-2} (43,560 lbs/acre) immediately before planting in the spring, it severely stunted cotton seedlings for the first 6 weeks after germination, particularly in treatments where the C:N ratio of the newsprint was not adjusted to 30:1 (*50*). A number of fungal organisms were found on the newsprint that require nutrients from the soil-nutrient pool; organisms detected were those generally found on organic material with high C:N ratios ($\geq$ 100:1).

Cotton yields were not affected when newsprint was surface-applied 4 to 6 weeks prior to planting, and C:N ratio was adjusted to 30:1 (*7*). When newsprint C:N ratio was adjusted to $\leq$ 30:1 in the fall, the following spring-planted cotton yields were increased 60% when compared to a control (standard cultural practices). With spring-applied newsprint and the C:N ratio adjusted $\leq$ 30:1, yields were increased 54% when compared to a control. When the C:N ratio of newsprint was not adjusted ($\geq$ 150:1), fall or spring application of newsprint decreased yields below the control.

Benefits from Land Application of Organic Wastes

Erosion Control. The 1985 Farm Bill recognized the danger of continued topsoil loss and mandated new approaches to erosion control. To achieve this goal, most farmers must plant cover crops during fallow periods, as well as leave crop residues on the soil surface. In most cases, yields are reduced because the cover crops use water during fallow periods that is needed to produce the subsequent cash crop.

Reduced Water Runoff. The effects of organic wastes and winter cover crop on sediment loss and runoff were evaluated in the Sand Mountain area in northern Alabama on agricultural land with a 9% slope (*33*). Cover was wheat (*Triticum aestivum* L.) and two organic wastes (ground newsprint and PL). The cover crop was established in the fall by direct-drilling wheat into the previous corn residue, along with application of two organic wastes. Newsprint was applied at 12.2 kg ha^{-1} (0.5 lbs ft^{-2}) and PL was applied at 8.3 kg ha^{-1} to adjust the C:N ratio of newsprint to 30:1.

Sample collections began in March and continued through August. Total rainfall for this period for 1994 was 793 mm. Accumulative runoff collected was 55 mm for wheat cover crop residue, 54 mm for the PL, and 42 mm for newsprint plus PL. The soil loss for the wheat cover was 3,584 kg ha^{-1}, PL was 3,360 kg ha^{-1}, and newsprint plus PL was 2,464 kg ha^{-1} (*33*). In a control plot with conventional tillage, the soil loss was 15,456 kg ha^{-1} as compared to 896 kg ha^{-1} from fescue sod. This soil loss from conventional tillage was 2 times higher than the acceptable level of 6,720 kg ha^{-1}. However, total rainfall for the same time period in 1993 was 300 mm. Newsprint cover reduced soil loss 30% more than the wheat cover crop.

Reduced Wind Erosion. Wind erosion can occur when soil aggregates, capable of being moved by wind are present on the surface soil, the wind velocity is sufficient to detach soil aggregates, and the soil surface is not protected with nonerodible material. The distance traveled by the airborne particles depends on the velocity of the wind and shape and density of the soil particles (*60*). In Big Spring, TX, a wind erosion event occurs an average of 26 days annually and the event averages 7.2 hours in duration.

Soil erodibility can be reduced by roughening the soil surface with tillage implements or placing vegetation on the soil surface. Covering even a small percentage of the soil surface with nonerodible material will reduce the loss of a highly erodible soil (*61*). The best means of controlling wind erosion may include a combination of various control methods.

Wind tunnel studies were conducted with 3, 8, and 19 mm diameter paper pellets under the following conditions: 1) dry pellets applied to a smooth, dry soil surface; 2) dry pellets applied to a dry soil, then wet with simulated rain. Soil loss from the trays covered with pellets were compared with soil loss from bare dry soil to evaluate the effect of rain on pellets alone or the combined effect of both rain and paper pellets on wind erosion. Dry pellets were applied so that 10, 20, or 30% of the soil surface was covered.

The 3 mm diameter pellet swelled to about 2 or 2.5 times their original diameter upon wetting. Without wetting, these pellets tended to blow on the smooth soil surface; wetting reduced this problem. The quantity of pellets required to cover different percentages of the soil surface will vary depending on the moisture the pellet can absorb when fully expanded. The moisture content of dry pellets at room temperature is approximately 10%. Simulated rainfall was applied at 16.6 mm hr^{-1} and then air-dried for 7 days. Dry pellets applied at a rate of 11,200 kg ha^{-1} (30% ground cover) reduced wind erosion by 95%.

Two sizes of pellets (8 or 19 mm diameter) were applied to a Miles (fine-loamy, mixed, thermic udic Paleustalfs) soil, a Olton (fine, mixed, thermic aridic Paleustolls) clay loam soil, or a Amarillo (fine-loamy, mixed, thermic aridic Paleustalfs) fine sandy loam soil to evaluate their effectiveness in reducing wind erosion. The dry pellets needed to obtained 30% ground cover required approximately 5600 kg ha^{-1}.

One advantage of applying pellets as ground cover is water conservation. However, improved water conservation was observed only with the 19 mm diameter pellet. If air-dried pellets are used for soil cover, less weight is required by the 5 mm diameter because of the increased surface area; however, if the surface area is estimated after the absorption of water, approximately 11,200 kg ha^{-1} of pellets are needed to obtain 30% soil cover. Cotton lint yields were increased as a result of increased water storage from rainfall. Even with only approximately 300 to 400 mm of rain each year, there are rainfall events where surface runoff occurs. Pellets reduced soil loss by surface runoff by 41%.

Application of pellets to the Miles soil increased stored water in the top 30 cm. This increase was evident in the period of reduced rainfall, i.e., July and August, where only 40 mm of rainfall was recorded. This increase in stored water was translated into a 95.2 kg ha^{-1} increase in cotton lint yield.

In a crop production system that is dependant on rainfall as the main source of water for plant growth, water deficiency is the major factor limiting crop yield *(62)*. Pellets can be used to provide an alternative mulch and/or residue cover to reduce wind erosion and nonpoint source pollution from cropland. When recommended soil management practices were utilized and pellets were applied to achieve a 30% soil cover, wind-blown sediment was reduced by 50%. However, when recommended soil management practices were not followed, pellets did not reduce wind-blown sediment.

Nitrogen Mineralization/Immobilization. The mineralization of N in soil-incorporated organic wastes does not necessarily coincide with the N demands of the plant. At equivalent rates, PL left 42% more N in the soil system at the end of the growing season than N from a commercial inorganic source *(63)*. This reduced efficiency in supplying nutrients has contributed to excessive application of organic wastes to land, resulting in leaching of NO$_3$-N to the groundwater *(64,65)*. The use of two different organic wastes, with C:N ratio balanced to about 30:1, may be a means to reduce the leaching of NO$_3$-N when an excessive amount has been applied to agricultural land.

Soil incorporation of plant residues or organic wastes with high C:N ratios (> 40 to 50:1) can result in the immobilization of N by the soil microorganisms. The extent of immobilization or mineralization of N depends on the composition of the organic waste or plant residue applied. Paper mill sludge with C:N ≥ 100:1 reduced plant-available N and the unrecovered N was assumed to be immobilized by incorporation into the microbial biomass *(66)*. When the mineralization of N is expressed as a percentage of total organic N contained in the residue, 3% was mineralized from soybean residue and 97% from alfalfa (*Medicago saiva* L.) residue; but corn (*Zea mays* L.) residue or sawdust resulted in immobilization of soil N ranging from 3% to 687% *(67)*.

When plant residue from *Lathyrus lingitanus* was incorporated into the surface soil from long-term spring wheat rotations, N mineralization increased when the following use was fallow, but decreased when spring wheat or canola (*Brassica napus*) was grown *(68)*. By following ^{15}N labeled residue from *Lathyrus lingitanus*, N mineralization of both the indigenous soil N and residue was found to be significantly increased when fallow was compared to residues from spring wheat or canola. The decrease in N mineralization in the spring wheat and canola crops was attributed to immobilization of N by the soil microorganisms.

Plant residue incorporated by moldboard plowing resulted in a uniform N mineralization potential to a soil depth of 15 cm. Chisel plowing and no-tillage resulted in higher N mineralization in the 0-5-cm depth than the 5-10 or 10-15-cm depths *(69)*. After 10 years of tillage and crop rotation management, surface soil

(0-5 and 5-10-cm depths) organic C was increased by 67%, and N concentration and potential mineralization was increased by 66% under no-tillage. Crop rotation with a higher frequency of corn in the rotation resulted inhigher organic C and N than continuous soybean with winter wheat cover (70).

Cotton yields were increased by applying ground newsprint in the fall as a surface residue cover when compared to fallow condition (cotton planted in previous year cotton residue). There were no differences in seed cotton yields when organic waste cover was compared to wheat cover. Cotton yields following application of organic waste residue for cover were approximately in the order: newsprint, wheat cover, and fallow.

Not only does the residue composition affect N mineralization or immobilization (71), but placement of the organic residues as influenced by tillage practices has an effect on N mineralization or denitrification. Denitrification was influenced more by the percent of water in the water-filled pore space (72) than by placement of residue or organic wastes (73). Surface application of organic wastes (feedlot manures) for a 20-year period showed that the percent of N mineralization from manures was independent of manure application rate and that all of the N applied could be accounted for by plant uptake, soil N, and soil NO_3-N levels (74). Land application of anaerobically digested biosolids increased N mineralization (75), and increased soil NO_3-N levels (76).

Because most of the total N and P in PL is in the organic form (77,78), and must be converted to the inorganic form, the rate of mineralization is governed largely by microbial-mediated processes (79) and becomes the limiting step in N and P availability for use by microorganisms as well as plants. Much research has been directed at laboratory N mineralization studies to determine quantity of N produced from the soil organic pool over a period of time (80-83). Other N mineralization studies have been conducted under field conditions to estimate the quantity of N available for plant uptake (84-86), and chemical extraction methods to estimate available-N content (67,83).

Phosphorus Mineralization/Immobilization. Phosphorus does not have a negative impact on agricultural land when it is applied, but it does adversely affect surface water if it is moved off-site by runoff or erosion (87). When PL is applied at recommended rates for N, P is often applied in excess of crop demands and has the potential of being removed in surface runoff (88). To minimize runoff losses of P while maintaining adequate N fertility, the readily-available P must be altered or the application rate should match plant uptake. One of the ways to accomplish this is by the addition of chemical amendments that immobilize P, or retard the mineralization of P (89). Thus, one must be aware of the concentration of all nutrients contained in the organic waste and alter their potential mineralization or immobilization to reduce the risk of environmental contamination.

Poultry litter or manure can be a source of P contamination to soil and water when it is applied at excessive rates. However, Beegle (90) reported the P in PL was mainly in the organic fraction, and only slowly available to the crop. The soil organic P pool is composed of inositol derivatives, lipids and nucleic acid (91); none of these organic forms of P are considered hard to mineralized and should be plant-available. In some states, consideration is being given to treating PL with Al, Ca and/or Fe amendments to precipitate P and reduce the risk of P contamination before it is applied to land (92). Phosphorus in plant residue can also be a source of P contamination. Phosphorus mineralization and movement through the soil was greater in plant residue left on the soil surface than when residue was incorporated, with less organic P leached from surface-applied compared to incorporated residue (93).

Biodegradable Mulch for Vegetable Crops. Vegetable growers use mulch systems because of there many advantages. Mulches are usually plastic, and are placed on top of a raised planting bed. The plastic mulches are not reusable, and disposal has become a problem. In many states landfilling is not an option, and researchers are examining alternative mulches for vegetable production, including degradable plastic, and latex.

In field trials using different mulches, there was no yield response in fall or spring-planted collards [*Brassica oleracea* L. (Acephala Group)] when the N rate was ≥ 201 kg N ha^{-1} (*94,95*). No mulch (bare soil) produced yields that were higher than black plastic and newsprint in the fall planting. However, in the spring planting, black plastic mulch produced yields that were 33% higher than newsprint or bare soil. The increase in yield may be attributed to increase in soil temperature with the black plastic when compared to the organic waste mulches.

A problem was the occurrence of plant root disease. Bare soil had a higher plant survival rate followed by newsprint and black plastic in the fall planting. However, newsprint had higher survival rates in the spring planting followed by bare soil and black plastic. From limited data, this organic waste (ground newsprint) can be used as biodegradable mulch.

Weed Suppression. Use of organic wastes on agricultural soils certainly shows potential for mitigating the effects of some weeds, thus hopefully reducing the amount of herbicides needed for crop production. However, the mechanisms involved must be identified; it is unclear whether they are physical and/or chemical in nature. Benefits observed with application of organic wastes were the control of summer and winter annual weeds in cotton production. Fall surface-applied ground newsprint reduced winter weeds. A reduction in summer weeds, particularly large crabgrass [*Digitaria sanguinalis* (L.) Scop.], has been documented for each of 3 annual applications; the populations of large crabgrass seedlings were significantly less the second and third year. When paper pellets were fall surface-applied, they were as effective in suppressing winter and spring annual weeds as the standard method of chemical weed control. Fall-applied newsprint had additional benefits, including a decrease of large crabgrass seedlings the following spring by an average of 65 to 70%, and complete control of winter annual weeds.

Enhanced Soybean Production. Reducing production costs and/or improving yield is needed to sustain soybean production in the southeast U.S. Research has identified cultural practices that have potential to accomplish these goals. For example, early drill-planting has improved yields and reduced the need for herbicides.

Organic wastes were fall-applied as a surface residue cover with no additional tillage. Standard row spacing (76 cm) and direct drill, 18-cm row spacings were used (*95*). Ground newsprint at 24,416 kg ha^{-1}/year suppressed winter annual grasses and broadleaf weeds, but newsprint had very little effect on suppressing summer annual broadleaf weeds. Since all tillage operations were conducted in the fall, soybean were planted in a stale seedbed. An unexpected result was the enhanced germination of soybean by four days in the fall-applied newsprint plots.

Increase in Soil Organic Matter Content. An improvement in the soil nutrient-storage capacity is the primary reason for the increase in yield of cotton, soybean, and corn when organic wastes were applied to the soil surface and incorporated. A 10-fold increase in soil N was observed during an 18 month period. The increased nutrient-storage capacity was reflected in the increase in soil pH, in cation exchange capacity, and in Mehlich-I extractable P, Ca, and Mg (*Entry et al., Bio. Fert. Soil, in press*).

Surface application of 4.9 kg m² of ground newsprint increased the SOM content by 5.3 g kg⁻¹ of soil without C:N adjustment in a seven-month period *(50)*. However, when PL was added to adjust the C:N ratio of the ground newsprint to 30:1, SOM was increased by 12.0 g kg⁻¹ of soil, with no visible evidence of any newsprint remaining to be decomposed. The use of a combination of organic wastes illustrates that the rate they are transformed into the stable SOM fraction can be controlled by the addition of nutrients, primarily N, used by microorganisms to aid decomposition.

On the average, four organic wastes increased SOM from 9.5 to 17 g kg⁻¹ after two annual applications of 2.44 kg m², plus inorganic N or PL used to adjust the C:N ratio to 30:1. Since woodchips contain a higher concentration of lignin, which is more resistant to decomposition, SOM content was increased from 9.5 to 14 g kg⁻¹ after two annual applications of woodchips. Pelletized paper increased SOM content from < 0.5 to 1.0 g kg⁻¹ and soil N content was increased from < 0.1 to 0.5 g kg⁻¹ in one crop year. Increasing the density of the pellets (5 mm diameter) reduced the rate of conversion of paper to SOM when compared to ground newsprint. Application of pellets to soil produces the desired improvement in the soil chemical properties, however, pellets have the advantage of being a controlled release organic C substrate for microbial reactions *(95)*.

Activity as Biological Control Agents. Although organics have been evaluated as soil amendments for suppression of nematodes and other soilborne plant pathogens *(96)*, their mode of action is not known. The increase in microbial activity following addition of organic amendments can result in increased parasitism of pathogens, increased competition for resources between pathogen and antagonist, and increases in the production of enzymes and secondary metabolites detrimental to pathogens *(97,98)*. The action of microorganisms on organic substrate during decomposition produces a wide range of compounds detrimental to pathogens including ammonia, nitrites, hydrogen sulphide, and a range of volatile organic compounds and organic acids *(99)*.

The efficacy of organic amendments for control of root-knot nematodes (*Meloidogyne* spp.) and other soilborne pathogens is dependent on the chemical properties of the organic wastes. Addition of chitin or chitinous organic materials generates ammonia, but also results in the stimulation of the activities of chitinolytic microflora *(98,100-103)*. The chitinolytic microorganisms are effective in the destruction of eggs of tylenchid nematodes, and mycelia and reproductive structures of phytopathogenic fungi *(104,105)*. Understanding the action of chitin and high N organics for suppression of nematodes permitted the preparation of an organic amendment that was effective at low application rates (5 to 8 ton ha⁻¹) *(101)*. The most effective organics were those with narrow C:N ratios and high protein or amine-type N contents *(106)*.

Degradation of organic amendments by microorganisms is mediated by enzymes; good correlations between enzyme activity and components of microflora have been established *(103,106-108)*. Pine bark powder used as an organic soil amendment altered the soil microflora and provided control of disease caused by the soybean cyst nematode, *H. glycines* *(109)*. Shifts in soil microflora were observed as increases in total fungal populations in species of *Penicillium*, and *Paecilomyces*.

Potential Environmental Problems from Land Application of Organic Wastes

When waste production is concentrated in a relatively small area such as the poultry-producing areas of Alabama, Arkansas, Georgia, North Carolina, Delmarva Peninsulas, and West Virginia, and in the beef cattle feedlots of the Midwest, the potential for environmental contamination from organic waste is enhanced when agricultural land is used as a disposal option. The National Research Council has stated that reducing nutrient loading of agricultural land will be difficult to achieve unless alternative means of using animal wastes are developed (*110*).

Increased Nitrate-N Concentration. An assessment of the extent of groundwater pollution in an area of intense poultry production was conducted by the USDA Soil Conservation Service, Alabama Department of Environmental Management, and the Geological Survey of Alabama in 1988 (Sand Mountain Groundwater Quality Assessment, first quarterly report, January, 1988). Concentrations of NO_3-N in 28 of 30 groundwater wells from the Sand Mountain area of northern Alabama exceeded the national primary drinking water standard of 10 mg L^{-1}. Fecal coliform or streptococci were detected in 28 out of 30 wells tested. In long-term PL versus non-PL studies on tall fescue (*Festuca arundinacea* Schreb.) pastures, Kingery et al. (*111*) found levels of soil NO_3-N ≥ 40 mg kg^{-1} at or near bedrock under pastures where litter was applied.

In a series of field trials conducted in Alabama to assess the use of ground newsprint on agricultural land, concentration of soil NO_3-N was ≤ 5 mg kg^{-1} at 2 m depth 8 months after newsprint plus PL was applied to the soil surface. When NH_4NO_3 was the N source, soil NO_3-N was ≥ 9 mg kg^{-1} at 2 m soil depth. Total N in the surface 15 cm was increased from ≤ 0.1 mg kg^{-1} to ≥ 1 mg kg^{-1} when newsprint was applied with PL. A higher portion of N in the PL and C from newsprint was being immobilized by increase in the soil microorganism biomass (*25,112*).

Buildup of Soil Phosphorus. In an experiment where PL was applied as the N source to adjust the C:N ratio of several organic wastes, its use resulted in a 3-fold increase in Mehlich-I extractable P in the surface 15 cm of soil after three annual applications (Table IV). Kingery et al. (*111*) found an increase of approximately 530% in extractable P to a depth of 60 cm under pastures where litter was applied as compared to pastures where litter was not applied. Elevated levels of P in the surface soils can lead to increased levels of biologically-available P with the potential to be transported to surface waters in runoff.

Long-term application of confined animal manures has resulted in an increase in extractable P in the top 30 cm of soil (*113*). However, the highest concentration has been found in the top 5 cm of soil (*113-115*), and this fraction has the greatest potential to be lost by soil transport. Also, with repeated application of manures, the capacity of soil to adsorb P is reduced (*113*). The decrease in P adsorption with repeated application suggests the potential for increased soil-P mobility via transport in surface runoff.

Increased Metal Loading from Waste Application. No adverse affects to date have been observed using newsprint with PL; micronutrient concentrations were found to be well within the ranges that can be tolerated by plants. Surface-applied newsprint did not affect the soil concentrations of Pb and Cr after four annual applications (*48*). Newsprint plus PL amendments increased soil Cu 100-fold, Mn two-fold, and Zn four-fold when compared to newsprint treatments where the C:N ratio was adjusted with NH_4NO_3. However, further research is needed to study the long-term effects of repeated applications of organic wastes on heavy metal accumulation in plant and soil systems.

Increased Aluminum Content from Waste Paper. Aluminum in newsprint is inherent to the manufacturing process of paper from the wood pulp industry. At least three forms of Al in paper have been identified and characterized. 1) The Al in the kaolin is a structural component of the mineral which contributes little to the soil Al activities. 2) The Al fraction that is a chelating ion in the cellulose fibers can be removed only by the decomposition of the paper and/or substituting another ion in the production of paper. 3) The adsorbed Al fraction, i.e., soluble in water, is probably the most active fraction and is the fraction inducing nutrient disorders in plants when ground newsprint is applied to soil without 4 to 6 weeks of *in situ* composting before planting (*Lu et al., Compost Sci. Utiliz., in press*).

Soil Nutrient Imbalances. A plant nutrient disorder is an indication of an improper nutrient balance in the soil, limited soil nutrient concentrations, or toxicities from excess soil nutrient levels. Since the uptake of nutrients depends partly on the volume of roots present, any condition that affects root growth will alter nutrient uptake and will be manifested as nutrient deficiency or toxicity by the plant (7). There are two possible explanations for the phytotoxic effect of uncomposted ground newsprint to plants. One may be the high demand for N by the microorganisms immediately after application. This condition would temporarily immobilize N and induce N deficiency in plants. The second may be related to the chemical composition of the paper. When advertisements were included in the waste paper, the Al content was 4525 mg kg^{-1} (Table III). This elevated Al content may have induced the phytotoxicity observed in cotton and corn seedings.

Nutrient imbalances were observed in corn seedlings when ground newsprint was surface-applied in the spring and the C:N ratio was adjusted to ≤ 30:1 with urea, NH$_3$, NH$_4$NO$_3$, or PL as N source. Evidence of P, Ca and S imbalances were observed 40 days after plant emergence (*48,113*). When compared to PL, all inorganic N sources stunted corn growth. When the N supply was from urea or NH$_4$NO$_3$, the corn plants were stunted, delayed in maturity, and had lower grain yield than when PL was the N source. Corn plants receiving NH$_3$ as the N source did not recover and no grain yield was obtained.

Table IV. Changes in Mehlich-I Extractable Nutrients after Three Annual Applications of Organic Wastes and Poultry Litter

Organic	Soil	Soil nutrient levels			
wastes	pH	P	K	Ca	Mg
		kg ha^{-1}			
Initial[1]	5.2[1]	32[1]	53[1]	306[1]	74[1]
Newsprint[2]	6.2	129	196	940	206
Yard wastes[2]	6.1	138	353	974	243
Wood chips[2]	6.6	155	233	1198	311
C gin waste[2]	6.2	177	355	1267	254
PL check[2]	6.1	132	228	874	206

[1] Analysis of soil collected from experimental area prior to initiation of experiments.

[2] Poultry litter was used to adjust C:N ratio to 30:1; PL check = Poultry litter check.

No nutrient imbalances were observed with corn seedlings when pape pellets paper (5 and 8 mm diameter) were surface-applied at a rate of 2.44 kg m^2 and incorporated in the fall. The C:N ratio was adjusted to ≤ 30:1 with different sources of N (urea, Ca(NO$_3$)$_2$, NH$_4$NO$_3$, or PL). No evidence of P, Ca and S deficiencies were observed during the first six weeks of growth. There was no effect from the different N sources when applied in combination with paper pellets on corn dry matter production, maturity, or grain yield.

Soil Microbial Dynamics

Soil microorganisms play an important role in improving soil nutrient levels and in the transformation of plant residues and organic wastes into the SOM fraction (*117*). Although they account for only 1 to 8% of the SOM fraction, they can have a dramatic effect on crop production (*118*). Soil microorganisms influence plant availability of N, P, S, and other elements by controlling the decomposition of plant residues and organic wastes, mineralization of N and P, and immobilization of N and S (*119,120*).

Recognizing the diverse activities of soil microorganisms and taking advantage of those processes to reduce fertilizer, herbicide, and pesticide inputs is an important aspect of maintaining or restoring soil quality (*121*). A better understanding of how to effectively work with soil microorganisms can be developed by using soil and crop management practices including preservation of plant residues on the soil surface, use of cover crops during fallow periods, use of reduced tillage, crediting animal manures as nutrient sources, and use of organic wastes.

Microbial Biomass. The influence of ground newsprint, combined with two sources of N, (NH_4NO_3 or PL) as a C source was evaluated on active bacterial, active fungal and total microbial biomass, cellulose decomposition, potential net mineralization of soil C and N and soil nutrient status in agricultural soils. Cotton gin waste promoted the highest potential net N mineralization and N turnover rates (*Entry et al., Bio. Fert. Soil, in press*). Organic wastes had no effect on active bacterial, active fungal or total microbial biomass, C turnover, or the ratio of net C:N mineralized. Organic wastes and N additions to soil did not consistently affect C turnover rates, active bacterial, active fungal or total microbial biomass. After 3, 6 or 9 weeks of laboratory incubation, soil amended with organic wastes plus PL resulted in higher cellulose degradation rates than soil amended with organic wastes plus NH_4NO_3. Cellulose degradation was highest when soil was amended with newsprint plus PL. When soil was amended with organic wastes plus NH_4NO_3, cellulose degradation did not differ from soil amended with only PL or unamended soil. Soil amended with organic wastes had higher concentrations of soil C than soil amended with only PL or unamended soil. Soil amended with organic wastes plus N as PL generally had higher Mehlich-I extractable P, K, Ca, and Mg concentrations than soil amended with PL or unamended soil. Since cotton gin waste plus PL resulted in higher cellulose degradation and net N mineralization, its use may result in faster increase in soil nutrient status.

Carbon Mineralization. El-Harris et al. (*69*) found that as soil organic C increased, net N mineralization increased in no-till systems, and the concentration of C in the 0-10 cm of soil correlated with net N mineralization. Campbell et al. (*122*) reported an increase in soil organic C, C mineralization, and microbial biomass C and N, especially in the 7.5 to 15-cm soil depth with increasing frequency of cropping, and with inclusion of legumes as green manure or hay crop in the rotation. Wood and Edwards (*70*) found that 10 years of conservation tillage increased soil organic C and N to a depth of 10 cm, compared to conventional tillage systems where residue was incorporated in the top 15 cm. Using stubble-mulch and no-till systems conserved up to 2% more SOM per year in surface soil than when conventional plowing was used (*123*). Crop rotation influenced soil organic C; corn two out of three years in the rotation promoted greater stability in the SOM fraction than did soybean (*124*). In a greenhouse study, Janzen and Radder (*82*) found that incorporation of green manure into agricultural soils growing wheat suppressed net N mineralization. The degree of N mineralization and/or immobilization was associated with the incorporation of crop residues and the level of labile organic matter in the surface soil and was not affected by different tillage practices (*125*).

Carbon mineralization of uncomposted newsprint and two aerobically digested paper mill sludges with either PL or urea as N sources used to adjust the C:N ratios were evaluated in a field study over a 14-month period (*126*). Most of the decrease in dry weight of each waste occurred in the first 60 days, for a total loss of 56% for paper pellets, 58% for ground paper, and 68% for cotton moats. The textile sludge and papermill sludge lost only about 20% of dry weight. In each of the uncomposted wastes the C:N ratio was approaching 20:1, from an initial C:N ratio of 140:1 for ground paper and paper pellets, and cotton moats of 40:1. The C:N ratio for the two sludges increased from approximately 10:1 to 20:1 at the end of 14 months.

Microbial Population Shifts. Fungal populations were increased by adjusting the C:N ratio of the applied waste paper (*112*). Bacterial populations were increased in soil amended with N, and populations in soil amended with PL were greater than populations in soil amended with NH_4NO_3-N. There was an inverse relationship between bacterial and fungal populations and C:N ratio. Actinomycetes populations were increased in soil amended with PL as the N source.

Bacterial populations after 9 weeks in ground newsprint-amended soil had greater diversity than populations in unamended soil (*25*). Gram positive bacterial species were increased in soil amended with PL as compared to soils amended with NH_4NO_3 or unamended soil. The dominance of Gram positive organisms in PL-amended soil may indicate that the soil environment is unsuitable for coliform bacteria; thus, coliform bacteria present in the PL are not present in the soil.

Many Gram negative strains of bacteria which colonize plant roots such as *Burkholdaria* sp. and *Pseudomonas* sp. have been associated with increased plant health, yield, and biological control of many plant pathogens (*127,128*). The shift to greater numbers of Gram negative organisms may indicate a potential for increase in plant growth due to the presence of some of these organisms.

Species diversity includes the number of species (richness) and the relative abundance of each species in the community (evenness) (*129*). Richness indices were higher in newsprint plus PL-amended soils compared to newsprint plus NH_4NO_3-amended soil. Evenness was greater for newsprint plus PL and PL-amended soils when compared to soils amended with NH_4NO_3. Newsprint alone was not greater than the unamended soil, which suggests that PL was the predominant influence on the increase in evenness.

Application of organic wastes into soil promoted microbial activity, and as a result increased soil enzyme activity (*130*). The effect appears to be influenced by the C:N ratio of the applied organic waste. When the C:N ratio of ground newsprint was set at 20:1, 40:1 or 60:1 and composted with soil for 9 weeks, total fungal population was greater at the 60:1 ratio. Similar results were observed with cotton growth during the first six weeks after newsprint was applied to the soil surface without adjusting the C:N ratio. Cotton plants had higher levels of plant death caused by *Sclerotium rolfsii* Sacc. (Sclerotium stem rot disease) and surviving plants were severely stunted by *Rhizoctonia solani* Kühn (Soreshin disease). Plants remained stunted, were delayed in maturity, and had lower lint yields. When C:N ratio of applied organic wastes was ≤ 30:1, total bacteria populations were increased in both the greenhouse and field studies. Cotton growth and yield were increased when the C:N ratio of the applied organic wastes were adjusted to 30:1 and was related to a shift from predominately fungal population to one that was mainly composed of either bacteria or actinomycetes (*95*).

Summary

The application of organic waste and animal manures to agricultural land creates dynamic soil ecosystems. The incorporation of C and N into the soil biomass and ultimately into the SOM fraction is an important pathway for C and N. This leads to improvement in soil chemical and physical properties and they have a positive beneficial effects on soil fertility. However, temporary soil conditions can exist that lead to increased leaching loss of NO_3-N, increased loss of soluble P, or they may cause shifts in microbial populations that lead to environmental degradation. Organic wastes-soil amendments-crop production systems are currently being developed to utilize them in an environmentally safe manner.

With the current restrictions on landfills and landfill space, land application of MSW may be a disposal option, but at the present time it is cost-prohibitive for wide use in agriculture. While land application of organic byproducts and MSW offer the potential to reduce MSW in landfills, and reduce fertilizer and herbicide inputs in crop production systems, the environmental impact on delicate soil and plant ecosystems remains largely unknown. We need to establish the loading rates of organic byproducts, in combination with different forms of N, to maximize crop production and minimize the potential for environmental pollution.

Disposal of organic wastes on agricultural soils have the potential for mitigating weed pressure and thus reducing the amount of herbicides needed for crop production. However, more research is needed to identify the mechanisms involved. Other pest-averting potentials for organic waste utilization may be discovered.

Surface-applied uncomposted ground newsprint should be composted *in situ* for an interval of 4 to 6 weeks, or care should be taken to adjust the C:N ratio of the waste paper before crops are planted in the area. When these two considerations are met, the results were satisfactory whether the organic waste was applied in the fall or the spring of the year.

Incorporation of organic wastes with high C:N ratios (> 100:1) may lead to N immobilization, but when the conditions are suitable the immobilized N may be mineralized and incorporated into microbial biomass or absorbed by the plant. The C:N ratio of the organic waste has an influence on the microbial populations and activity. At C:N ratios of > 100:1, fungal populations are enhanced, but C:N ratios < 30:1 enhanced bacteria and actinomycete populations. High nutrient demands occurred in environments where water deficiency is not the major yield-limiting factor.

The 3 mm diameter paper pellet was blown by high velocity wind on the soil surface when dry. However, the wet 3 mm diameter paper pellet eliminated the problem. Paper pellets did not have any adverse affects on cotton grown the following season. The concentrations of extractable soil macro- or micronutrients were not affected. One advantage of applying paper pellets is water conservation. However, improved water conservation was observed only with the 19 mm diameter pellet.

Organic wastes (ground paper and paper pellets) can be used to provide an alternative mulch and/or residue cover for agricultural row crops to reduce wind erosion and nonpoint source pollution from cropland. When recommended soil management practices were utilized and organic wastes were applied to achieved a 30% soil cover, wind-blown sediment was reduced by 50%. Cotton lint yields were increased as a result of increased water storage from rainfall. Paper pellets reduced soil loss by surface runoff by 41%.

Acknowledgments

Contribution of USDA-ARS, National Soil Dynamics Laboratory, P.O. Box, 3439, Auburn, AL 36831-3439; Department of Agronomy and Soils, 202 Funchess, Hall, Auburn University, AL 36849-5412; and the Alabama Agriculture Experiment Station. Journal Series No. 3-965750. Special thanks to Ms. S.A. Morey, Agricultural Research Technician for her invaluable help in searching the literature and editing of this manuscript.

Literature Cited

1. Parr, J.F.; Papendick, R.I.; Colocicco, D. *Biol. Agric. Hort.* **1986**, *3*, 115-130.
2. Edwards, J.H. *Nat. Conserv. Tillage Dig.* **1995**, *2*(5), 2.
3. Alter, H. *Waste Manage. Res.* **1991**, *3*, 3-20.
4. Finstein, M.S. In *Environmental Microbiology;* Mitchell, R., Ed.; Wiley-Liss, NY, 1992, pp 355-374.
5. Steuteville, R. *BioCycle* **1995**, *36*(5), 30-37.
6. U.S. Environmental Protection Agency (USEPA). The solid waste dilemma: An agenda for action; U.S. Gov. Print. Office, Washington, DC; 1989, pp 1-70.
7. Edwards, J.H.; Walker, R.H.; Lu, N.; Bannon, J.S. *BioCycle* **1993**, *34*(10), 48-50.
8. He, X.T.; Traina, S.J.; Logan, T. *J. Environ. Qual.* **1992**, *21*, 318-329.
9. Goldstein, N. *BioCycle* **1989**, *30*(4), 32-37.
10. Kashmanian, R.M. *BioCycle* **1995**, *36*(1), 55-57.
11. Kashmanian, R.M.; Rynk, R.F. *Compost Sci. Utiliz.* **1995**, *3*(3), 84-88.
12. Kashmanian, R.M.; Rynk, R.F. *J. Soil Water Conserv.* **1996**, *51*(3), 194-201.
13. Rynk, R. *Proc. of the 3rd International Dairy Housing Confer.* American Society of Agricultural Engineers. February 2-5, 1994.
14. Maynard, A.A.; Hill, D.E. *BioCycle* **1994**, *35*(3), 66-67.
15. Buchanan, M.; Gliessman, S.R. *BioCycle* **1991**, *32*(12), 72-77.
16. Campbell, C.A.; Schnitzer, M.; Stewart, J.W.B.; Biederbeck, V.O.; Selles, F. *Can. J. Soil Sci.* **1986**, *66*, 601-613.
17. Chefetz, B.; Hatcher, P.G.; Hadar, Y.; Chen, Y. *J. Environ. Qual.* **1996**, *25*, 776-785.
18. Chanyasak, V.; Mirai, M.; Kubota, H. *J. Ferment. Technol.* **1982**, *60*, 439-446.
19. Chanyasak, V.; Katayama, A.; Hirai, M.; Mori, S.; Kubota, H. *Soil Sci. Plant Nutr.* **1983**, *29*, 251-259.
20. Jimenez, E.I.; Garcia, V.P. *Agric. Ecosyst. Environ.* **1992**, *38*, 331-343.
21. Harada, Y.; Inoko, A.; Tadaki, M.; Izadaki, T. *Soil Sci. Plant Nutr.* **1981**, *27*, 357-364.
22. Iannotti, D.A.; Gregbus, M.E; Toth, B.L.; Madden, L.V.; Hoitink, H.A.J. *J. Environ. Qual.* **1994**, *23*, 1177-1183.
23. Chen, Y.; Inbar, Y. In *Science and Engineering of Composting: Design, Environmental, Microbiological and Utilization Aspects;* Hoitink, H.A.J; Keener, H.M., Eds.; Renaissance Publ., Worthington, OH, 1993, pp 551-600.
24. Edwards, J.H. *BioCycle* **1994**, *35*(4), 69-72.
25. Press, C.M.; Mahaffee, W.F.; Edwards, J.H,; Kloepper, J.W. *Compost Sci. Utiliz.* **1996**, *4*(2), 70-80.
26. Havlin, J.L.; Kissel, D.E.; Maddux, L.D.; Claasen, M.M.; Long, J.H. *Soil Sci. Soc. Am. J.* **1990**, *54*, 448-452.

27. Tester, C.F. *Soil Sci. Soc. Am. J.* **1990**, *54*, 827-831.
28. Pagliai, P.; De Nobili, M. *Geoderma* **1993**, *56*, 243-256.
29. Blevins, R.L.; Thomas, G.W.; Cornelius, P. *Agron. J.* **1977**, *69*, 383-386.
30. Skidmore, E.L.; Layton, J.B.; Armbrust, D.V.; Hooker, M.L. *Soil Sci. Soc. Am. J.* **1986**, *50*, 415-419.
31. Metzger, L.; Levanon, D.; Mingelgrin, U. *Soil Sci. Soc. Am. J.* **1987**, *51*, 346-351.
32. Raper, R.L.; Edwards, J.H.; Way, T.R.; Washington, B.H.; Burt, E.C.; Hill, D.T. *ASAE Paper No.* 961044. ASAE, St. Joseph, MI, 1996.
33. Edwards, J.H.; Norton, L.D.; Walker, R.H. *88th Annual Meeting & Exhibition* (95-mp12.02), San Antonio, TX, June 18-23, 1995, pp 1-15.
34. Edwards, J.H.; Walker, R.H.; Burt, E.C.; Raper, R.L. In *Agricultural Utilization of Urban and Industrial By-products;* Karlen, D.L., et al., Eds., Agron. Special Publication No. 58, ASA, CSSA, SSSA, Madison, WI, 1995, pp 225-249.
35. Chen, Y.; Inbar, Y.; Hadar, Y.; Malcolm, R.L. *Sci. Total Environ.* **1989**, *81/82*, 201-208.
36. Hoitink, H.A.J.; Boehn, M.J.; Hadar, Y. In *Science and Engineering of Composting: Design, Environmental Microbiological and Utilization Aspects.* Hoitink, H.A.J.; Keener, H.M., Eds.; Renaissance Publ., Worthington, OH, 1993, pp 601-621.
37. Giusquiani, P.L; Pagaliai, M.; Gigliotti, G.; Businelli, D.; Benetti, A. *J. Environ. Qual.* **1995**, *24*, 175-182.
38. Sims, J.T. *Agron. J.* **1987**, *79*, 563-570.
39. Flynn, R.P.; Wood, C.W.; Touchton, J.T. *Bioresource Tech.* **1993**, *44*, 165-173.
40. Wood, C.W.; Torbert, H.A.; Delaney, D.P. *J. Sustain. Agric.* **1993**, *3*(2), 21-36.
41. Ritter, W.F.; Chirnside, A.E.M. *Ground Water* **1984**, *22*, 38-47.
42. Chandler, K.; Brookes, P.C. *Soil Biol. Biochem.* **1993**, *25*, 1231-1239.
43. Christie, P.; Beattie, J.A.M. *J. Appl. Ecol.* **1989**, *26*, 597-612.
44. Bache, C.A.; Lisk, D.J. *J. Agric. Food Chem.* **1990**, *38*, 190-194.
45. Christie, P.; Kilpatrick, D.J. *Soil Biol. Biochem.* **1992**, *24*, 325-330.
46. del Castilho, P.; Chardon, W.J.; Salomons, W. *J. Environ. Qual.* **1993**, *22*, 689-697.
47. Wood, C.W.; Hattey, J.A. In *Animal Waste and the Land-Water Interface*; Steel, K., Ed.; Lewis Publishers, Boca Raton, FL, 1995, pp 334-352.
48. Lu, N.; Edwards, J.H.; Walker, R.H. *Compost Sci. Utiliz.* **1995**, *3*(1), 6-18.
49. Jawson, M.D.; Elliot, L.F. *Soil Biol. Biochem.* **1986**, *18*(1), 15-22.
50. Edwards, J.H.; Burt, E.C.; Raper, R.L.; Hill, D.T. *Compost Sci. Utiliz.* **1993**, *1*(2):79-92.
51. Chen, Y.; Inbar, Y.; Hadar, Y. *Soil Sci.* **1988**, *145*, 298-303.
52. McKenna, I.M.; Chaney, R.L.; Williams, F.M. *Environ. Pollution* **1991**, *79*, 113-120.
53. Walker, C.D.; Welch., R.M. *J. Agr. Food Chem.* **1987**, *38*, 721-727.
54. Chaney, R.L.; Munns, J.B.; Cathey, H.M. *J. Amer. Soc. Hort. Sci.* **1980**, *105*, 485-492.
55. Edwards, J.H. *BioCycle* **1992**, *33*(7), 71-72.
56. Entry, J.A.; Wood, B.H; Edwards, J.H.; Wood, C.W. Alabama Agricultural Experiment Station, *Highlights Agricultural Research* **1996**, 43(1), 8-9.
57. Burt, E.C.; Edwards, J.H.; Raper, R.L.; Hill, D.T. *ASAE Paper No.* 921560. ASAE, St. Joseph, MI, 1992.
58. Raper, R.L.; Way, T.R.; Burt, E.C.; Hill, D.T.; Edwards, J.H.; Reeves, D.W.; Trotman, A. In *Agricultural Utilization of Urban and Industrial By-*

products. Karlen, D.L., et al., Eds.; Agron. Special Publication No. 58, ASA, CSSA, and SSSA, Madison, WI., 1995, pp 251-259.
59. Rhoton, F.E.; Tyler, D.D. *Soil Sci. Soc. Am. J.* **1990**, *54*, 223-228.
60. Fryrear, D.W. In *Adv. Soil Science: Dryland Agriculture: Strategies for Sustainability,* Singh, R.P., et al., Eds.; Springer-Verlag: New York, NY, 1990, Vol. 13, pp 187-197.
61. Fryrear, D.W. *Tran. ASAE* **1985**, *28*(3), 781-784.
62. Stewart, B.A.; Steiner, J.L. *Adv. Soil Sci.* **1990**, *13*, 151-173.
63. Wood, C.W.; Cotton, C.D.; Edwards, J.H. In *Proc. Conf. Environmentally Sound Agriculture*; Bottcher, A.B., et al., Eds.; April 11-18, Orlando, FL, 1991, pp. 568-575.
64. Liebhardt, W.C.; Golt, C.; Tupin, J. *J. Environ. Qual.* **1979**, *8*, 211-215.
65. Copper, J.R.;Reneav, R.B.; Kroontje, W.; Jones, G.D. *J. Environ. Qual.* **1984**, *13*, 189-193.
66. King, L.D. *J. Environ. Qual.* **1984**, *13*, 609-612.
67. Chae, Y.M.; Tabatabai, M.A. *J. Environ. Qual.* **1986**, *15*, 193-198.
68. Janzen, H.H.; Radder, G.D. *Plant Soil* **1989**, *120*, 125-131.
69. El-Harris, M.K.; Cochran, V.L.; Elliott, L.F.; Bezdicek, D.F. *Soil Sci. Soc. Am. J.* **1983**, *47*, 1157-1161.
70. Wood, C.W.; Edwards, J.H. *Agric. Ecosystems Environ.* **1992**, *39*, 123-138.
71. Entry, J.A.; Backman, C.B. *Can. J. For. Res.* **1995**, *25*, 1231-1236.
72. Aulakh, M.S.; Doran, J.W.; Walter, D.T.; Moiser, A.R.; Francis, D.D. *Soil Sci. Soc. Am. J.* **1991**, *55*, 1020-1025.
73. Aulakh, M.S.; Rennie, D.A.; Paul, E.A. *Soil Sci. Soc. Am. J.* **1984**, *48*, 790-794.
74. Chang, C.; Janzen, H.H. *J. Environ. Qual.* **1996**, *25*, 785-790.
75. Boyle, M.; Paul, E.A. *Soil Sci. Soc. Am. J.* **1989**, *53*, 99-103.
76. Lindemann, W.C.; Cardenas, M. *Soil Sci. Soc. Am. J.* **1984**, *48*, 1072-1077.
77. Peperzak, P.; Caldwell, A.G.; Hunziker, R.R.; Black, C.A. *Soil Sci.* **1959**, *87*, 293-302.
78. Wood, C.W.; Hall, B.M. *Comm. Soil Sci. Plant Anal.* **1991**, *22*, 1677-1688.
79. Stevenson, F.J. In *Cycles of Soil*. John Wiley and Sons, New York, NY, 1986.
80. Bitzer, C.C.; Sims, J.T. *J. Environ. Qual.* **1988**, *17*, 47-54.
81. Deans, J.R.; Molina, J.A.E.; Clapp, C.E. *Soil Sci. Soc. Am. J.* **1986**, *50*, 323-326.
82. Fine, P.; Mingelgrin, U.; Feigin, A. *Soil Sci. Soc. Am. J.* **1989**, *53*, 444-450.
83. Douglas, B.F.; Magdoff, F.R. *J. Environ. Qual.* **1991**, *20*, 368-372.
84. Sims, J.T. *J. Environ. Qual.* **1986**, *15*, 59-63.
85. Barbarika, A., Jr.; Sikora, L.J.; Colacicco, D. *Soil Sci. Soc. Am. J.* **1985**, *49*, 1403-1406.
86. Wiseman, J.T.; Zibilske, L.M. *J. Environ. Qual.* **1988**, *17*, 334-339.
87. Sharpley, A.N.; Menzel, R.G. *Adv. Agron.* **1987**, *41*,297-324.
88. Edwards, D.R.; Daniel, T.C. *Tran. ASAE* **1992**, *35*, 1827-1832.
89. Shreve, B.R.; More, P.A., Jr.; Daniel, T.C.; Edwards, D.R.; Miller, D.M. *J. Environ. Qual.* **1995**, *24*, 106-111.
90. Beegle, D.B. In *Proceedings of the National Poultry Waste Management Symposium,* Columbus, OH 18-19 April, 1988. The Ohio State University, Columbus, OH, 1988, pp 120-124.
91. Anderson, G. In *The Role of Phosphorus in Agriculture;* Sample; Kamprath, Eds; ASA, CSSA, SSAJ, Madison, WI, 1980, pp 411-431.
92. Moore, P.A., Jr; Miller, D.M. *J. Environ. Qual.* **1994**, *23*, 325-330.

93. Sharpley, A.N.; Smith, S.J. *J. Environ. Qual.* **1989**, *18*, 101-105.
94. Guertal, E.A.; Edwards, J.H. *HortSci.* **1996**, 31(5):823-826.
95. Edwards, J.H.; Walker, R.H.; Guertal, E.A.; Eason, J.T. *BioCycle* **1994**, 35(11), 66-68.
96. Ramirez-Villapuda, J.; Munnecke, D.E. *Phytopath.* **1988**, *78*, 289-295.
97. Mian, I.H.; Godoy, G.; Shelby, R.A.; Rodríguez-Kábana, R.; Morgan-Jones, G. *Nematrópica* **1982**, *12*, 71-84.
98. Godoy, G.; Rodríguez-Kábana, R., Shelby, R.A.; G. Morgan-Jones, G. *Nematrópica*, **1983**, *13*, 63-74.
99. Stirling, G.R. In *Biological Control of Plant Parasitic Nematodes: Progress, Problems, and Prospects.* C.A.B. International, Wallingford, U.K., 1991, p. 282.
100. Rodríguez-Kábana, R.; Boubé, D.; Young, R.W. *Nematrópica* **1989**, *19*, 53-74.
101. Rodríguez-Kábana, R.; Boubé, D.; Young, R.W. *Nematrópica* **1990**, *20*, 153-168.
102. Culbreath, A.K.; Rodríguez-Kábana, R.; Morgan-Jones, G. *Nematrópica* **1985**, *15*, 49-75.
103. Rodríguez-Kábana, R. *J. Nematology* **1986**, *18*, 129-135.
104. Cook, R.J.; Baker, K.F. In *The Nature and Practice of Biological Control of Plant Pathogens.* Amer. Phytopath. Soc., St. Paul, MN, 1983.
105. Hoitink, H.A.J. *Ann. Rev. Phytopathol.* **1988**, *24*, 93-114.
106. Rodríguez-Kábana, R.; Morgan-Jones, G.; Chet, I. *Plant Soil* **1987**, *100*, 237-247.
107. Kokalis-Burelle, N.; Rodriguez-Kábana, R. *Biol. Control* **1994**, *4*, 269-276.
108. Kokalis-Burelle, N.; Rodríguez-Kábana, R.; Weaver, C.F.; King, P.S. *Plant Soil* **1994**, *162*, 163-168.
109. Kokalis-Burelle, N.; Rodriguez-Kábana, R. *Plant Soil* **1994**, *162*, 169-175.
110. National Research Council. *Opportunities to Improve Soil and Water Quality: an Agenda for Agriculture.* Natl. Acad.
111. Kingery, W.L.; Wood, C.W.; Delaney, D.P.; Williams, J.C.; Mullins, G.L.; van Santen, E. *J. Prod. Agric.* **1992**, *6*(3), 390-395.
112. Edwards, J.H.; Press, C.M.; Mahaffee, W.F.; Kloepper, J.W. In *Proc. of Workshop on Management of the Soil Biota in Sustainable Farming Systems.* Pankhurst, C.E. (ed.) CSIRO, Australia, 1994, pp 91-93.
113. Sharpley, A.N.; Smith, S.J.; Bain, W.R. *Soil Sci. Soc. Amer. J.* **1993**, *57*, 1131-1137.
114. Christie, P. *J. Agric. Sci. Camb.* **1987**, *108*, 529-541.
115. Kingery, W.L.; Wood, C.W.; Delaney, D.P.; Williams, J.C.; Mullins, G.L. *J. Environ. Qual.* **1994**, *23*, 139-147.
116. Lu, N.; Edwards, J.H.; Walker, R.H.; Bannon, J.S. In *Proc. 2nd Conf. on Environmentally Sound Agriculture*; Bottcher, A.B., et al., Eds.; April 20-22, 1994, Orlando, FL, 1994, pp 431-438.
117. Parkinson, D.; Coleman, D.C. *Agri. Ecosystems Envir.* **1991**, *34*, 3-33.
118. Roder, W.; Mason, S.C.; Cleff, M.D.; Doran, J.W.; Kniep, K.R. *Soil Sci. Soc. Am. J.* **1988**, *52*, 1337-1342.
119. Saranthchandra, S.U.; Perrott, K.W.; Boase; M.R.; Waller, J.E. *Biol. Fertil. Soils* **1988**, *6*, 328-335.
120. Perucci, P. *Biol. Fertil. Soils* **1992**, *14*, 54-60.
121. Doran, J.W.; Parkin, T.B. In *Defining Soil Quality for a Sustainable Environment*, Doran, J.W.; Coleman, D.C.; Bezdicek, D.F.; Steward, B.A., Eds, SSSA Special Publ. No. 35. Am. Soc. Agron. and Soil Sci. Soc. Am., Madison, WI, 1994, pp.3-22.
122. Campbell, C.A.; Biederbeck, V.O.; Zentner, R.P.; Lafond, G.P. *Can. J. Soil Sci.* **1991**, *71*, 363-376.

123. Rasmussen, P.E.; Collins, H.P. *Adv. Agron.* **1991**, *45*:93-134.
124. Edwards, J.H.; Wood, C.W.; Thurlow, D.L.; Ruf, M.E. *Soil Sci. Soc. Am. J.* **1992**, *56*, 1577-1582.
125. Carter, M.R.; Rennie, D.A. *Soil Sci. Soc. Am. J.* **1984**, *48*, 1077-1081.
126. Arriaga, F.J. MS Thesis, Auburn Univeristy, AL, 1996.
127. Burr, T.J.; Schroth, M.N.; Suslow, T. *Phytopath.* **1978**, *68*, 1377-1383.
128. Kloepper, J.W.; Hume, D.J.; Scher, F.M.; Singleton, D.; Tipping, B.; Laliberté, M.; Frauley, K.; Kutchaw, T.; Simonson, C.; Lifshitz, R.; Zaleska I.; Lee, L. *Plant Dis.* **1988**, *72*, 42-46.
129. Ludwig, J.A.; Reynolds, J.F. In *Statistical Ecology: A Primer on Methods and Computing*. John Wiley and Sons, New York, NY, 1988, pp 71-103.
130. Martens, D.A.; Johanson, J.B.; Frankenberger, W.T. *Soil Sci.* **1992**, *153*, 53-61.

Chapter 12

Composition and Land Application of Paper Manufacturing Residuals

J. J. Camberato[1], E. D. Vance[2], and A. V. Someshwar[2]

[1]Faculty of Soils and Land Resources, Clemson University, 2200 Pocket Road, Florence, SC 29506-9706
[2]National Council of the Paper Industry for Air and Stream Improvement, Box 141020, Gainesville, FL 32614-4501

Paper manufacturing generates sludge and causticizing residuals that can be used as soil amendments to enhance productivity of agricultural systems while maintaining high environmental quality standards. Crop response to land-applied sludge depends largely on its N content, and supplemental N is sometimes required to improve its effectiveness as a soil amendment. Slaker grits, green liquor dregs, and lime mud have high alkalinity and are effective agricultural limestone substitutes. Paper manufacturing residuals are generally low in metals and organic compounds of environmental concern. Successful utilization of residuals necessitates an accurate measure of nutrient content and alkalinity and a uniform, timely, and appropriate application to a suitable, well-managed soil-plant system.

Residuals from paper manufacturing include waste treatment sludges, bark and combination bark boiler ashes, lime mud, slaker grits, and green liquor dregs. Estimates of solid waste generation for 26 USA Kraft pulp and paper mills were obtained by the National Council of the Paper Industry for Air and Stream Improvement (NCASI) during two recent life cycle inventory surveys (Table I). Since nearly 80% of the chemical pulp produced in the USA is from the Kraft process, these data are a good representation of current pulp and paper industry residuals from chemical pulping. Boiler ash and sludges constitute the majority of pulp and paper mill residuals. The causticizing operations involved in the Kraft pulping and recovery process generate alkaline residuals such as lime mud, slaker grits, and green liquor dregs. These residuals are not produced in other forms of chemical or non-chemical pulping. Estimates of sludges produced in other forms of chemical pulping (e.g., sulfite and semi-chemical) and non-chemical pulping (e.g., direct recycled, deinking, groundwood, and non-integrated) were obtained during a 1989 survey (Table II).

Table I. Summary from NCASI Surveys of Residuals Generated at Thirteen Unbleached Kraft and Thirteen Bleached Kraft USA Pulp Mills

Residual	Units	No. Mills	Mean	Median	Max.	Min.
Wastewater Sludges[a]	kg wet/ODMTP[b]	22	64.2	49.2	268.8	0.5
Combustion Ashes	kg/ODMTP	19	41.2	33.8	107.3	2.2
Lime Mud or Sludge	kg/ODMTP	12	71.9	46.0	284.2	0.9
Slaker Grits and Dregs	kg/ODMTP	18	19.2	11.1	66.9	0.4
Woodwaste, Bark Grit	kg/ODMTP	10	26.9	13.7	90.2	0.6
Other Solid Residuals[c]	kg/ODMTP	20	29.9	20.3	204.1	0.5

[a]Note that some mills burn their sludge in boilers, thus the amount reported may be less than the total generated.
[b]ODMTP - oven-dry metric ton of pulp.
[c]May include waste paper/paperboard, mill trash, construction debris, and cinders.
Source: NCASI, unpublished data, 1993-1995.

Table II. Pulp and Paper Mill Sludge Generation Data from 1989 NCASI Survey

Production Category	No. of Mills Reporting	Sludge Produced	
		Mean	Median
		----- lb/ADMTP[a]-----	
Non-Integrated	25	48.5	34.5
Waste Paper	4	48.5	17.5
Deinking	8	380.0	390.0
Groundwood	10	52.0	40.5
Sulfite	10	85.5	62.5
Semi-chemical	7	23.0	13.5
Unbleached Kraft and Cross Recovery	21	24.5	18.5
Bleached Kraft	33	58.0	51.5

[a]ADMTP - air-dry metric ton of pulp.
Source: Adapted from ref. 1.

Paper manufacturing residuals have traditionally been landfilled but stringent regulations on the construction of new landfills has greatly increased the cost of disposal. Land application of these residuals to crop or forest lands represents a beneficial alternative to disposal. Based on their composition, mill residuals are classified as non-hazardous under the United States Environmental Protection Agency (USEPA) Resource Conservation and Recovery Act (RCRA). Under this Act, only non-hazardous materials can be land applied. Beneficial use through land application is based on their ability to favorably alter soil properties such as plant nutrient availability, soil reaction, or properties related to enhanced soil organic matter status such as cation exchange capacity, water holding capacity, tilth (physical condition of soil related to tillage, seedbed, and rooting media), and soil strength.

This paper will review the composition and land application of wastewater sludges and causticizing residuals generated by paper manufacturing. Land application of wastewater from sludge treatment and boiler ashes are not discussed in this paper, but information on these topics can be obtained from other sources [wastewater-(2), boiler ashes (3-5)].

Waste Treatment Sludges

Plant Nutrient, Metal, and Organic Content of Sludges. Essential plant nutrient concentrations for pulp and paper mill sludges from surveys across the USA and typical nutrient levels for municipal sewage sludges are presented in Table III. These data show that pulp and paper sludges vary substantially in plant nutrient composition and typically have lower levels of essential plant nutrients than municipal sewage sludges.

Metal concentrations of paper manufacturing sludges and municipal sewage sludges are presented in Table IV. Median levels of As, Cd, Co, Hg, Pb, Sn, and Se are substantially lower in paper manufacturing sludges than in municipal sludges. Aluminum is perhaps the only metal that may be present in paper manufacturing sludges at higher levels than in municipal sludges. Thacker and Vriesman (6) report that the results of Extraction Procedure (EP) toxicity tests conducted by NCASI (14) and individual paper companies indicate that paper manufacturing sludges would rarely be classified as hazardous materials due to heavy metal content.

The characterization data presented in Tables III and IV pertains primarily to sludges from virgin chemical pulping mills. Someshwar et al. (15) presented a detailed analysis of the composition of several deinking sludges, addressing chemical constituents and the results of EP toxicity and Toxicity Chemical Leaching Procedure (TCLP) characterizations. Their data showed that the composition of deinking sludges were comparable to those from chemical pulping operations and were also comparable or superior in quality to municipal sewage sludges. The EP toxicity and TCLP characterizations showed that all measured concentrations were less than 5% of hazardous waste thresholds. TCLP characterizations of sludges from virgin pulping mills have also generally been found to be non-hazardous (e.g., Table V).

Besides the TCLP analyses, only limited information on organic compounds in sludges is available. Only chloroform was found in detectable quantities (between 0.26 and 2.4 mg kg^{-1} dry weight--well below TCLP regulatory levels) in analyses of wastewater sludges from an integrated bleached Kraft/fine paper mill and an integrated

Table III. Macronutrient and Micronutrient Concentrations in Pulp and Paper Manufacturing Wastewater Sludges

Nutrient	Sludge Type [a]	NCASI 54 Mill Sludge [a] Survey (6)		7 Combined Sludges [a] from Wisconsin (7)		12 Sludges [a] from Maine (8)		16 Sludges [a] from Alabama (9)		Municipal Sewage Sludges		
		Range	Median	Range	Mean	Range	Median	Range	Mean	Range	Median	Source
Macronutrient (g/Kg)												
N	All	0.51-87.5	8.98			0.1-61	1.1	3.0-15.5	8.76	<1.0-210	32.0	(10)
	Combined	1.1-59	8.5	1.4-41	29							
	Primary	0.51-9.0	2.7									
	Secondary	6.2-87.5	23.3									
P	All	0.01-25.4	2.35			0.01-3.1	0.85	0.19-2.57	1.2	<1.0-150	14.0	(10)
	Combined	0.1-25.4	0.67	1.3-5.8	3.4							
	Primary	0.01-4.0	1.6									
	Secondary	0.42-16.7	4.2									
K		0.12-10	2.2	0.37-6.0	1.37	0.08-3.4	0.46	0.15-2.65	0.96	0.2-650	2.3	(10)
Ca		0.28-210	14.0	2.7-90.2	29.5	0.001-81	4.2	13.0-148	60.0	1.0-250	27.0	(10)
Mg		0.2-19.0	1.55	0.11-1.0	0.65	0.05-3.7	1.8	0.72-5.07	2.30	0.3-25.0	4.0	(10)
S		0.2-20.0	4.68			0.3-6.5	0.50	0.1-10.5	4.1	6.0-15	11.0	(11)
Micronutrient (mg/Kg)												
B		<1-491	25.0	8.5-21.1	13.3			5.11-28.2	13.5	4-1,000	33	(12)
Cl		<0.06-8,500	383									-
Cu		3.9-1,590	52.0	2.2-90.9	67.4	2.0-204	48.0	10.3-69.6	34.3	6.8-3,120	463	(13)
Fe		97.1-10,800	1,540	675-7,162	2,930	162-5,120	1,670	790-12,800	3,040	1,000-15,400	1,700	(12)
Mn		13-2,200	155.0	22.9-286.4	119.0	9.8-1,500	99.0	40.0-2,555	900	32-9,870	260	(12)
Mo		<2.5-14.0	--	5.2-8.9	6.7			2.44-32.2	11.9	2.0-67.9	11	(13)
Zn		13-3,780	188.0	41.3-214	127.0	9.8-843	62.0	50.2-695	215	38-68,000	725	(13)

[a] An assortment of primary, secondary, and combined sludges unless otherwise noted.

Table IV. Metal Concentrations in Pulp and Paper Manufacturing Wastewater Sludges (excluding essential plant nutrients)

	NCASI 54 Mill Sludge Survey [a] (6)		7 Combined Sludges [a] from Wisconsin (7)			12 Sludges [a] from Maine (8)			16 Sludges [a] from Alabama (9)			Municipal Sewage Sludges		
	Range	Median	Range	Mean		Range		Median	Range		Mean	Range	Median	Source
Heavy Metals (mg/Kg)														
Cd	<0.09-56	1.2	1.6-9.4	3.9		0.008-7.7		1.16	0.1-3.69		1.01	0.7-8,220	7	(13)
Cr	3.0-2,250	42.0	29.2-55.8	39.6		0.02-172		11.4	17.9-789		145	2.0-3,750	40	(13)
Co	ND-9.7	--	1.8-4.5	2.5		--		--	1.6-9.7		4.36	1-260	10	(12)
Pb	<0.05-880	28	37.8-129.3	81.9		0.05-1,000		14.3	9.5-116		48.6	9.4-1,670	106	(13)
Hg	0.0009-3.52	0.35	3.9	3.9		0.003-<2		0.17	--		--	0.2-47.0	4	(13)
Ni	1.3-133	18.3	--	--		1.7-97		10.6	0-41.0		15.2	2.0-976	29	(13)
Ag	<0.1-<11	0.55	--	--		0.003-10		1.06	--		--	--	--	--
Sn	<70.6	--	--	--		5.07-135		10.7	--		--	40-700	150	(12)
Other Metals (mg/Kg)														
Al	590-59,000	13,400	15 to 22,000	19,500		120-16,000		6,190	790-40,700		13,100	1,000-13,500	4,000	(11)
As	<0.07-8.3	1.2	--	--		--		--	--		--	0.3-315.6	6	(13)
Ba	17.9-1,800	160.0	--	--		--		--	0.0-289		144	<0.01-9,000	200	(11)
Se	<0.01-<31	0.21	--	--		--		--	--		--	0.5-70.0	5	(13)
Na	300-66,700	2,200	490-4,120	1,730		219-14,600		585	440-48,000		7,750	100-30,700	2,400	(11)
Ti	3,100-76,000	--	--	--		--		--	--		--	--	--	--

[a] An assortment of primary, secondary, and combined sludges unless otherwise noted.

Table V. TCLP Analyses of a Typical Unbleached Kraft Mill Sludge, Slaker Grit, and Green Liquor Dreg, and Compared to Regulatory Levels

Parameter	Sludge	Grit	Dreg	Reg. Level
		mg/L		
Arsenic	<1.0	<1.0	<1.0	5.0
Barium	<1.0	<1.0	<1.0	100.0
Cadmium	<0.5	<0.5	<0.5	1.0
Chromium	<1.0	<1.0	<1.0	5.0
Lead	<1.0	<1.0	<1.0	5.0
Mercury	<0.2	<0.2	<0.2	0.2
Selenium	<0.5	<0.5	<0.5	1.0
Silver	<1.0	<1.0	<1.0	5.0
Benzene	<0.01	<0.01	<0.01	0.5
Carbon Tetrachloride	<0.01	<0.01	<0.01	0.5
Chlorobenzene	<0.01	<0.01	<0.01	100.0
Chloroform	<0.01	<0.01	<0.01	6.0
1,4-Dichlorobenzene	<0.01	<0.01	<0.01	7.5
1,2-Dichloroethane	<0.01	<0.01	<0.01	0.5
1,1-Dichloroethylene	<0.01	<0.01	<0.01	0.7
Methyl Ethyl Ketone	0.24	<0.01	<0.01	200.0
Tetrachloroethylene	<0.01	<0.01	<0.01	0.7
Trichloroethylene	<0.01	<0.01	<0.01	0.5
Vinyl Chloride	<0.01	<0.01	<0.01	0.2
o-Cresol	<0.01	<2.0	<0.01	200
m-Cresol	<0.01	<2.0	<0.01	200
p-Cresol	<0.01	<2.0	<0.01	200
Pentachlorophenol	<0.01	<2.0	<0.01	100.0
2,4,5-Trichlorophenol	<0.01	<2.0	<0.01	400.0
2,4,6-Trichlorophenol	<0.01	<2.0	<0.01	2.0
2,4-Dinitrotoluene	<0.01	<0.13	<0.01	0.13
Hexachlorobenzene	<0.01	<0.13	<0.01	0.13
Hexachlorobutadiene	<0.01	<0.13	<0.01	0.50
Hexachloroethane	<0.01	<0.13	<0.01	3.0
Nitrobenzene	<0.01	<0.13	<0.01	2.0
Pyridine	<0.01	<0.13	<0.01	5.0
Chlordane	<0.03	<0.03	<0.03	0.03
Endrin	<0.02	<0.02	<0.02	0.02
Heptachlor and hydroxide	<0.008	<0.008	<0.008	0.008
Lindane	<0.005	<0.005	<0.005	0.4
Methoxychlor	<0.025	<0.025	<0.025	10.0
Toxaphene	<0.005	<0.005	<0.005	0.5
2,4-D	<1.0	<1.0	<1.0	10.0
2,4,5-TP (Silvex)	<1.0	<1.0	<1.0	1.0

Sources: Data for sludge, grit, and dreg from NCASI member company testing during May 1990. Regulatory levels from Federal Register 55(61)11804.

bleached sulfite/fine paper mill (*16*). Thacker and Vriesman (*6*) reported on sludge composition data for several aromatic hydrocarbons, phthalates, chlorinated pesticides, volatiles, semi-volatiles, phenols, and triaryl aryl phosphate esters for two deinking, one groundwood/fine paper, and two neutral sulfite semi-chemical pulping mills. The concentrations of most compounds were below detection limits. Naphthalene and three of the phthalates were detected in some sludges at concentrations >10 mg kg^{-1}. The source(s) of naphthalene are unknown. The phthalates may have originated in coatings and adhesives used in paper manufacturing.

Dioxin and Furan Content of Sludges. USEPA's Tier IV National Dioxin Study revealed trace levels of chlorinated dioxins and furans in the effluents and sludges of several bleached Kraft pulp mills. In early 1988, a joint USEPA/Paper Industry Cooperative Dioxin Study was conducted and results from 104 bleached Kraft mills were summarized by Whittemore (*17*). Since that time, NCASI has conducted a number of study updates to document tetrachlorodibenzo-dioxin and -furan (TCDD/TCDF) concentrations of bleached Kraft mill effluents, sludges, and pulps. A summary comparison of statistical parameters from the "104 Mill Study" and the most recent update (*18*) is presented in Table VI. It should be noted that 27 different isomers of polychlorinated dibenzo-dioxins (PCDDs) and -furans (PCDFs) are typically included in deriving a 2,3,7,8 TCDD Toxic Equivalent (TEQ). However, analyses of PCDDs and PCDFs from sludges from a number of bleached Kraft pulp and paper mills processing primarily virgin fiber have uniformly shown the 2,3,7,8 TCDD and 2,3,7,8 TCDF isomers to be the only principal isomers present (*17*).

In an effort to minimize the formation of PCDD/Fs during bleaching of pulp, the paper industry has made extensive changes in the bleaching process over the last decade, the key one being the substitution of chlorine dioxide for chlorine. These changes have dramatically reduced the levels of PCDD/Fs in sludge, effluent, and pulp. The reduction in sludge PCDD/F levels is evident from Table VI.

Land Application of Sludges. Primary sludges are typically low in plant nutrients, especially N, and have high C:N ratios (Table III). Secondary sludges have higher concentrations of N and P and lower C:N ratios than primary sludges (Table III), because N and P are commonly added to the waste treatment system to enhance biological degradation. Mixtures of primary and secondary sludges are also generated, with properties dependent on the proportion of each sludge in the mix.

Crop responses to land-applied paper manufacturing sludges have been variable, dependent on the sludge N concentration, C:N ratio, and amount applied. Increased crop yields resulting from application of low C:N ratio sludges have been obtained in some studies (*19,20*), whereas other studies have shown decreased crop productivity from high C:N ratio sludges (*21-23*). Plant N deficiencies in high C:N ratio sludges result from N immobilization, which occurs when the N concentration of the sludge is insufficient to meet the demands of the soil microbial community. Nitrogen from the sludge and soil is then immobilized into microbial tissues, rendering it unavailable for plant uptake. As the sludge decomposes, C is evolved as CO_2, resulting in a gradual decline in C:N ratio and an increase in N availability. Strategies to overcome this limitation include; (a) applying sludge well in advance of crop planting so that the C:N ratio of the sludge

Table VI. Comparison of 2,3,7,8 TCDD/F Concentrations in Bleached Kraft Mill Sludges - '104 Mill Study' versus the 1994 NCASI Update

Data from the USEPA/Industry '104 Mill Study' *(17)*

All Concentrations in ng/Kg

	Max.	90th% [a]	Mean	Median	% < ML [b]	% < 10*ML [c]
TCDD	1,390	161	72	16	na	--
TCDF	17,100	1,300	607	78	--	22

1994 NCASI Survey Data (99 Mills Responding) *(18)*

All Concentrations in ng/Kg

	Max.	90th %	Mean	Median	% < ML	% < 10*ML
TCDD	92	17	6	1	39	--
TCDF	735	76	31	5	--	54

[a] 90th% - ninetieth percentile.
[b] %<ML - percent of TCDD measurements in each data set that are less than the nominal "Minimum Level' of EPA's proposed Method 1613A (1 ng/Kg for sludges).
[c] %<10*ML - percent of TCDF measurements in each data set that are less than 10 times the ML because the toxicity equivalence factor of 2,3,7,8-TCDF is one tenth that of 2,3,7,8-TCDD.
na = not available.
Sources: Adapted from ref. 17, 18.

has been reduced to the point that immobilization no longer occurs, (b) adding additional N to satisfy microbial demand for N necessary to decompose the sludge, or (c) planting legumes so that soil N is not required by the crop.

The duration of N immobilization that occurs when sludge is applied to soil determines when N will be available to the crop, and dictates the timing of crop planting relative to sludge application. Unfortunately, the immobilization period may be variable and unpredictable. In one study, immobilization of soil N occurred for four and two weeks after soil incorporation of primary and secondary sludges, respectively (24). Subsequently, microbial mineralization of N exceeded immobilization. However, in another study of three sludges with C:N ratios of 50:1, 142:1, and 151:1 incorporated in five different soils, complete immobilization of N continued through a 90 day incubation period (25). Hatch and Pepin (26) reported a 60 day N immobilization period following land application of primary and secondary sludges from an integrated Kraft mill. Mineralization studies with mixed sludges consisting of 50 and 90% primary sludge suggested sludge applications should precede crop planting by several months (27). Timing of N availability from sludge is critical, with sludge applied far enough in advance of planting so that immobilization of N does not occur when the crop requires it. However, if excessive mineralization occurs prior to crop growth, there is a potential for leaching losses of N to occur.

Adding fertilizer N to soil amended with high C:N ratio sludge is also an effective method of eliminating the effects of N immobilization on crop productivity. In one study, sludge applied at 168 dry MT ha^{-1} (C:N of 169:1) decreased winter wheat (*Triticum aestivum* L.) grain yield 60% compared to the standard fertilizer treatment (22). However, when 56 kg N ha^{-1} or more additional N was supplied with the sludge, grain yield was equivalent to the standard fertilizer treatment. In another study, sludge with a C:N ratio >100:1 applied at 12 and 24 dry MT ha^{-1} decreased corn (*Zea mays* L.) grain yield when no fertilizer N was applied, but increased grain yield at 12 dry MT ha^{-1} when 100 kg N ha^{-1} was added and at both sludge rates when 200 kg N ha^{-1} was applied (21). Potato (*Solanum tuberosum* L.) yield was decreased when a deinking sludge (C:N of 344:1) was applied to a silty loam soil in Canada at 12 dry MT ha^{-1} unless an additional 90 kg N ha^{-1} was applied (23). Another study found that primary sludge with a C:N ratio of 169:1 required 0.6 kg N per dry MT sludge to satisfy microbial demand for N during decomposition in initial research studies, but experience in commercial application of the sludge suggested that about 0.9 kg N per dry MT was actually required (22).

Composting residuals may be an alternative method of increasing N availability of high C:N ratio sludges. Bowen et al. (19) reported that N immobilization increased with fresh sludge rate (45, 90, and 180 dry MT ha^{-1}) and persisted for six to eight weeks, but N immobilization did not occur when the same sludge was composted prior to soil incorporation. Subsequent studies showed that composting reduced the C:N ratio of the sludge from 23:1 to 10:1. In another study, a primary sludge, tailings, ash, and N source mixture with an initial C:N ratio>270:1 was composted for 14 weeks and cured for 4 weeks, which, depending on the amount of the ash in the mixture, resulted in final compost C:N ratios ranging from 14:1 to 67:1 (28). Nitrogen immobilization would be considerably less with the composted mixture than with the initial sludge mixture.

It has been suggested that N-fixing legumes could be effectively grown on soils recently amended with high C:N sludges (25). Since legumes do not require soil N to grow, N immobilization would not affect plant productivity and supplemental N may not be necessary. Although this is a promising concept, it has not been thoroughly tested.

Nitrogen mineralization from land-applied sludge occurs subsequent to N immobilization. Predicting the quantity of N mineralized from sludges is necessary to provide sufficient N to enhance crop productivity without providing excess, which may contaminate ground water. Nitrogen mineralization was estimated in one study at 14 and 35% after 16 weeks of a laboratory incubation for primary and secondary sludges, respectively (24). Eleven and 21% of sludge-applied N (C:N=22:1, 17 g N kg^{-1}, 2.5:1 primary: secondary sludge mixture) was mineralized in 16 weeks when applied to a red pine (*Pinus resinosa* Ait.) plantation at 94 and 32 MT ha^{-1} (29). In a greenhouse study, N mineralization occurred in soil amended with three sludges having C:N ratios less than 10:1, while N immobilization occurred with a sludge having a C:N of 22:1 (27). The sludges releasing N were composed of >70% secondary sludge and <30% primary sludge; whereas, the N immobilizing sludge contained 70% primary sludge and 30% secondary sludge. The exact proportion of N mineralized during incubation could not be calculated from the data presented. A considerably greater understanding of N mineralization from low C:N paper manufacturing sludges is needed to maximize their effectiveness for use in crop production systems.

Paper manufacturing sludges may also have positive effects on soil physical properties. High application rates (448 and 672 MT ha^{-1}) of primary clarifier sludge to a sandy soil increased soil cation exchange capacity and available moisture content as much as two to five fold (30). In this case, both organic matter and kaolinite clay in the sludge were likely responsible for the increase in these parameters. Simulated weathering treatments reduced the increase in cation exchange capacity and available moisture, probably due to oxidation of the added organic matter. In another example, deinking sludge added to a sandy soil at 50 and 112 MT ha^{-1} increased cation exchange capacity from 1.6 to 2.5 cmol kg^{-1} but did not affect soil porosity or bulk density (31).

Causticizing Residuals

Chemical Composition of Causticizing Residuals. Causticizing residuals include lime mud, slaker grits or rejects and green liquor dregs. Typical quantities of these residuals generated at a Kraft mill are presented in Table I. Lime mud refers to residuals arising from the reaction of CaO with green liquor (solution of Na_2CO_3 and Na_2S), yielding $CaCO_3$, NaOH, and Na_2S. The majority of the NaOH and Na_2S are recovered as white liquor and reused in the process. The $CaCO_3$ (lime mud) can become a residual at this point, although it is usually introduced into a lime kiln to be converted back to CaO. If operation of the kiln is disrupted, the materials must be removed from the kiln and become residuals requiring utilization or disposal. The amount of lime mud generated is highly variable and is dependent on limitations in lime kiln capacity and frequency of kiln malfunction. Slaker grits comprise mainly large, unreactive lime particles and insoluble impurities remaining after lime is mixed in with green liquor in a chemical reactor or slaker. Green liquor dregs are undissolved substances in the green liquor

comprised mainly of C (50% or more) and foreign materials (mainly insoluble metal carbonates, sulfates, sulfides, hydroxides, and silicates).
Published information on the chemical characteristics of causticizing residuals is limited. Muse (9) provided the only published detailed characterization of these materials, reporting data for six lime residuals, five slaker grits, and three green liquor dregs from Kraft pulp mills in Alabama. These data, in addition to those reported by Thacker and Vriesman (6) and from NCASI surveys, are summarized in Table VII. TCLP analyses for slaker grits and green liquor dregs are presented in Table V.

Land Application of Causticizing Residuals. Relatively few studies have reported crop yields from field trials where causticizing residuals have been substituted for agricultural limestone to increase soil pH. Those conducted have found that residuals increased crop yields to a similar extent as that from commercial limestone. For example, lime mud and limestone increased forage yield of alfalfa (*Medicago sativa* L.) and birdsfoot trefoil (*Lotus corniculatus* L.) 31 and 27% and 6 and 7%, respectively (32). In another study (33), lime mud, grits, and limestone increased forage yield of a dallisgrass-fescue (*Paspalum dilatatum* Poir.-*Festuca arundinacea* Schreb.) pasture 42, 47, and 51%, respectively.

Causticizing residuals are utilized as a replacement for agricultural limestone to increase soil pH. Soil pH is an important chemical characteristic because it affects the availability of many plant nutrients and toxic elements. The soil pH desirable for crop production is dependent both on the soil type and the crop species, but in general is in the range of 5.8 to 7.0 (34). Soil pH levels below or above the optimum range can be detrimental to crop growth (35,36). The land grant university and Cooperative Extension Service in each state publish recommendations for desirable pH ranges for most crops and soils.

Soils acidify over time. In intensive agricultural cropping systems, acidity is predominantly generated by the oxidation of reduced-N compounds commonly applied in nutrient sources (fertilizers, plant residues, animal manures, and organic residuals) (37). In forests and natural systems, acidification is generally more gradual and due to the leaching of bases from the soil by rainfall and acidity and from soil organic matter decomposition and root respiration. In many soils, periodic liming is required so that conditions are favorable for plant rooting and nutrient acquisition. The frequency of lime additions is dependent on the soil type, N fertilization rate, and crop removal and leaching of bases from the soil. The quantity of lime applied is based on the initial soil pH, the desired final pH, and the buffer capacity of the soil. Limestone has been the material of choice for increasing soil pH for centuries. Recent estimates indicate 11 million Mg of agricultural limestone are sold annually in the USA at a cost of $58 million (38). Industrial residuals, particularly those from the paper manufacturing industry, provide potential alternatives for adjusting soil pH.

Calcium carbonate equivalence (CCE) is a chemical measurement of the relative acid neutralizing ability of a liming materials. Pure $CaCO_3$ is defined as having a CCE of 100%. Agricultural limestones typically have CCEs of 90-98% (39). Analysis of a limited number of lime mud and slaker grit samples show that the CCE of these residuals is often similar to or greater than that of agricultural limestone (Table VII).

Table VII. Macronutrient, Micronutrient, and Metal Concentrations of Kraft Causticizing Residuals

Var.	No.	Lime Muds			Green Liquor Dregs				Slaker Grits			
		Mean	Median	Range	Sample	Mean	Median	Range	Sample	Mean	Median	Range
pH	7	11.4	11.7	8.4-13.0	4	10.8	10.8	9.5-12.1	6	12.5	12.5	12.2-12.7
CCE	7	101	101	91-109	3	78	77	77-79	6	104	103	97-118
Macronutrient (g/Kg)												
N	6	1.78	1.60	2.0-3.0	3	2.70	2.30	1.6-4.2	5	2.22	1.50	1.3-5.0
P	11	2.83	1.80	1.07-5.7	5	0.67	0.66	0.13-1.17	6	0.81	0.54	0.23-1.68
K	13	0.44	0.26	<0.2-1.16	5	4.63	1.90	0.33-11.95	6	2.70	1.93	1.32-5.98
Ca	13	379	374	281-499	5	194	209	35-322	6	364	363	340-391
Mg	13	6.23	6.78	2.35-10.4	5	22.65	20.10	5.85-42.7	6	7.17	7.36	3.7-10
S	7	2.0	2.0	0.5-4.0	4	14.2	14.4	6.2-21.7	6	1.2	0.8	0.2-2.9
Micronutrient (mg/Kg)												
B	8	5.9	7.3	<20-10.1	3	20.6	25.4	9.2-27.1	5	7.3	6.9	5.5-9.9
Cl	1	180	180	180	1	260	260	260	1	4,900	4,900	4,900
Cu	12	15.8	10.0	<5.0-46.8	5	239.6	206.0	65-387	6	16.0	14.5	11.7-28
Fe	13	1,512	1,000	500-3,147	5	8,977	4,590	1,000-28,000	6	3,300	3,015	1,640-6,020
Mn	12	310.3	235.0	52-1,070	5	9,611	4,710	930-21,600	6	192	125	30-440
Mo	8	3.6	4.6	<5.0-5.2	3	8.3	7.6	4.7-12.7	5	11.4	7.2	5.3-26.8
Zn	12	255.0	24.2	<3.0-2,700	5	1,052.3	678.0	18.7-2,800	6	487.7	52.7	15.7-2,700

Sources: Adapted from ref. 6 and 9 and data from reports of tests conducted by NCASI member companies.

Table VII, Continued. Macronutrient, Micronutrient, and Metal Concentrations of Kraft Causticizing Residuals

		Lime Muds				Green Liquor Dregs				Slaker Grits		
Var.	No.	Mean	Median	Range	Sample	Mean	Median	Range	Sample	Mean	Median	Range
Heavy Metal (mg/Kg)												
Cd	12	0.2	0.1	<0.01-0.6	4	6.0	7.3	0.1-9.5	6	0.04	<0.25	<0.01-0.2
Cr	12	44.2	44.3	8.0-94.8	5	143.8	130.0	18-342	6	47.5	48.5	12-72
Co	9	4.4	4.9	<5.0-9.5	3	26.1	24.9	16.1-37.3	5	7.4	8.4	4.7-9.7
Pb	12	83.8	42.4	<1.0-574	5	57.2	80.9	18-83.2	6	107.5	47.0	<10.0-439
Hg	2	<0.05	<0.05	<0.05	1	<0.05	<0.05	<0.05	1	<5.0	<5.0	<5.0
Ni	12	37.6	23.3	1.2-158	5	183.1	27.5	9-780	6	46.6	13.5	0.0-124.6
Ag	2	<1.0	<1.0	<1.0	1	4.3	4.3	4.3	1	<1.0	<1.0	<1.0
Sn	2	<20.0	<20.0	<15-<25	0	--	--	--	0	--	--	--
Other Metal (mg/Kg)												
Al	13	1,163	1,000	630-2,220	4	3,137	3,024	820-5,680	5	10,600	4,860	1,420-33,400
As	2	0.19	0.19	<10.0-0.37	1	<10.0	<10.0	<10.0	1	<10.0	<10.0	<10.0
Ba	8	241.9	206.5	110-518	3	147.1	133.0	90.3-218	5	256.0	240.0	9.9-719
Se	1	<10.0	<10.0	<10.0	1	<10.0	<10.0	<10.0	1	<10.0	<10.0	<10.0
Na	13	9,175	7,000	3,680-38,000	5	37,380	25,500	20,401-87,500	6	12,742	8,010	170-50,000
Ti	1	346	346	346	1	259	259	259	0	--	--	--

Sources: Adapted from ref. 6 and 9 and data from reports of tests conducted by NCASI member companies.

The average CCE of three green liquor dregs was reported to be 78%, somewhat lower than the CCE of typical agricultural limestone (Table VII). The reaction rate of a liming material is at least as important as the CCE in determining the effectiveness of the material in increasing soil pH *(40)*. Particle size is the main factor determining reaction rate. Causticizing residuals generally have smaller particle sizes than agricultural limestone and therefore tend to react faster *(24,32,33)*. For example, one study showed that maximum soil pH in lime mud-amended soil occurred within four weeks of application whereas limestone continued to increase soil pH through 24 weeks *(32)*. The fast reaction of causticizing residuals may also be related to the presence of oxides and hydroxides *(24)*. The rapid rate of reaction of these materials compared to limestone may be an advantage if soils are planted shortly following amendment application.

Amending soils with dolomitic agricultural limestone provides Ca and Mg in addition to increasing soil pH. In many soils, Mg deficiency of crop plants can occur if calcitic agricultural limestone (0-0.6% Mg) is repeatedly used to adjust soil pH. Dolomitic limestone contains 1.3 to 6.5% Mg *(40)*. Lime mud and slaker grits are low in Mg with maximum values <1% Mg (Table VII). Magnesium deficiencies are thus a concern with repeated use of lime mud and slaker grits, although tests to date have not revealed Mg deficiencies *(32,33)*. The Mg concentrations of green liquor dregs (Table VII) are generally comparable to those of dolomitic limestone and Mg availability in dregs may be greater than that in agricultural dolomitic limestone *(33)*. Thus, land application of dregs is unlikely to result in crop Mg deficiency.

Potential Effects of Land Application of Paper Manufacturing Residuals on Water Quality and Wildlife

As with any soil amendment, appropriate management practices are needed in order to insure that high environmental quality standards are met when land-applying sludges and other mill residuals. High-N secondary sludges or sludges mixed with fertilizers should be applied at rates consistent with fertilizer recommendations for the crop in question, with some consideration for the rate at which the N contained in the sludge will mineralize, nitrify, and become susceptible to leaching losses. Temporary, elevated levels of nitrate in soil water have been observed following secondary sludge application to land used to grow corn *(27)*, and in soil water and groundwater following very high application rates (equivalent to 1120 and 1670 kg N ha^{-1}) of a primary/secondary sludge mix to a red pine plantation *(29)*. Other studies have shown no detectable change in soil water or groundwater quality following application of secondary sludge (rates as high as 22 dry MT ha^{-1}) and deinking mill sludges mixed with N fertilizer (rates as high as 18 dry MT ha^{-1} and 135 kg N ha^{-1}) *(23,41)*.

No adverse effects of land-applied paper mill sludges on wildlife have been documented in studies of forest stands on strip-mined land *(42-46)*. In a study of a Wisconsin red pine plantation, no effects of land-applied paper mill sludge on reproduction or histopathological symptoms for deer mouse populations and several bird species were observed *(43,44)*. In that study, earthworms, deer mice, and insectivorous bird populations were generally higher in areas where sludge had been applied, while litter invertebrate diversity and density were not affected and soil invertebrate diversity and density were reduced by sludge application (*44*). The authors of the study concluded that the increase in wildlife and decrease in soil invertebrates were due to the physical effects of the sludge and the resulting changes in food availability. Another study of wildlife populations in regenerating spruce-fir stands in Maine over two years following paper mill sludge application showed no significant effects on breeding bird densities, small mammals, and foliar and ground invertebrates (*46*). Although some shifts in several bird species were apparent on treated plots, this was attributed to increases in ground vegetation.

There have been concerns about the environmental implications of trace concentrations of TCDD and TCDF in sludges from mills using chlorine bleaching processes. Studies of wildlife exposed to land-applied sludges from mills using chlorine bleaching have shown no adverse effects, however (*42-47*). As previously noted, concentrations of TCDDs and TCDFs in bleaching mill sludges have been reduced dramatically in recent years due to the implementation of new bleaching technologies (Table VI). These and other chlorinated compounds should continue to decrease in significance. Studies have also shown that chlorolignin compounds formed during chlorine bleaching processes are rapidly immobilized in soil and are slowly mineralized to inorganic chloride (*42,48,49*). According to these studies, low molecular weight chlorinated degradation products appear to rapidly decompose in soil, and do not accumulate, leach or create a toxic environment for soil bacteria.

Federal and State Regulation of Land Application of Paper Manufacturing Residuals

Land application of paper mill sludges and other residuals are regulated primarily at the state level, although they are potentially subject to regulation under several federal statutes. Since mill residuals are not defined as hazardous wastes, they are not regulated under the Resource Conservation and Recovery Act (RCRA). Analyses of organic compounds using TCLP characterization, heavy metal concentrations and pH are generally needed to confirm this fact. As with any soil amendment, water quality standards for nutrients and heavy metals developed under the Clean Water Act must not be exceeded.

In March 1994, a Memorandum of Understanding between the USEPA and the American Forest and Paper Association established voluntary TCDD/TCDF concentration limits, application rates, site management practices, monitoring, record keeping, and reporting requirements for the land application, distribution and marketing of residuals from Kraft and sulfite pulp and paper mills using chlorine and chlorine-derivative bleaching processes. The Memorandum applies to residuals with TCDD/TCDF concentrations greater than or equal to 10 ppt TEQ. Residuals with concentrations below 10 ppt TEQ are excluded from the Memorandum, except for monitoring, testing, distribution and reporting requirements. Maximum residuals TCDD/TCDF concentrations of 50 ppt TEQ (or temporarily up to 75 ppt TEQ) and maximum soil concentrations up to 10 ppt TEQ are permitted. For agricultural application, sludge may be applied at rates up to 68 dry MT ha^{-1}, unless greater application rates are permitted by the individual state.

Current state regulations for land application of paper mill sludges and other residuals vary widely. Only a few states, including Maine, Ohio, and Wisconsin, have provisions which specifically regulate paper mill residuals such as sludges. As long as analyses (e.g. TCLP) show the materials to be applied are not hazardous, they are most often regulated under general state solid waste requirements or under "Beneficial Use" provisions. In the latter case, regulatory burdens and permitting requirements may be reduced if the benefits of the materials to the site can be demonstrated. Many states use the guidelines for heavy metals and management practices defined in USEPA 503 standards for land application of municipal sewage sludge biosolids as a baseline for land application of paper mill residuals. Paper mill residuals easily meet the USEPA 503 composition standards in most cases. Some states have more stringent standards, however, which can limit land application in some situations. Typical requirements include information on site and soil characteristics, set-back distances from surface water and wells, depth to groundwater, slope, vegetative cover, and proximity to floodplains or wetlands. A major regulatory issue for the generators and users of mill residuals is whether a general permit for residuals, site requirements and management practices is sufficient, or whether each site and practice must be individually permitted.

Summary

Land application of wastewater sludges and causticizing residuals from paper manufacturing is an alternative to landfilling that can provide nutrients and organic matter beneficial to crop growth. Mill residuals have also been typically found to be non-hazardous when assessed by elemental analysis and toxicity tests. These characteristics make them suitable as beneficial soil amendments. Sludges are generally applied based on N content. Sludges with high C:N ratios (e.g., primary sludges) can result in N immobilization when incorporated into the soil. So as not to hinder crop and forest plant growth these sludges should be applied well in advance of plant need for N, or in conjunction with supplemental N. Low C:N ratio sludges (e.g., secondary sludges) may supply N to the crop shortly after application. For all sludges, predicting the quantity and timing of N mineralization is critical to satisfactory crop production. Additional benefits to crop plant growth from sludge application may accrue from increased soil organic matter and associated changes in soil water and nutrient holding capacities. Causticizing residuals can be used satisfactorily as substitutes for agricultural limestone to increase soil pH and provide Ca. The CCE of these residuals is comparable to that of agricultural limestone and the reaction rate is generally higher. When applied at reasonable rates and under appropriate management guidelines, paper manufacturing sludges and residuals can enhance crop growth and site productivity while maintaining high environmental quality standards.

Literature Cited

1. Miner, R. A.; Unwin, J. P. *Progress in Reducing Water Use and Wastewater Loads in the U.S. Paper Industry;* NCASI Tech. Bull. 603; NCASI: Research Triangle Park, NC, 1991.
2. Thacker, W. E. In *The Forest Alternative for Treatment and Utilization of*

Municipal and Industrial Wastes; Cole, D. W., et al. Eds.; Univ. of Washington Press: Seattle, WA, 1986; 41-54.
3. Mitchell, C. C.; Black, E. D. In *Uses of By-products and Wastes in Agriculture;* 1997; Chapter 13.
4. Someshwar, A. V. *J. Environ. Qual.* **1996,** *25,* 962.
5. Vance, E. D. *J. Environ. Qual.* **1996,** *25,* 937.
6. Thacker, W. E.; Vriesman, R. *The Land Application and Related Utilization of Pulp and Paper Mill Sludges;* NCASI Tech. Bull. 439; NCASI: Research Triangle Park, NC, 1984.
7. McGovern, J. N.; Berbee, J. G.; Bockheim, J. G.; Baker, A. J. *TAPPI J.* **1983,** *66,* 115.
8. Rock, C. A.; Byer, B. S. In *Proc. Symposium on Long Range Disposal Alternatives From Pulp and Paper Sludges;* Rock, C. A.; Alexander, J. A., Eds.; Univ. of Maine: Orono, ME, 1982.
9. Muse, J. K. Inventory and evaluation of paper mill by-products for land application. M.S. thesis, Dep. of Agronomy and Soils, Auburn Univ., Auburn, AL, 1993.
10. Hue, N. V. In *Soil Amendments and Environmental Quality;* Rechcigl, J. E., Ed.; Lewis Publ.: Boca Raton, FL, 1995; 199-239.
11. Sommers, L. E. *J. Environ. Qual.* **1977,** *6,* 225.
12. Chaney, R. L. In *Sludge - Health Risks of Land Application;* Bitton, G., et al. Eds.; Ann Arbor Science Publ.: Ann Arbor, MI, 1980; 59-83.
13. Kuchenrither, R. D.; McMillan, S. J. *Biocycle* **1991,** *32,* 60.
14. Marshall, D. W.; Fisher, R. P. *Nature and Environmental Behavior of Manufacturing - Derived Solid Wastes of Pulp and Paper Origin;* NCASI Stream Improvement Tech. Bull. 319; NCASI: Research Triangle Park, NC, 1979.
15. Someshwar, A. V.; Wiegand, P.; Borton, D.; Miner, R.; Fisher, R.; Berger, H.; Unwin, J. *Characterization of Wastes and Emissions from Mills Using Recycled Fiber;* NCASI Tech. Bull. 613; NCASI: Research Triangle Park, NC, 1991.
16. Thiel, D. A. *Proc. of TAPPI Environ. Conf.;* TAPPI Press: Atlanta, GA, 1984; 93-102.
17. Whittemore, R. C. *USEPA/Paper Industry Cooperative Dioxin Study: The 104 Mill Study;* NCASI Tech. Bull. 590; NCASI: Research Triangle Park, NC, 1990.
18. Gillespie, W. J. *Progress in Reducing the TCDD/TCDF Content of Effluents, Pulps and Wastewater Treatment Sludges from the Manufacturing of Bleached Chemical Pulp;* NCASI Spec. Rep. 95-12; NCASI: Research Triangle Park, NC, 1995.
19. Bowen, B. D.; Wolkowski, R.; Hansen, G. *Proc. of TAPPI Environ. Conf.;* TAPPI Press: Atlanta, GA, 1996; 523-527.
20. Thiel, D.A. *Proc. of TAPPI Environ. Conf.;* TAPPI Press: Atlanta, GA, 1985; 261-278.
21. Bellamy, K. L.; Chong, C.; Cline, R. A. *J. Environ. Qual.* **1995,** *24,* 1074.
22. Field, G.; Carroll, B.; Fasth, W. *Proc. of TAPPI Environ. Conf.;* TAPPI Press: Atlanta, GA, 1996; 539-544.
23. Trepanier, L.; Theriault, G.; Caron, J.; Gallichand, J.; Yelle, S.; Beauchamp, C. J. *Proc. of TAPPI Environ. Conf.;* TAPPI Press: Atlanta, GA, 1996; 529-537.

24. Simpson, G. G.; King, L. D.; Carlile, B. L.; Blickensderfer, P. S. *TAPPI J.* **1983**, *66*, 71.
25. Dolar, S. G.; Boyle, J. R.; Keeney, D. R. *J. Environ. Quality* **1972**, *1*, 405.
26. Hatch, C. J.; Pepin, R. G. *Proc. of TAPPI Environ. Conf.;* TAPPI Press: Atlanta, GA, 1985; 287-291.
27. Shimek, S.; Nessman, M.; Charles, T.; Ulrich, D. *TAPPI J.* **1988**, *71*, 101.
28. Campbell, A. G.; Zhang, X.; Tripepi, R. R. *Compost Science & Utilization*, **1995**, Winter, 84.
29. Bockenheim, J. G.; Benzel, T. C.; Lu, Rui-Lin; Thiel, D. A. *J. Environ. Qual.* **1988**, *17*, 729.
30. Einspahr, D.; Fiscus, M. H.; Gargan, K. *Proc. of TAPPI Environ. Conf.;* TAPPI Press: Atlanta, GA, 1984; 253-257.
31. Diehn, K. *Proc. of TAPPI Environ. Conf.;* TAPPI Press: Atlanta, GA, 1991; 739-746.
32. Simson, C. R.; Kelling, K. A.; Liegel, E. A. *Agron. J.* **1981**, *73*, 1003.
33. Muse, J. K.; Mitchell C. C. *Agron. J.* **1995**, *87*, 432.
34. Brady, N. C. *The Nature and Properties of Soils, 8th ed.;* Macmillan: New York, NY, 1974.
35. Kamprath, E. J. *Soil and Crop Sci. Soc. Florida Proc.* **1971**, *31*, 200.
36. Kamprath, E. J.; Foy C. D. In *Fertilizer Technology and Use;* Engelstad O. P. Ed.; SSSA: Madison, WI, 1985; 91-151.
37. Heylar, K. R. *J. Aust. Inst. Agric. Sci.* **1976**, *42*, 217.
38. Tepordei, V. V. *Crushed Stone. Annual Report;* U.S. Dep. of the Interior, Bureau of Mines: Washington, DC, 1993.
39. Tisdale, S. L.; Nelson, W. L. *Soil Fertility and Fertilizers, 3rd ed.;* Macmillan: New York, NY, 1975.
40. Barber, S. A. In *Soil Acidity and Liming, 2nd ed.* Adams, F., Ed.; Agronomy; ASA: Madison, WI, 1984; Vol. 12; 171-209.
41. Bailey, S. L.; O'Neill, D. C.; Smithe, R.E. *Proc. of TAPPI Environ. Conf.;* TAPPI Press: Atlanta, GA, 1995; 735-750.
42. Sherman, W. R. *TAPPI J.* **1995**, *78*, 135.
43. Thiel, D. A.; Martin, S. G.; Duncan, J. W.; Lemke, M. J.; Lance, W. R.; Peterson, R. E. *Proc. of TAPPI Environ. Conf.;* TAPPI Press: Atlanta, GA, 1988; 487-507.
44. Thiel, D. A.; Martin, S. G.; Duncan, J. W.; Lance, W. R. *TAPPI J.* **1989**, *72*, 94.
45. Krouskop, D. J.; Ayers, K. C.; Proctor, J. L. *Proc. of TAPPI Environ. Conf.;* TAPPI Press: Atlanta, GA, 1991; 761-775.
46. Vera, C. J.; Servello, F. A. *J. Wildl. Manage.* **1994**, *58*, 719.
47. McFadden, D. P.; Proctor, J. L.; Krouskop, D. J.; Ayers, K. C. *Water Environ. Res.* **1995**, *67*, 792.
48. Dargitz, P.; Brezny, R.; Joyce, T. W.; Overcash, M. R. *Proc. of TAPPI Environ. Conf.;* TAPPI Press: Atlanta, GA, 1991; 339-345.
49. Brezny, R.; Joyce, T. W.; Gonzalez, B.; Slimak, M. *Environ. Sci. Technol.* **1993**, *27*, 1880.

INORGANIC BY-PRODUCTS AND WASTES

Chapter 13

Land Application of Boiler Wood Ash in the Southeastern United States

C. C. Mitchell[1] and E. D. Black[2]

[1]Department of Agronomy and Soils, 202 Funchess Hall,
Auburn University, AL 36849-5412
[2]122 West Lochwood Drive, Cary, NC 27511

Most agricultural soils in the southeastern U.S. require periodic application of ground limestone in order to maintain productivity. Using boiler wood ash and combination ash as an alternative to ground limestone is agronomically productive, environmentally safe, and fiscally sound for both the ash producer and the landowner/farmer. While plant nutrient content of ash is variable, it should be considered as an incidental source of plant nutrients, especially K and Mg, for field crops. An analysis of boiler ash from 14 Alabama pulp and paper mills averaged 38% $CaCO_3$ equivalent (CCE) with a dry density of 500 kg m^{-3}. Although popular concerns are often expressed about land application of metals in boiler wood ash and in combination ash, levels are well within EPA's guidelines for land application of biosolids. There are no published reports of metals being an environmental or crop production/crop quality problem, especially when ash is used at recommended rates as a soil liming material. An analysis of the CCE along with a routine analysis of plant nutrients and selected total metals is needed to utilize ash as a soil amendment. If growers follow a conscientious soil testing program and apply ash as a liming material, some variability in the ash can be tolerated, especially at the rates generally used. Because boiler wood ash is considered a non-hazardous waste, it is regulated by individual states. Where it is being utilized as an agricultural lime or plant nutrient source or soil amendment, the state departments of agriculture regulate its licensing. Hauling and spreading can be a logistical problem because of the physical condition of ash and collection and handling practices by the generator. Nevertheless, research and experience from Maine to Alabama suggest that land application is a safe and practical approach to utilization of boiler wood ash and mixed ash by-products.

Every industry produces wastes, by-products or co-products that must be either disposed or recycled. Economics and government regulations usually determine how they are handled. With increased regulations on public and private landfills, the cost of disposal has increased dramatically over the past 10 years. In 1994, the costs of disposal in public

landfills ranged from $8 per U.S. ton in New Mexico to $175 per U.S. ton in New Jersey (*1*). Industrial and private landfills would probably be more expensive. This has made the economics of alternative uses for some by-products more attractive.
Tisdale et al. (*2*) made the following historical observations:
The Bible records the value of wood ashes in its reference to the burning of briars and bushes by the Jews, and Xenophon and Virgil both report the burning of stubble to clear fields and destroy weeds. Cato advised the vine keeper to burn prunings on the spot and to plow in the ashes to enrich the soil. Pliny states that . . . some farmers burned manure and applied the ashes to their fields. Columella also suggested the spreading of ashes or lime on lowland soils to destroy acidity.

There are over 90 wood-fired, electrical cogenerating facilities consuming over 40 million metric tons of fuelwood each year in the 13 southeastern United States (*3*). Sometimes combinations of wood, coal, oil, or biosolids may be burned. The pulp and paper industry is one of the largest users of renewable, biomass energy sources in the United States. Much of the energy used in the pulping process is derived from the burning of bark, small sized trees, logging slash, and fibrous by-products in boilers (*4*). More wood may be consumed in the biomass boiler than in the paper making process itself (*5*). The ash generated from these facilities is bottom ash from the base of the boiler, fly ash from cyclones, ash from wet scrubbers, or, more commonly, a mixture of two or more sources. All are referred to as "boiler wood ash". The U.S. pulp and paper industry is probably the largest generator of producing an estimated 3.6 million metric tons of boiler ash annually which is about evenly divided between coal ash and bark/wood ash (*6*). In addition, many public and private utilities and wood-using industries burn wood or wood wastes in boilers.

The average pulp and paper mill in the southeastern U.S. can produce an average of approximately 18 metric tons pulp and paper products per day (*7*) 43 metric tons boiler ash per day from their boilers (*8*). A 1991 survey of 88 pulp and paper mills (*9*) indicated that most of this ash is from the burning of hardwoods, a wood-coal mixture or a wood-sludge mixture (Figure 1). Ash from mixtures of wood, coal, and/or sludge is often referred to as mixed or combination ash. Most of the mills in the southeastern U.S. place their boiler ash in company-owned landfills. Only 5% are land-applying some of their boiler ash and much of the land application activity involves wood ash. Both the survey by Muse in the southeastern U.S. and a national survey by Miner and Unwin (*6*) report similar figures for ash disposal. Other alternative disposal practices include lagoons, stockpiles, or reuse as construction materials. However, almost 60 percent of the mills contacted by Muse indicated that the company either approved or strongly approved a land application program. In the northeastern U.S., up to 80% of the boiler ash is land-applied (*10*).

Land application of many by-products often seems to be an inexpensive alternative to rising disposal costs (*11*). However, there are some very practical considerations before committing to such a program whether the by-product is boiler wood ash or some other material. These are the types of considerations the authors have to address frequently when dealing with land application of by-products. Boiler wood ashes have a long history of successful, beneficial use when land applied. Extensive research has been conducted throughout North America and Europe, yet obstacles still are encountered whenever land application of boiler wood ashes are proposed.

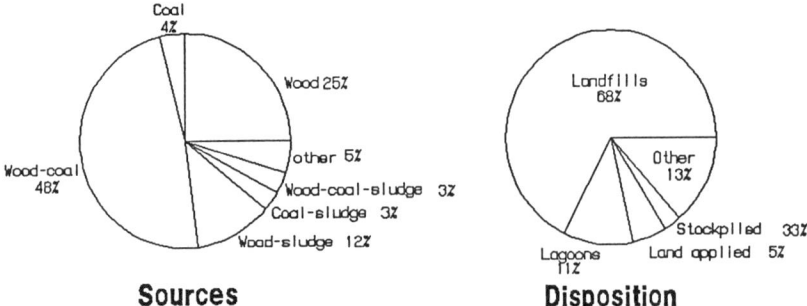

Figure 1. Sources of boiler ash and methods of ash disposal estimated from a 1991 survey of 88 paper mills in the southeastern U.S. (Adapted from ref. 9).

Environmentally Harmful Properties. The first consideration is, *"Does my by-product have any obvious properties that could be potentially harmful to the environment (soil, plants, animals, water) if land applied?"* This may be a difficult one to answer, particularly if it contains traces of metals, carcinogens, or organics that would be rate dependent. However, unless the answer is clearly, "Yes, it does", then it may be a candidate for land application and one can proceed. Most boiler wood ashes are low in metals, organics, and potential carcinogens. There are no cases of boiler wood ashes harming the environment when applied at reasonable rates (*12, 13,14*). Magdoff et al. (*14*) reported overliming injury in alfalfa when excess limestone was applied but not when excess ash was used. Research in Oregon has reported poor herbicide efficacy in perennial ryegrass or fescue fields previously treated with "boiler fly ash" as a liming material (unpublished report from G.W. Mueller-Warrant, S. Aldrich-Markham, and M.E. Mellbye, USDA-ARS and Oregon State Univ.)

Value of Ash. The second consideration is, *"Does it have any value if land applied?"* If it is inert and has no value as a source of plant nutrients, alternative soil liming material, source of soil organic matter, or other environmentally or agronomically advantageous properties, then one will have a difficult time selling the landowner, a regulatory agency or the public on a land application program. Boiler wood ash, on the other hand, has demonstrated properties of benefit to the land, primarily its neutralizing value as an alternative soil liming material (*8, 15,16, 17*) and its plant nutrient content (*12, 13, 15, 18*). Research is being conducted by the authors to evaluate its effect on soil physical properties such as bulk density and aggregate stability.

Logistics of Application. A third consideration is, *"Can we logistically and economically apply it to either our own land or offer it for public use?"* Some materials may be too wet, too dusty, too bulky, or too odorous to move and spread economically. If the logistics of handling the materials cannot be economically overcome, then a land application program may not be the best alternative. Boiler wood ash is a product that can be handled and spread successfully (*4, 6*), but some companies are not willing to make the commitment. The bottom line is economics.

Public Objections. The final question to consider is, *"Will there be any public objection to a land application program?"* Often, there are some public or political objections to almost any activity that involves land application of a by-product. This has led to complicated regulations by many states. However, research and extensive practical experience has failed to identify negative environmental effects from application of boiler wood ash when used at reasonable agronomic rates. Sometimes, this last issue may be overcome by a good public relations campaign. However, if public objections and potential litigations seem insurmountable, then an alternative to land application may be appropriate.

The following is a review of published research reports, journal articles, abstracts, and reports of practical experiences involving the beneficial use and recycling of boiler wood ash and combination ash as an agricultural and silvicultural soil amendment along with unpublished data from the authors.

Ash Use as Lime

From a land application standpoint, the most important property of ash is the total alkalinity, total neutralizing value or potential liming value most often reported as $CaCO_3$ equivalent (CCE). Although there have been alternatives to CCE proposed for wood

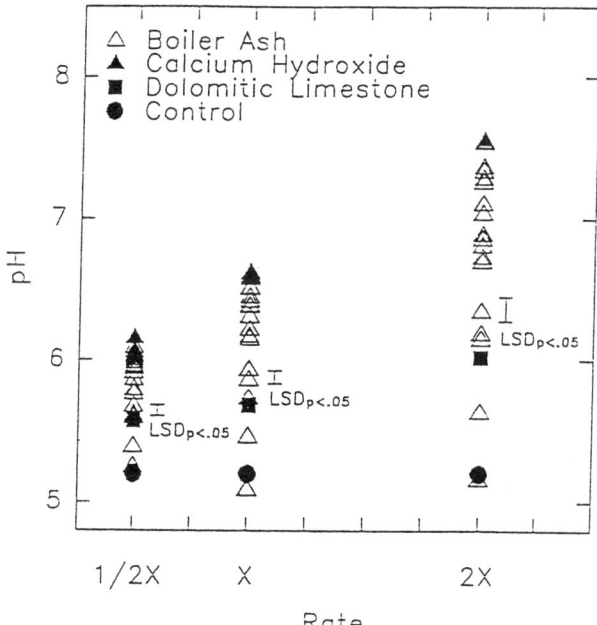

Figure 2. Effectiveness of 19 boiler ash samples, regent grade calcium hydroxide, and a commercial, dolomitic limestone when applied to an acid surface soil based only upon CCE and the recommended rate to achieve a soil pH of 6.5 (x rate) (unpublished data by the authors).

ash, the method suggested by the Association of Official Analytical Chemists (AOAC) for the determination of the CCE of agricultural limestone is the one routinely used for most alkaline by-products to be land applied. Erich and Ohno (*18*) found that CCE values obtained by this procedure were similar to those obtained from a soil incubation method.

Muse and Mitchell (*17*) reported an average CCE of 38% from 19 boiler ash samples from Alabama pulp and paper mills. Values ranged from 0 for a coal ash sample to 70% for a pure hardwood ash sample. Most samples were combination ash from the burning of wood with some coal, sludge, or other fibrous wastes. Lerner and Utzinger (*19*) reported values that ranged from 88% for black cherry ash to 115% for red oak ash. However, they generated their own ash in the laboratory for the purposes of the study. A sample of mixed ash from a homeowner's stove had a CCE of 83%. Most industrial boiler ash has a much lower CCE in the range of 25 to 70% for dry ash (Table I). Values in the lower range are probably ash from mostly fossil fuels or wood ash that has been slaked through exposure to weather and water.

Boiler wood ash is generally more effective at neutralizing soil acidity compared to commercial, ground, agricultural limestone when both are compared at equivalent rates based only on CCE (*8*). This is due to the lower effective neutralizing value of ground agricultural limestone due to its relatively coarse particle size. Most boiler ash samples in Figure 2 were almost as effective as calcium hydroxide in neutralizing soil acidity when applied at equivalent rates based upon CCE in an 84-d incubation study. Similar results were found in other field and incubation studies by the authors (Figures 3-4).

In humid regions of the U.S. where soils tend to be acidic and become more acid under cultivation, the accepted practice is for the grower or his consultant to take a soil sample according to directions by each state's extension service recommendations (*20*). It is then tested by a state supported (e.g. university) or private soil testing laboratory which makes an agricultural lime recommendation based upon the soil pH, soil buffering capacity, and crop to be grown. For example, in Alabama, 43% of all soil samples tested annually have a soil pH below 5.8 and would receive an agricultural lime recommendation for optimum production of most crops (unpublished data from the Auburn University Soil Testing Laboratory, Auburn, AL). The grower has the responsibility of acquiring a suitable liming material, applying it at recommended rates, and using other best management practices in his cropping systems. Boiler wood and combination ash would provide another option for the crop or forestry manager.

Plant nutrients

Most of the N and S that may have been present in wood, coal, or biomass is volatilized upon combustion in the boilers. At very high temperatures in most modern boilers, some P and K may also be lost. Oxides, hydroxides, and carbonates of Ca, Mg, and K make up a significant fraction of wood ash (*18*). These along with minute quantities of micronutrients (Zn, Cu, Fe, Mn) are the plant nutrients of most interest in boiler wood ash. When ash is used as an alternative to agricultural lime at rates based upon the CCE, significant plant nutrients are also added to the soil (Table I). A detailed review of the macro and micro element concentrations in wood and combination boiler ashes has been prepared by Someshwar (*21*).

Table I. Important properties of representative boiler wood ash and mixed ash samples for consideration when land applying ash as a beneficial soil amendment

Property	Mean of 6 wood ash, Vermont (14)	Ave. of wood/ coal ash, Maine (37)	Wood ash, New York (13)	Wood ash, New York (12)	Mean of 19 mixed ash, Alabama (8)	Range of 6 wood ash, Maine (16)	Range of fly, bottom & mixed ash, Georgia (47)
Neutralizing Value and Physical Properties							
CCE, %	39.2	--	50	35	37.5	25 to 59	5 to 61
Solids, %	--	--	99.7	72.4	74	--	2 to 98
Ash, %	93.6	--	--	--	70	--	57 to 100
Density, g/cm³	--	--	--	--	0.51	--	--
pH	--	11.7	12.7	11.9	9.9	--	10.6 to 13.4
Primary and Secondary Plant Nutrients							
N, g/kg	0.8	--	0.5	0.5	<0.1	--	0.15 to 2.64
P, g/kg	9	<0.1	7.9	3.3	<0.1	5.1 to 14.3	0.48 to 5.30
K, g/kg	30.3	26	30.8	16.6	13.3	26.8 to 73.8	--
Mg, g/kg	14	7.1	15	8.1	7.3	9 to 22	--
Ca, g/kg	135	73.5	270	128	120	74 to 280	--
S, g/kg	--	9.5 (as SO$_4$)	--	--	0.4	--	--
Micronutrients							
B, mg/kg	292	--	--	--	95	--	--
Cu, mg/kg	73	120	90	40	67	--	1 to 165
Fe, g/kg	21.3	17.0	11.1	13.2	6.3	--	--
Mo, mg/kg	--	--	--	--	14.6	--	0.1 to 11
Mn, g/kg	9.1	3.3	12.7	6.6	2.6	--	0.2 to 4.4
Zn, mg/kg	590	370	381	200	183	--	45 to 1560
Other elements							
Al, g/kg	19.7	32.0	15.9	15.9	12.5	2.4 to 17.0	2.7 to 20.0
Ba, mg/kg	--	--	--	--	588	--	74 to 1590
Cd, mg/kg	--	--	--	4.2	1.5	--	bd to 1.2*
Co, mg/kg	--	--	--	--	14	--	bd to 22.7
Cr, mg/kg	--	--	--	9.1	75	--	bd to 97
Na, mg/kg	--	--	--	2000	1410	430 to 5600	--
Ni, mg/kg	--	--	--	11.6	15.8	--	1 to 128
Pb, mg/kg	54	59	72	38	72	--	1 to 94

*bd=below detectable limit

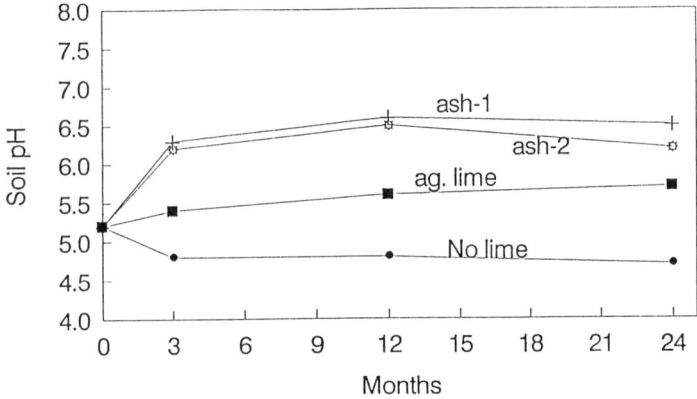

Figure 3. The effect of ground, dolomitic agricultural lime, a boiler wood ash (ash-1) and a mixed wood-coal-sludge ash (ash-2) on surface soil pH (0-5 cm depth) of an acid, Vaiden clay in a field experiment over 2-yr period. All materials were applied based on CCE of the material and the recommended lime necessary to achieve a pH approximately 6.5 (unpublished data by the authors).

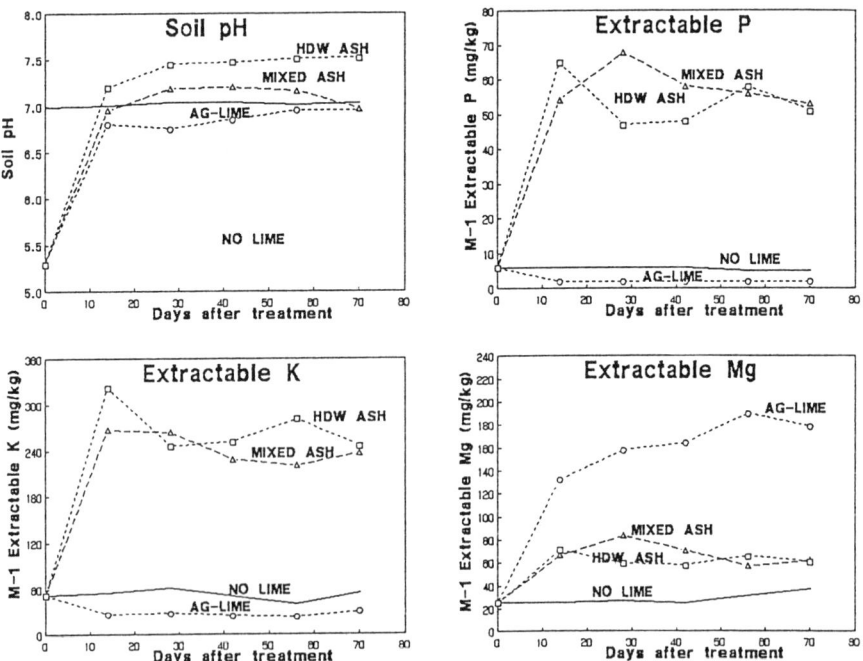

Figure 4. Effect of time after treatment on soil pH and Mehlich-1 extractable P, K, and Mg when three liming materials were applied to a Dothan f.s.l. at approximately equivalent rates based upon the CCE of each material (adapted from ref. 42).

Phosphorus. The P concentration in boiler wood ash is generally low compared to inorganic fertilizers and organic soil amendments. Total P in 19 boiler ash samples ranged from 1 to 7 g/kg with a mean of 3 g P/kg (8). Ammonium citrate-extractable P (available P) in 6 ash samples from Maine ranged between 43 and 53% of total P (16) and averaged 48% of total P in 19 Alabama ash samples (8). As a result, total P in wood ash applied to corn was determined to be less available than P in monocalcium phosphate (22). Otherwise, soil P reactions appear to be similar regardless of P source (15, 16, 22).

At a typical application rate to supply the equivalent of 2 metric tons of CCE lime per hectare (6 t ash/ha assuming 33% CCE), one would apply approximately 18 kg total P/ha (37 pounds P_2O_5 per acre). For forage crops on soils that test "medium" in P, typical fertilizer P recommendations would range between 20 and 40 kg P/ha (23). Provided the P in ash is plant available, moderate applications would provide some P to crops but generally not enough for field crops on soils already low in P. However, ash used in Fig. 4 by the author increased Mehlich-1 (dilute, double acid) extractable P in an unfertilized soil from less than 10 mg P/kg (medium level) to over 50 mg P/kg (very high level). The mixed ash had a CCE of 26% and the hardwood ash had a CCE of 67%. They were applied to the soil at rates based upon their CCE (11.5 g/kg and 4.5 g/kg, respectively) to be approximately equivalent to recommended ag. lime (3.0 g/kg) for this soil.

Potassium. The term "potash" was derived from the practice of burning wood for the salts. According to Collins (24), a seemingly inexhaustible source of wood for making potash was ". . . one of the incentives of Great Britain for establishing colonies along the Atlantic Seaboard." For two centuries the U.S. cut the eastern forests for potash to meet the growing industrial needs of England. Early potash was produced by leaching wood ashes and evaporating the solution to dryness. The first patent approved by the U.S. government was for a process for making potash from wood in 1790. Today, potash refers principally to K fertilizers or K as a plant nutrient expressed as percentage K_2O.

The K content of wood ash from high temperature boilers is lower than the K content of wood ashes from open incinerators or wood burning stoves and fireplaces. Nevertheless, K is the most significant primary plant nutrient applied in boiler wood ash (Table I). Ash is also high in the secondary nutrients Ca and Mg. Ash K is comparable to fertilizer K sources in its plant availability (15, 16). When ash is applied at soil liming rates, adequate K is available for most field and forage crops (Figure 4). Excessive K application to soils is not an environmental concern in humid regions of the world. Rates as high as 280 kg K/ha may be recommended for productive hay crops on soils low in extractable K. In the southeastern U.S., loss of stand of hay crops such as alfalfa and bermudagrass is often attributed to inadequate K fertilization (23). Coleman (25) reported that ". . . ash successfully raised soil pH as well as levels of calcium, phosphorus, magnesium and potassium. Crop response was excellent, with dramatic results seen when ash was applied to 'run-out' hay fields which were in a low pH and low fertility state prior to ash application." In Alabama, 57% of all soil samples test "medium" or "low" in K and would warrant a fertilizer K recommendation (unpublished data from the Auburn University Soil Testing Laboratory, Auburn, AL).

Micronutrients and Metals. In spite of the fact that all analyses of boiler wood ash and most analyses of combination ash indicate relatively low levels of most plant micronutrients and non-nutrient metals from an agricultural and silvicultural standpoint

(Table I), the most frequently expressed environmental concern is with heavy metals. Shetron et al. (*26*) reported that ". . . objections to large scale application of wood ash have been raised in Michigan. The major concern is the potential leaching of heavy metals in the ash into ground waters." However, this concern is not supported by any known research with wood ash or combination ash. Metal leaching into groundwaters is almost unknown in productive, agricultural soils. Other concerns are with metals entering the food chain (*27*). However, Mullins and Mitchell (*28*) stated, "*Widespread public concern about their (wastes/by-products) potential negative effects on crops is largely unsupported by research. These concerns, particularly regarding heavy metal applications, may be based upon past government regulations and limited research. Only 5 of 74 (research) summaries . . . indicate negative effects of soil amendments on crop yield, quality, or metal concentrations (and) . . . were predictable based upon known properties of the amendments and intentionally gross over applications.*"

A symposium by the National Council of the Pulp and Paper Industry for Air and Stream Improvement (NCASI) on "Land Application of Wood-fired and Combination Boiler Ashes" in 1995 had 17 speakers from North America and Europe address the topic. None of the papers presented indicated any environmentally harmful property of the ash studied beyond what would be normally expected from adding nutrients or a liming material to a soil. Morris et al.(*29*) noted that ". . . *ash from 10 southern pulp and paper mills indicates low potential to leach compounds of regulatory concern. With the exception of the metal barium, metal, organic and pesticide concentrations determined by the EPA Toxicity Characteristic Leaching Procedure (TCLP) are below detectable limits.*" Someshwar (*21*) concluded that burning other fuels such as coal, sludges, non-recyclable paper, etc. along with wood residue has little impact on ash metal composition. He also found that the ". . . trace and heavy metal composition in combustion ashes are comparable to those in other potential soil amendments such as coal ashes, waste treatment sludges, and limestone but lower than those in sewage sludges." Comparisons of mean values for ash (*8*) with some other common materials and EPA standards emphasize this point (Table II).

Other Environmental Concerns

Organics of critical environmental concern have not been identified in boiler wood and combination ash. Chlorobenzenes, chlorophenols, and polychlorinated biphenyls were below detectable levels (*21*). However, a frequently mentioned concern when disposing of dewatered pulp and paper mill sludge in wood-fired boilers is the presence of "dioxins" when chlorine bleaching processes are used. In a study to investigate the burning of dioxin-laden sludges in bark-fired boilers, NCASI (*30*) reported that less than one percent of the PCDD/Fs (polychlorinated dibenzodioxins and polychlorinated dibenzofurans) in sludge fed to boilers were found in the combustion ash. The higher boiler heat from fossil fuels increased the PCDD/F destruction efficiencies of boilers. In addition, the dioxin of main concern, TCDD (2,3,7,8-tetrachlorodibenzo-p-dioxin), is highly photo degradable and possibly easily volatilized when initially applied to soils (*31*). It is strongly sorbed once it moves into soils or sediments, and is unlikely to be taken up by plants. Lawrence et al. (*32*) reported that land spreading of waste water sludges containing low levels of TCDD (up to 190,000 parts per trillion) presents no significant threat to human health. Someshwar (*21*) concluded that dioxins and furans in ash should be of no concern as long

Table II. A comparison of metals in boiler ash with metals in other materials and US-EPA's Section 503 regulations for metals applied in municipal biosolids

Metal	Mean of 19 boiler ash (8)	Mississippi surface soils (38)	Poultry manure/litter (39, 40)	Fertilizer phosphate rock (41)	Municipal biosolids (42)	EPA's maximum concentration for class A biosolids	EPA's cumulative soil loading rate (33)
			---mg/kg---				---kg/ha---
As	--	--	281	12	10	41	41
Cd	1.5	0.5	2.4	11	10	39	39
Cr	75	--	8.5	109	500	1200	3000
Cu	67	10.4	332	23	800	1500	1500
Fe	6260	11600	1950	--	--	--	--
Mo	14.6	--	5.7	--	4	18	420
Mn	2600	700	277	--	--	--	--
Ni	15.8	15.2	7.6	37	80	420	17
Pb	72	20.8	14.6	123	50	300	300
Se	--	--	--	--	5	36	100
Zn	183	48	252	19	1700	2800	2800

as extraneous sources of chloride such as from salt water or pulp bleaching are either kept out of the wood fuel mix or kept to a minimum (<0.03% of the fuel).

Variability of Ash

Wood ash characteristics important for the use in a land application or agricultural by-product program can vary. The wood ash calcium carbonate equivalence (CCE) has been observed to vary from boiler to boiler even between operations that used similar type of boilers. In a sampling of fresh wood ashes by the authors during a one month period from Southern plants that used spreader stokers and similar wood, the wood ash CCE ranged from a low of 10% to a high of 94%. Wood ash CCE also varied at each plant. Wood ash and combination ash samples were collected periodically by the authors during a one month period from boilers at mills in North Carolina and Alabama, respectively. The ash was characterized as being normal for the boiler. Calcium carbonate equivalent values ranged from 52 to 95% in the wood ash and 13 to 33% in the combination ash (Figure 5). Total K was also determined in the combination ash; values ranged from 0.3 to 1.4%. Factors that affected this variability may include boiler operation especially the reburning of the wood ash and the O_2 levels within the boiler and the source of the fuel although these factors could not be confirmed during this sampling period.

Regulatory

US-EPA. Since wood ash has qualified as a non-hazardous waste under subtitle C of the Resource Conservation and Recovery Act, its regulation at operating facilities has been traditionally conducted by the individual states within the solid waste regulations.

State Regulations. For most wood ash generators, the wood ash is regulated as a solid waste. Using the wood ash in a land application type program allows the wood ash to be used in a beneficial manner. Developing the wood ash land application program further into an agricultural by-product program provides the opportunity for the wood ash waste to be transformed into a by-product that can be sold as a liming material or fertilizer. By developing a by-product program, the wood ash producer can gain a waste minimization program, income to offset the cost of the program and in many by-product operations less complex regulations.

Environmental. Experience by the authors indicate that when used as an agricultural by-product, wood ash is either exempted from solid waste regulations or is regulated as a solid waste material in a beneficial use manner. Each state has specific policy. Alabama solid waste regulations exempt wood ash from solid waste regulations when the ash is solely derived from wood. Other states such as Georgia, Florida, Louisiana and Virginia (Table III) can exempt wood ash from solid waste regulations if the wood ash qualifies as a recovered material.

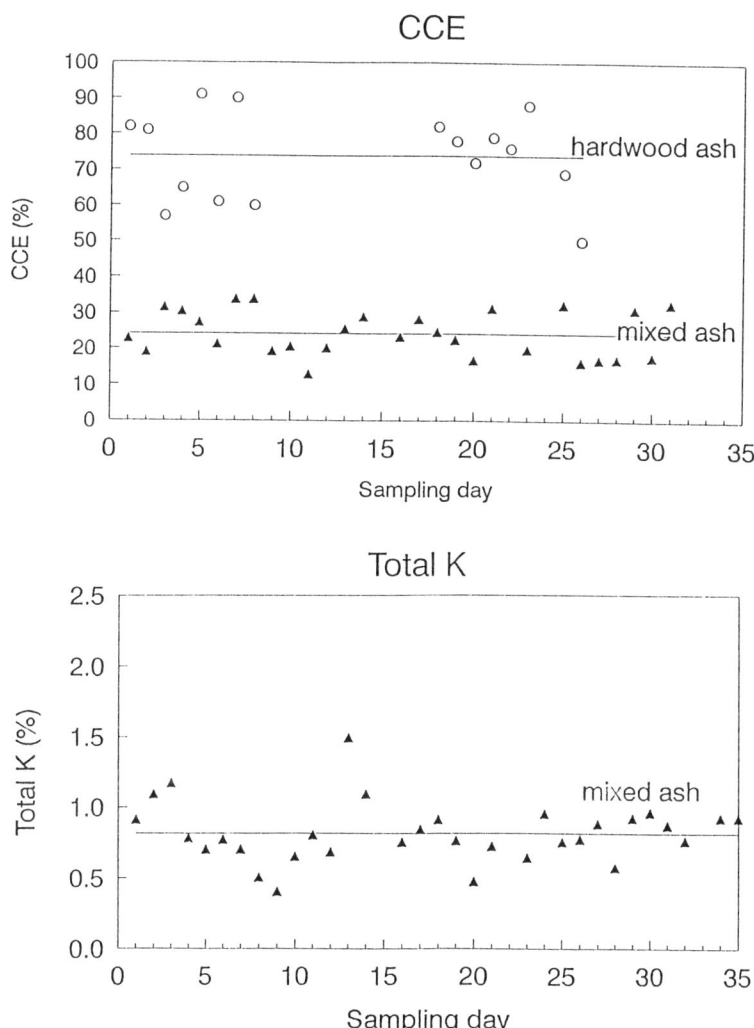

Figure 5. Variability of neutralizing value expressed as calcium carbonate equivalent (CCE) and total K in ash from two sources (unpublished data by the authors).

Table III. General regulatory policy for the use of wood ash as an agricultural or forestry by-product in selected Southern states

Regulatory policy	States						
	AL	GA	FL	LA	MS	NC	SC	VA
Exempted from regulation	*							
Exempted as qualifying recovered material		*	*	*			*	*
Regulated as a solid waste					*	*		

To qualify as a recovered material, the wood ash must be documented as safe and beneficial for use in agriculture. Typically states that issue a recovered material exemption will require documentation in the form of TCLP (US-EPA's Toxicity Characteristic Leaching Procedure) metals content (33) and annual pollutant loading rates; calcium carbonate equivalence, K and solids content. In addition, a management plan that describes the intended use, monitoring, QA/QC, basis of application rate, records and spreading operations is also required as part of the regulatory review for a recovered material status.

The regulatory emphasis for recovered materials is focused on the characteristics and the use of the wood ash and does not involve site investigation or characterization as is required for land application programs. Some of the wood ashes that receive the recovered material exemption are sold. These must be licensed or registered with the state department of agriculture. For some states, the recovered material status is conditional on obtaining a department of agriculture licensing or registration as a lime by-product, lime material or soil amendment. In South Carolina wood ash can be exempted from solid waste regulations, if the wood ash is licensed or registered with the South Carolina Department of Agriculture.

In Mississippi and North Carolina, wood ash is regulated as a solid waste with established metal loading limits, site requirements and investigation, monitoring, reporting and management requirements. In North Carolina, the North Carolina Department of Environment, Health and Natural Resources (NCDEHNR) requires analyses of As, Ba, Cd, Cr, Fe, Pb, Se, Ag, Zn, CCE, P and K. The wood ash must also be registered or licensed with the North Carolina Department of Agriculture, and it must be used in a manner that follows the guidelines of the North Carolina State University Cooperative Extension Service Best Management Practices (BMPs). In addition, the field that is to receive the wood ash must be sampled and analyzed to determine the appropriate wood ash spreading rate. The NCDEHNR also requires that the management plan include buffers from surface waters and the depth to groundwater for each specific field that receives the wood ash.

Agricultural. Individual state departments of agriculture regulate wood ash that is marketed or sold for use in agriculture as either a liming material, fertilizer, by-product, soil additive or soil amendment. The advantage of licensing or reregistering the wood ash with the state department of agriculture is that the wood ash is considered a by-product that can be sold for use in agriculture rather than a waste. If the manufacturer/supplier claims that the wood ash has liming value or the wood ash can neutralize soil acidity, the wood ash must be licensed as either a by-product or liming material. Or if the wood ash is described as containing K fertilizer or plant nutrients, it must be registered as a fertilizer or by-product.

The state departments of agriculture regulations are developed to maintain a standard lime or fertilizer quality and the exclusion of materials that could be damaging to the crop or the soil. Most state departments of agriculture require notification from the solid waste section that the wood ash is non-hazardous based on TCLP results and EPA limits for regulated parameters. In addition, wood ash total As, Cd, Cr, Cu, Pb, Hg, Mo, Ni, Se, Zn and Na content is required as part of the review by some departments of agriculture.

To license a wood ash as an agricultural liming material, the wood ash must contain a minimum content of lime expressed on a calcium carbonate equivalence (CCE) basis. The specific amount varies from state to state. Georgia has a minimum CCE of 85%, in Florida the minimum CCE is 90% for standard and 75% for non-standard, and in Mississippi and Alabama, the minimum CCE is 90%. Failure to meet the minimum CCE values can result in suspension of license, sales and fines. Because of this high CCE requirement, wood ash is usually licensed as a lime by-product rather than as agricultural lime. Several states have established policy for the licensing or registration of the wood ash. Alabama, Florida, Georgia, Louisiana, Mississippi and North Carolina have established by-product categories for materials such as wood ash. South Carolina and Virginia establish standards on a case by case basis.

Although the state departments of agriculture require minimum CCE standards, the wood ash producer can declare that minimum value at the time the material is registered as a lime by-product. And like agricultural lime, lime by-products can be marketed and sold as liming materials.

Georgia also allows the producer to select the minimum moisture and particle size (mesh) to be guaranteed for the wood ash. Alabama requires that the by-product meet established particle size requirements such as 90% passing a #10 mesh screen (<2.5mm) and 50% passing a #60 (<0.4mm) mesh screen. For wood ashes produced in states with established particle size requirements, the particle size requirements can be the most limiting factor for the licensing of the wood ash as a by-product. Failure to meet the guaranteed limits for particle size or moisture can result in the loss of the by-product license, sales stoppage and fines. The state departments of agriculture also require that lime by-product invoices and labels acknowledge that the material is a by-product and state the equivalent amount of lime by-product required to equal one ton of agricultural lime.

Because boiler wood ash also contains K, ash may also be registered as a potash fertilizer. If the wood ash is registered as a fertilizer by-product, the producer selects the minimum guaranteed amount expressed as percentage $N-P_2O_5-K_2O$. Several ash by-products have been registered with state departments of agriculture as 0-0-1, 0-0-2 or higher analysis potash fertilizers. Like the lime material and lime by-products, failure to maintain the minimum guarantee, in this case K_2O concentration, can result in the loss of the by-product registration, sales stoppage or fines.

Although wood ash can be registered and used as potash fertilizers, the liming characteristics of the wood ash and the amount to be applied to a field as potash fertilizer by-product must be considered by the user. When the wood ash is applied to an

Table IV. Summary of some published research with land application of boiler wood and combination ash

Location	Type of ash	CCE	Treatments/ rate(s) applied	Nature of research; crops/plants involved	POSITIVE effects of ash	NEGATIVE effects of ash	References
				Domestic research			
AL	wood	58-69%	4.5 and 9.0 Mg/ha	pasture; effects on soils	raised soil pH above equivalent rate of lime based on CCE; higher forage yield than equivalent lime rate; no effect on forage quality.	none	8
AL	mixed	38%	–	summary of research and on-farm demonstrations	recommends use as liming material; field and greenhouse research confirmed safety and practicality.	none	8, 43
AL	mixed	20%	4.5 to 27 Mg/ha	field research/forages	confirmed safety, effectiveness, and predictability of ash; no harmful effect on plant growth; no significant uptake of metals by forages.	ash variability	44
AL	wood and mixed	26-67%	6.7 to 51.3 Mg/ha	soil incubation with legumes	ash comparable to ag. lime when use based on CCE; soil reaction rapid; increases soil P and K levels.	none	45
AL	mixed	30%	based on CCE and lime recommendation	report of on-farm demonstrations	Farmers accepted and used ash in a county-wide demonstration project based on CCE and lime recommendations based on soil tests	handling/ spreading	35
AL	hardwood	43%	10.3 Mg/ha	soybean in field	more effective than ag. lime on soil pH; increased yield and soil nutrients	none	46
GA	fly, bottom, & mixed	5-60%	to soil pH=6.5 and 7.5	residue characterization and application on pine seedlings	no detrimental effects on pine trees even when soil pH is above what is thought to be optimum	none	47
ID	mixed wood	92%	0-30% of soil mass	wheat & poplar trees in greenhouse	increased biomass and no detrimental effects when used at rates <2% of soil; recommended use as lime or source of nutrients.	none at rates <2% of soil	48
ID	mixed wood	92%	10 Mg/ha	bush beans in greenhouse	Soil pH and K concentration are rate limiting components of ash; rate should be based on CCE.	yield reduced when pH>6.5 or >2662 kg K/ha applied	49

Table IV (continued).

Location	Type of ash	CCE	Treatments/ rate(s) applied	Nature of research; crops/plants involved	POSITIVE effects of ash	NEGATIVE effects of ash	References
ID	wood	36%	0-36 Mg/ha	wheat	growth and protein unaffected by ash; no accumulation of heavy metals; pH increased	none	50
ME	4 sources of wood ash	18-48%	based upon P or K	nutrient uptake by corn	ash effective as source of both P and K.	none	35
ME	6 sources of wood ash	26-59%	4.1 to 12.3 g/kg soil	soil liming and fertilizer source	ash acceptable alternative for lime; lab CCE more variable than incubation CCE values; ash provides modest amounts of P and K to soils.	potential K:Mg balance in soils; Na saturation for repeated ash applications.	16
ME	wood	48%	incubation study	P and K release	soil acidity neutralization rates pH dependent; P sorption rapid at low pH values; solubility of wood ash K very high.	none	15
ME	wood	?	greenhouse study	spinach	soil pH reacted faster to ash than to lime; yields increased; no significant uptake of Cu or Cd.	none	51
ME	wood	?	industry example	summary of industry experience	local landowners acceptance exceeded expectations; costs savings to landowners;	none	37
ME	papermill sludge ash	?	?	soil solution study; ecosystem of clearcut forests	soil solution concentration of cations and anions increased; ash had less impact on soil solution than timber harvesting	none	52
ME, NH, VT	wood ash	~25%	–	summary of experiences	crop response "excellent" on old, acid; "enthusiastic" acceptance by farmers; high C ash useful in biosolid composting.	spreading; combustion; dust	25

Continued on next page

Table IV (continued).

Location	Type of ash	CCE	Treatments/ rate(s) applied	Nature of research; crops/plants involved	POSITIVE effects of ash	NEGATIVE effects of ash	References
MI	5 sources of wood ash	varies	11 and 33 Mg/ha	field soil leaching study	raised soil pH; no adverse effect on soil or groundwater	suggested not use ash on low CEC soils with high leaching potential	26
NH	wood	?	11, 22, 44 Mg ash/ha	white pine	raised soil pH; increased soil cations; maintained "acceptable" environmental quality	none	53
NY	wood	35%	0 to 50 Mg/ha	alfalfa hay; soil quality	ash increased soil pH and exchangeable K and Mn; decreased exchangeable Al and Fe; increased hay yields and hay quality.	excessive rates not detrimental to crop or soil.	12
NY	willow ash	?	10-20 Mg ash/ha	willow (*Salix purpurea*)	replaced nutrients removed from intensive, annual harvests	none	54
OH	several wood ash	83-116%	variable	characterization of ash; effect on snapbeans	rapid increase in soil pH; good alternative to lime	none	19
SC	bark bottom ash	?	11 to 44 Mg/ha	field and column soil leaching study	increased soil cations; no adverse impact on ground water.	none	55
SC	bark	?	10 Mg/ha	metals in fauna	no impact	none	56
VT	wood	39%	0 to 200% of lime requirement	greenhouse study with corn and alfalfa	plants grew better with ash than lime; no overliming effect with ash; ash should be used based upon lime recommendations	none	14

Table IV (continued).

Location	Type of ash	CCE	Treatments/rate(s) applied	Nature of research; crops/plants involved	POSITIVE effects of ash	NEGATIVE effects of ash	References
WA	mixed	29%	3, 4, & 5 equivalent tons/acre compared to ag. lime	oat & bean	raised soil pH; increased biomass yield; increased oat grain yield; increased nutrient uptake by plants	none	57
WA	wood	46%	3 to 6 equivalent tons/acre compared to ag. lime	green pea & wheat	wheat yield increased 15%, green pea yield increased 63%, raised soil pH	none	58
International research and Reviews							
Finland	wood	?	variable	ash on agricultural and forestry lands	positive aspects of spreading ash on forests in Finland for both lime and plant nutrients	spreading	4
Finland	wood	?	10 Mg/ha	microbial activity in organic soils	soil pH, microbial activities and cellulose decomposition increased;	mineralization of soil	59
Review	wood	--	--	application to agricultural and forestry lands	Ash can be used as excellent substitute for lime; Zn may be rate-limiting metal but doil pH and K threshold would be exceeded first.	none	10
Review	wood and mixed	--	--	characterization of ash	Trace and heavy metals in ash low and not impacted by other fuels; organics not an evironmental concern; PCDD/Fs of no concern from inland woods.	PCDD/Fs only when Cl burned	21

agronomic crop as a K fertilizer, excess lime may also be applied to the field. This will likely result in elevated soil pH and potential crop micronutrient deficiencies.

The CCE, moisture, particle size or K_2O content variation may be sufficiently large that it is not possible to effectively guarantee a minimum CCE, moisture, particle size or K_2O content. This ash may be registered as "soil amendments" or "soil additives". The requirements for soil additives or soil amendments do not include minimum CCE, moisture, particle size or K2O contents. Soil additives and soil amendments are materials that enhance soil physical properties, condition or biological activity such as peat, compost, pine bark, perlite, or vermiculite. However unlike liming materials, lime by-products or fertilizer by-products; no nutrient or lime claims can be made in regard to the wood ash.

Land application of boiler wood ash in the U.S.

Wood ash is commonly applied to fields as a lime material or potash fertilizer without further treatment, although in some cases the wood ash is mixed with agricultural lime or screened to obtain a desired mesh size. Wood ash as a liming material and nutrient source has been applied to both agricultural land and woodlands (*34*). Many pulp and paper mills and wood products companies have plants that apply wood ash either to agricultural land or forest lands. About 80% of the wood ash generated in the Northeast US is applied to agricultural lands (*10*). Erich (*35*) notes that during 1986 through 1991 100,000 to 150,000 dry tons of wood ash was used as a lime by-product in the US.

In Alabama, Kimberly-Clark at Mobile has distributed a wood ash under the name of BioAsh. The BioAsh is delivered to Baldwin County farmers as a lime by-product to neutralize the soil acidity. Mitchell and Dunn (*36*) report that in 1988 over 50 farmers used a total of 5,000 tons of the BioAsh in on-farm demonstrations. In Louisiana, the International Paper Mill in Mansfield also distributes a wood ash to local farms. In Georgia, the Stone-Savannah Pulp and Paper Mill distributes to agricultural land a wood ash mixture.

Many Southern pulp and paper mills spread wood ash onto woodlands. The Georgia-Pacific mill in Palatka, Florida, spreads the mill wood ash onto area woodlands. The Union Camp Savannah Mill in Georgia has spread a wood ash to the company woodlands. In South Carolina the International Paper Mill at Georgetown spreads the mill wood ash onto company owned woodlands. Also the Champion International Mill at Texarkana, Texas, spreads the mill wood ash onto the company woodlands. In Arkansas an International Paper Mill at Camden is spreading wood ash to the company woodlands. In addition to the pulp and paper mills, several wood products companies and saw mills spread their wood ash onto woodlands.

Summary of Land Application Research and Experiences

Land application of boiler wood ash is primarily for its value as a soil liming material and/or a source of plant nutrients. Ash with a low CCE, high C content, and low density potentially could be used to improve soil physical characteristics such as to decrease bulk density, improve water infiltration, improve drainage, etc. but these uses are not documented. Summaries in Table IV clearly indicate few if any problems with land application that would not have been predicted based upon an analysis of the ash and acceptable agronomic practices. Major problems appear to be in handling and distributing ash. However, equipment and hauling/spreading logistics can be overcome (*4, 35*); cost of transportation is usually the prohibitive factor.

Wood ash and, more recently, combination boiler ash have been used successfully in the U.S. and Europe as an alternative to ground agricultural limestone, as a minor source of plant nutrients, and as a soil conditioner. Although popular concerns are often expressed about land application of metals in boiler wood ash and in combination ash, levels are well within EPA's guidelines for land application of biosolids. There are no published reports of metals being an environmental or crop production/crop quality problem, especially when ash is used at recommended rates as a soil liming material. An analysis of the CCE and a routine analysis of plant nutrients and selected total metals are needed to justify ash as a soil amendment. Significant plant nutrients, principally K and Ca, are applied to soil when ash is used as a liming material. Variability in the quality of ash is a concern, but if a grower follows a conscientious soil testing program and applies ash as a liming material, some variability in the ash can be tolerated, especially at the rates generally used. Ash is regulated by either the state environmental agencies or the departments of agriculture or both. For beneficial use as an alternative to agricultural lime or as a soil amendment, its regulation as an industrial by-product used as lime or as an agricultural soil amendment is preferable.

Literature Cited

1. Steuteville, R. 1995. The state of garbage in America. Part I. BioCycle 36(11):54-63.
2. Tisdale, S.L., W.L. Nelson, and J.D. Beaton. 1985. Soil fertility and fertilizers. 4th ed. 751 pp. Macmillan Pub. Co., New York, NY
3. Meridian Corp. 1994 (revised). Economic impact of industrial wood energy use in the southeastern region of the U.S. Vol. II. prepared for the Southeastern Reg. Biomass Energy Prog. TVA. Muscle Shoals, AL., by Merdian Corp., 4300 King St., Alexandria, VA 22302
4. Hakkila, P. and Hikalaja. 1983. The technique of recycling wood and bark ash. Folia Forestalia 552:35-37 (Finnish with summary in English).
5. Coleman, P. 1984. Keeping it down on the farm. The American Tree Farmer. p. 21.
6. Miner, R., and J. Unwin. 1992. Solid waste management and disposal practices in the U.S. paper industry. NCASI Tech. Bul. no. 641. National Council of the Paper Industry for Air and Stream Improvement, Inc. 260 Madison Ave., New York, NY 10016.
7. Forest Farmers Association. 1991. Twenty eighth manual edition. 50(3):106-113.
8. Muse, J.K., and C.C. Mitchell. 1995. Paper mill boiler ash and lime by-products as soil liming materials. Agron. J. 87:432-438.
9. Muse, J.K. 1993. Inventory and evaluation of paper mill by-products for land application. M.S. Thesis. 140 pp. Auburn University, Alabama.
10. Campbell, A.G. 1990. Recycling and disposing of wood ash. TAPPI J. 73(vol. 9):141-146.
11. Hatch, C.J., and R.G. Pepin. 1985. Recycling mill wastes for agricultural use. TAPPI J. 68:70-73.
12. Naylor, L.M., and E. Schmidt. 1989. Paper mill wood ash as a fertilizer and liming material: Field trials. TAPPI J. 72:199-206.
13. Naylor, L.M., and J.A. Johnson. 1985. Papermill wood-derived boiler ash as a fertilizer. I. Available nutrients and liming value. *Abstracted in* Summary of research on woodash utilization. Resource Conservation Services, Inc., Yarmouth, ME.
14. Magdoff, F., R. Bartlett, and D. Ross. 1983. Final report: Wood ash research project, Burlinton/Electric Dep. Univ. of Vermont, Burlinton, VT. 23 p. *Abstracted in* Summary of research on woodash utilization. Resource Conserv. Services, Yarmouth, ME.
15. Ohno, T. 1992. Neutralization of soil acidity and release of phosphorus and potassium by wood ash. J. Environ. Qual. 21:433-438.

16. Ohno, T., and M.S. Erich. 1990. Effect of wood ash application on soil pH and soil test nutrient levels. Agric. Ecosyst. Environ. 32:223-239.
17. Muse, J.K., and C.C. Mitchell 1993. Paper mill byproducts find a home on the farm. Highlights of Agric. Res. 40(3):4. Alabama Agric. Exp. Stn., Auburn University, AL.
18. Erich, M.S., and T. Ohno. 1992. Titrimetric determination of calcium carbonate equivalence of wood ash. Analyst 117:993-995.
19. Lerner, B.R., and J.D. Utzinger. 1986. Wood ash as soil liming material. HortScience 21(1):76-78.
20. Thom. W.O., and W. Sabbe. 1994. Soil sampling procedures for the southern region of the United States. Southern Coop. Ser. Bul. 377. Univ. of Kentucky, Lexington, KY.
21. Someshwar, A.V. 1996. Wood and combination wood-fired boiler ash characterization. J. Environ. Qual. 25(5): (in press).
22. Erich, M.S., and T. Ohno. 1992. Phosphorus availability to corn from wood ash-amended soils. Water, Air, and Soil Pollution. 64:475-485.
23. Adams, J.F., C.C. Mitchell, and H.H. Bryant. 1994. Soil test fertilizer recommendations for Alabama crops. Ala. Agric. Exp. Stn. Agron. and Soils Dep. Ser. no. 178. Auburn University, AL.
24. Collins, G.H. 1947. Commercial fertilizers, their sources and use. 4th edition. 522 pp. The Blakiston Co., Philadelphia, PA
25. Coleman, P.M. 1995. Land application of wood ash: 12 years in northern New England. Abstracts of Presentations p. 4-6.. Symposium on land application of wood-fired and combination boiler ashes. National Council for Air and Stream Improvement. Gainesville, FL
26. Shetron, S.G., S.C. Defour, C. Brecken, J. Rughani, J. Diebel, G. McGinnis, and M.F. Jergensen. 1995. The environmental fate of wood ash applied to selected soils in northern Michigan. Abstracts of Presentations p. 34-36.. Symposium on land application of wood-fired and combination boiler ashes. National Council for Air and Stream Improvement. Gainesville, FL
27. McBride, M.B. 1995. Toxic metal accumulation from agricultural use of sludge: Are USEPA regulations protective? J. Environ. Qual. 24:5-18.
28. Mullins, G.L., and C.C. Mitchell. 1994. Crops *In* J.E. Rechcigl (ed.) Soil amendments: Impacts on biotic systems (pp. 1-40). CRC Press, Inc./Lewis Publishers, Boca Raton, FL.
29. Morris, L.A., W.E. Miller, F.M. Sounders, and J.F. Sanders. 1995. Characteristics of ash and alkaline pulp mill residues affecting their use as lime substitutes. Abstracts of Presentations p. 9-10.. Symposium on land application of wood-fired and combination boiler ashes. National Council for Air and Stream Improvement. Gainesville, FL
30. NCASI. 1990. Burning sludge from bleached kraft mill in bark-fired boilers: A study of polychlorinated dibenzodioxins and dibenzofurans (PCDD/F) concentrations in the combustion ashes. National Council of the Paper Industry for Air and Stream Improvement, Inc. Tech. Bull. no. 580.
31. US-EPA. 1990. Integrated risk assessment for dioxins and furans from chlorine bleaching in pulp and paper mills. Office of Pesticides and Toxic Substances. EPA 560/5-90-011.
32. Lawrence, F.H., R.E. Kennan, D.W. Crawford, and M.M. Sauer. 1987. Human health risk assessment related to exposure to dioxin from land application of waste water sludge in Maine. *In* Proc. 1986 National Council of the Paper Industry for Air and Stream Improvement, Inc. NE Regional Meeting. NCASI Spec. Rep. no. 87-03.
33. US-EPA. 1993. 40 CFR Parts 257, 403, and 503. Standards for the use or disposal of sewage sludge. Fed. Reg. 58:9248-9415.
34. Vance, E.D. 1996. Land application of wood-fired and combination boiler ashes: An overview. J. Environ. Qual. 25:937-944.
35. Erich, M.S. 1991. Agronomic effectiveness of wood ash as a source of phosphorus and potassium. J. Environ. Qual. 20:576-581.
36. Mitchell, C.C., and D.E. Dunn. 1989. Evaluation of Scott's Bioash in Baldwin County. Ala. Coop. Ext. Serv. New Technology Demonstration Rep. no S-4-89. Auburn University, AL.
37. Pepin, R.G., and P. Coleman. 1984. Paper mill sludge and ash as soil conditioner. BioCycle 25(4):52-55.
38. Pettry, D.E., and R.E. Switzer. 1993. Heavy metal concentration in selected soils and parent materials in Mississippi. Miss. Agric. & Forestry Exp. Stn. Bul. 998. Mississippi State, MS.
39. Mitchell, C.C., and J.O. Donald. 1995. The value and use of poultry manures as fertilizer. Ala. Coop. Ext. Ser. Cir. ANR-244. Auburn University, AL

40. Edwards, J.H., E.C. Burt, and R.L. Raper. 1995. Issues affecting application of noncomposted organic waste to agricultural land. *p.* 225-249. *In* Agricultural utilization of urban and industrial by-products. ASA Spec. Pub. no. 58. Amer. Soc. Agron., Madison, WI.
41. Mortvedt, J.J., and J.D. Beaton. 1995. 6 heavy metal and radionuclide contaminants in phosphate fertilizers. p. 93-105. *In* H. Tiessen (ed.) Phosphorus in the Global Environment. John Wiley & Sons, Ltd.
42. Chaney, R.L. 1983. Potential effects of waste constituents on the food chain. p. 50-76. *In* J.F. Parr (ed.) Land treatment of hazardous wastes. Noyes Data Corp. Park Ridge, NJ
43. Mitchell, C.C., Jr. 1995. Boiler ashes as an alternative soil liming material in Alabama. Abstracts of Presentations p. 19-20.. Symposium on land application of wood-fired and combination boiler ashes. National Council for Air and Stream Improvement. Gainesville, FL
44. Mitchell, C.C., and J.K. Muse. 1993. Boiler wood ash as an alternative soil liming material. Ala. Coop. Ext. Serv. New Technology Demonstration Rep. no S-5-93. Auburn University, AL.
45. Mitchell, C.C., and J. Cotten. 1989. The effect of boiler ashes on soil pH and extractable nutrients. Ala. Coop. Ext. Serv. New Technology Demonstration Rep. no S-5-89. Auburn University, AL.
46. Mitchell, C.C., and S.D. Carroll. 1989. Effect of papermill boiler ash as an alternative liming material in Dallas County. Ala. Coop. Ext. Serv. New Technology Demonstration Rep. no. S-2-89. Auburn University, AL.
47. Morris, L.A., W.L. Nutter, J.F. Sanders, W.P. Miller, M.E. Sumner, and F.M. Saunders. 1996. Use of pulp and paper mill residues for forest production in the U.S. Southeast. 13pp. *In* Proc. of the South. Sec. Air & Waste Mgmt. Assoc., Biloxi, MS. South. Sec. Air & Waste Mgmt Assoc., 3988 Flowers Rd., Atlanta, GA 30360.
48. Etiegni, L., A.G. Campbell, and R.L. Mahler. 1991a. Evaluation of wood ash disposal on agricultural land. I. Potential as a soil additive and liming agent. Commun. Soil Sci. Plant Anal. 22:243-256.
49. Etiegni, L., R. L. Mahler, A.G. Campbell, and B. Shafii. 1991b. Evaluation of wood ash disposal on agricultural land. II. Potential toxic effects on plant growth. Commun. Soil Sci. Plant Anal. 22:257-267.
50. Huang. H., A.G. Campbell, R. Folk and R.L. Mahler. 1992. Wood ash as a soil additive and liming agent for wheat: field studies. Commun. Soil Sci. Plant Anal. 23:25-33.
51. Zibilske, L., and W. Clapham. 1986. Evaluation of two papermill sludges and bioash for plant production. Dep. Plant and Soil Sci., Maine Agric. Exp. Stn. and USDA, Orono, ME.
52. Boyer, J.A., and I.J. Fernandez. 1995. Biogeochemical effects of papermill ash in a western Maine forest clearcut. Abstracts of Presentations p.32-33.. Symposium on land application of wood-fired and combination boiler ashes. National Council for Air and Stream Improvement. Gainesville, FL
53. Smith, C.T., and W.B. Bowden. 1995. Soil chemical changes after forest utilization of wood ash. Abstracts of Presentations p. 28-31. Symposium on land application of wood-fired and combination boiler ashes. National Council for Air and Stream Improvement. Gainesville, FL
54. Sahm, J.M., E.H. White, and L.P. Abrahamson. 1995. Wood ash applications in willow bioenergy systems--an example. Abstracts of Presentations p.24-25.. Symposium on land application of wood-fired and combination boiler ashes. National Council for Air and Stream Improvement. Gainesville, FL
55. Williams, T.M., B.R. Smith, and C.A. Hollis. 1995. Soil and water chemistry following boiler bottom ash application. Abstracts of Presentations p. 37-38.. Symposium on land application of wood-fired and combination boiler ashes. National Council for Air and Stream Improvement. Gainesville, FL.
56. Sweeney, J.R., and P.D. Jones. 1995. Heavy metal whole body burdens of fauna inhabiting mill ash recycling sites in southeastern Coastal Plain forests. Abstracts of Presentations p. 39-40.. Symposium on land application of wood-fired and combination boiler ashes. National Council for Air and Stream Improvement. Gainesville, FL
57. Krejsl, J. 1995a. Greenhouse evaluation of the beneficial use of boiler ash on oats and beans. Abstracts of Presentations p. 21.. Symposium on land application of wood-fired and combination boiler ashes. National Council for Air and Stream Improvement. Gainesville, FL
58. Krejsl, J. 1995b. The effect of wood ash on wheat and green pea yields. Abstracts of Presentations p. 22. Symposium on land application of wood-fired and combination boiler ashes. National Council for Air and Stream Improvement. Gainesville, FL
59. Weber, A. M. Karsisto, R. Leppanen, V. Sundman, and J. Skujins. 1985. Microbial activities in a Histosol: Effects of wood ash and NPK fertilizers. Soil Biol. Biochem. 17:291-296.

Chapter 14

Agricultural and Industrial Uses of By-Product Gypsums

W. P. Miller and M. E. Sumner

Department of Crop and Soil Sciences, University of Georgia, Athens, GA 30602

By-product gypsums are produced by the phosphate refining and electric power industries, as well as from other sources. The large amounts produced annually cannot be accomodated in traditional uses, and new applications in the building, cement, road-building, and chemical processing industries are being examined. Agricultural uses on acid soils also look promising as a way to boost productivity of acid, dispersive soils of the Southeastern U.S. The radionuclide content of phosphate-derived gypsums is a road-block to such uses, unless further research demonstrates more limited environmental impacts of uses of this material.

Gypsum ($CaSO_4 \cdot 2H_2O$) is a widely occurring geologic mineral that is found on nearly every continent. Extensive deposits occur world-wide as evaporite beds, formed over geologic time by precipitation from brines concentrated from evaporating seawater. It is one of the first minerals to precipitate from such solutions due to its limited solubility (about 2.5 g per L) in water.

In the modern world gypsum is used in a number of industrial and agricultural applications. Gypsum mined from geologic deposits is used in manufacture of wallboard, plaster and cement products, as an additive in a range of industrial formulations, and as a fertilizer and soil conditioner in agriculture.

Gypsum is also produced as a by-product by several important manufacturing processes. In addition to wastes generated from wallboard manufacture and use, processing of mineral phosphate ores and other industrial processes involving neutralization of sulfuric acid solutions with lime are responsible for most of the by-product gypsum produced in the U.S. and world-wide. Chemical removal of sulfur dioxides from coal-fired electric utility stack gases is a growing source of by-product gypsum. Currently most of these "waste" materials are not recycled or used in other applications, but are either stockpiled on-site or landfilled.

The increased environmental awareness of the past several decades has placed a greater emphasis on waste recycling, reduction of landfill volumes, and reduced contaminant emissions, and this has had an impact on producers of by-product gypsum materials. Considerable efforts are being made to substitute by-product gypusms for mined gypsum in established markets such as cement additives, wallboard manufacture, and traditional agricultural uses. Problems such as excess soluble salt content, poor handling characteristics, and trace element contamination in by-product gypsums have hindered these efforts to some extent, although continuing research is proposing solutions. However, the greater issue is the large supply of by-product material relative to current demand. New uses of by-product materials that have economic benefit over alternative technologies, and are proven to be environmentally sound, need to be developed to provide an expanded market for the large and growing annual supply of these materials.

By-product Gypsum Production and Potential Uses

Sources and Production of By-product Gypsum. Historically, the most significant producer of by-product gypsum has been the phosphate fertilizer industry, concentrated in Florida and North Carolina in the U.S. Sedimentary phosphatic rock formations are surface-mined in these areas, and the ground rock containing the mineral fluorapatite is treated with sulfuric acid to produce phosphoric acid; gypsum is a product of the reaction, as follows:

$$Ca_{10}(PO_4)_6F_2 + 10H_2SO_4 + 20H_2O \rightarrow 6H_3PO_4 + 10CaSO_4 \cdot 2H_2O + 2HF$$

The stoichiometry and masses involved in the reaction result in production of hydrated $CaSO_4$ in amounts equal to 1.0-1.5 times the mass of flourapatite reacted. Thus, in 1995 annual production of approximately 40 million t of phosphogypsum (as it is referred to) resulted from the processing of 45 million t of raw rock phosphate mined. In central Florida nearly 1 billion t of this material is stockpiled in stacks that cover hundreds of acres and range up to 100 feet tall; total inventory in the U.S. may be as high as 7 billion t (*1*). World-wide production is estimated at 150 million t annually (*2*).

Phosphogypsum (PG) is typically 85-95% gypsum; the major impurity is residual quartz sand (SiO_2) carried through the process stream, which may range from 3-17% (*3*). Residual P and F may make up 0.1-1.0% of PG as well. The material is typically acidic (pH 4.5-5.5) due to free acid remaining in the pore fluids, but it is not strongly buffered; soluble salts are not present. Gypsum crystals vary in size and morphology due to process conditions, but are often small (50-200 μm) needles or plates.

Another major source of by-product gypsum is removal of SO_2 from exhaust gases of coal-fired power plants, prompted by the 1990 amendments to the Clean Air Act aimed at reducing emissions that contribute to potential acidification of rainfall. Numerous desulfurization technologies have been developed over the past decade, primarily in Europe and Japan where stringent air quality standards have been in place since the 1970's. Both "dry" scrubbers (fluidized bed systems, where lime is injected into the boiler or stack directly) and "wet" systems (where the stack gas is directed through a lime slurry) are in service in the U.S. and abroad, and are effective at SO_2 removal. Most of these systems do not oxidize the sulfur, but rather produce calcium sulfite ($CaSO_3$) after solution of the SO_2:

$$SO_2(g) + H_2O \rightarrow H_2SO_3$$
$$H_2SO_3 + CaCO_3 \rightarrow CaSO_3 + CO_2(g) + H_2O$$

Such by-products may also contain considerable ash and unreacted lime, as well as some gypsum. The $CaSO_3$ itself is not a useful product, and nearly all this material (15 million t in 1984) is landfilled (1).

Newer designs in wet scrubbing technology use a "forced oxidation" process, pumping air through the slurry or generating surface foams to enhance the oxidation of sulfite to sulfate:

$$SO_3^{2-} + \tfrac{1}{2} O_2(g) \rightarrow SO_4^{2-}$$

which then precipitates and hydrates as $CaSO_4 \cdot 2H_2O$. An important advantage of this technology is that it produces a potentially valuable by-product that can be marketed by the utility.

As of 1989, only about 16 forced-air scrubbers were producing flue-gas desulfurization gypsum (FDG) in the U.S., with an annual output of about 0.7 million t (1). Based on regulatory pressures, projections of up to 30 million t per year have been made in order to reach mandated reductions in SO_2 emissions. However, the high capital and operating costs of scrubbers have caused utilities to look to alternative strategies, one being the use of low-sulfur coals. Several Eastern U.S. utilities are beginning to substitute coals of the Powder River Basin of Montana (0.5-1.0% S) for higher sulfur Kentucky-West Virginia coals (2-3% S), and coupled with favorable rail charges, have been able to reduce S emissions to regulatory levels. Future FDG production obviously depends on the balance between relative costs of various coal types, transportation costs, capital costs of scrubber construction, and future regulatory changes. In addition, the market value of the FDG produced is a consideration; if the by-product can be reliably marketed at a reasonable price, scrubber operation becomes a more attractive option. In Europe and Japan, low-S coals are less available, and FDG production is appreciable, with much of it being recycled to the wallboard and plaster industries.

Flue-gas desulfurization gypsum is nearly pure $CaSO_4 \cdot 2H_2O$; some FDG may have 0.2-0.5% Mg derived from the use of more dolomitic lime in the scrubber. Depending on the installation, some fly ash may enter the scrubber past the electrostatic precipitator (EP), but this is typically < 1% of the by-product mass. An experimental forced-air scrubber near Atlanta was operated with the EP shut down, thereby using the scrubber to remove both SO_2 and fly ash; this saves the utility the considerable cost of running the EP. This system operated satisfactorily, but did cause increased wear and abrasion inside the fiberglass scrubber vessel, and obviously produced a mixed fly ash-FDG product (about 50% each) that is probably less marketable for many uses (4). Typical FDG materials have a near-neutral to slightly acidic pH, as lime feed to the scrubber is carefully controlled, but some by-product has appreciable soluble salt content in the pore fluids, due either to soluble Mg salts (from the lime used) or from salts accumulated in recycled process water used in closed-loop systems. Crystal morphologies of FDG materials vary widely based on scrubber design and operating conditions, but are typically 40-400 μm needles (4); smaller crystal size enhances solubility, but makes the material more difficult to handle and is undesirable for use in wallboard manufacture.

Other sources of by-product gypsums include industries that neutralize waste or spent sulfuric acid solutions using lime; these include electro-plating, pigment and mineral processing (especially titanium ores), and metal etching industries. Amounts of material produced are unknown, and locations of production and stockpiles are widely scattered. These materials are likely to contain metal contaminants co-precipitated during lime addition, depending on the particular industry; these may be more-or-less inert phases (Fe, Al oxides) making up an appreciable portion of the product, or trace levels of more hazardous metals (Cd, Cr) that may constitute an environmental concern.

Wallboard waste may be listed as a final source of "by-product" material. Many wallboard manufacturing plants are unable to re-process out-of-spec product, resulting in large piles of waste near the plant. Several plants in the Southeast are looking at grinding these materials for agricultural use. In new home construction, approximately 0.5 kg wallboard waste (cut-off) is generated per square foot of floor space, resulting in about 1 million t of such waste produced annually in the U.S. (Yost, P., Natl. Assoc. Homebuilders, Washington DC). This material is essentially mined gypsum, with 2-3% backing paper and minor additives used in the manufacturing process. On-site grinding and application to the often disturbed soil on the building lot may be a feasible alternative to landfilling for this material.

Markets and Demand for By-product Gypsum. Given that much of the by-product gypsum produced is relatively pure, its use will depend on economic variables, as long as other critical properties are satisfactory and it does not contain environmental contaminants at significant levels (which will be addressed below). A summary of current gypsum supply and demand is given in Table I. It shows that most of the current supply is mined domestically, with some imports (especially on the East coast) from Canada; only small amounts of by-product are currently in the marketplace. About three-quarters of this production is used in wallboard and plaster manufacture, 13% as a cement additive (to retard setting), and 10% in agriculture.

Table I. U.S. Supply and Demand for Gypsum, 1995

SUPPLY:	million t/yr
Mined gypsum (domestic)	17
Mined gypsum (imported, Canada & Mexico)	8
By-product gypsum	1
Total:	26
DEMAND:	
Wallboard and plaster manufacture	20
Cement additive	3.5
Soil amendment (agriculture)	2.5
Total:	26

Source: *(20)*

Thus, the total annual gypsum market is considerably less than the yearly PG production in the Southeastern U.S. (about 40 million t). Including expanded FDG production in the future, by-product gypsum could clearly swamp the market and drive prices to a low level. Florida PG has not been extensively marketed to date due to ongoing debate with EPA over radioactivity levels in the material (discussed below); however, the industry is interested in reducing stockpiles. Utilities are actively seeking uses for FDG, some even cooperating directly with wallboard industries to co-locate facilities.

A possible solution is to develop new markets for by-product gypsum in order to create increased demand for use of this material. Possibilities exist in the engineering (structural/chemical recovery/construction) arena, in the form of substitution of gypsum for other materials, and in agriculture and horticulture, where research indicates that gypsum may be of much wider benefit on a range of crops and soils than previously thought.

Engineering Markets for By-product Gypsum

Traditional Markets. Wallboard and plaster manufacture, along with use as a retarder in cement, are the current primary uses of gypsum. While little by-product gypsum has entered this market in the U.S., efforts are being made following the lead of Europe and Japan, where substantial amounts of both PG and FDG are re-processed to these markets. Problems with some by-product gypsums include fine particle size (which modifies the hydration/dehydration processes during calcination), excessive soluble salts, and offcolor due to pigmented impurities. The later detracts from the necessary white color of plaster products. Most wallboard manufacturers seem to be able to adapt their process to account for differences in crystallite size, and thereby utilize the by-products after some experimentation. Other problems may require washing of the gypsum to remove porefluid salts, or further processing to remove color; alternatively, small changes made inplant may be able to minimize such contaminants.

If by-product gypsum meets the specifications of the industries, it will compete strongly with mined gypsum, as it has overseas. In Japan, where all gypsum is imported, about 2 million t/yr of domestic by-product is used internally in these markets. Transport costs will be a major consideration in moving large stockpiles of PG in Florida to market.

Structural Components. The phosphate industry has a long history of exploring new markets for their by-products, particularly in Europe where environmental constraints have been in place for a longer time. In this country the Florida Institute of Phosphate Research (FIPR) has sponsored research for a number of years on the use of PG in novel cement formulations, structural block and panel construction, aggregate manufacture, and as compacted fill. Reports on the progress of this research are given in various FIPR publications, including symposia they have sponsored.

The current opinion appears to be that large additions (up to 50%) of PG directly to portland cement mixtures tends to reduce strength and stability (corrosion-resistance) of the mix, although in fabrication of pre-cast concrete beams and panels using PG-concrete-sand mixes with compactive pressure, good strength was obtained. Under very high compaction, mixes of 90% PG/10% cement gave very high strength even when wet

(5). Other research is examining mixtures of PG with fly ash in the production of portland-type cements, where the gypsum acts as a source of Ca in the formation of calcium silicate hydrate and other pozzolanic, cementitious compounds. Strengths of such cements were adequate, and in one study where PG was calcined at 950°C, exceeded that of traditional portland cement (6). Specialized products such as corrosion-resistant blocks can be made of PG-sand-elemental S mixtures, compacted and heat-cured, for use environments requiring salt and acid resistance. One particular problem of many PG-based building materials is that they show poor water resistance, wetting readily and slowly dissolving; this problem is under continued investigation. Overall, a wide range of potential products has been identified, largely aimed at substituting gypsum for the more expensive portland cement, but are yet to be accepted widely in the building industry.

Road Construction. Related to the above applications, it has been observed that mixtures of gypsum and soil can be compacted into strong, durable sub-grades for road construction. In Florida, where clay must be incorporated into sandy soils to create an adequate base, work has shown that a mix of 30-50% PG in the subgrade, compacted according to standard practices, gives a suitably strong base for asphalt roads at reduced cost (7). The same may be true of very clayey soils, where gypsum additions to sub-bases may decrease shrink-swell potential and prolong pavement life.

Research on "compaction concrete" has suggested that mixtures of 80-90% by-product gypsum + 10-20% cement may be compacted under high pressures to produce very strong subgrade or pavement surfaces (8). However, strength is significantly reduced when the material is water-saturated, potentially reducing the range of applications.

Chemical Recovery. Phosphate producers would be very happy to be able to recover S in the form of sulfuric acid from by-product PG if economically possible. A number of aproaches have been taken that appear promising in this regard. In one method, a high-temperature, reducing environment is used to reduce SO_4^{-2} to H_2S, which can then be readily converted to H_2SO_4. A pilot-scale fluidized bed system has been demonstrated (9) to do this very efficiently, using pulzerized coal as a fuel in the bed. Up to 90% recovery of the S was obtained, with the other product being relatively pure CaO, which is also a useful by-product. Using a different approach, a South African phosphate firm has an operational coal-fired rotary kiln that produces SO_2 and CaO under mildly reducing conditions at 900°C (10):

$$4CaSO_4 + 2C \rightarrow 4\ CaO\ +\ 2CO_2\ + 4SO_2 \quad \Delta H = 62.3\ \text{Kcal/mole } CaSO_4$$

The kiln is also fed with a clay soil, which reacts with the CaO at the high temperature to produce a cement clinker, which is then ground and blended on-site to produce bagged cement. This operation is profitable under their local conditions, where both external cement and acid prices are relatively high. Other related experimental technologies generate CaS in a similar fashion, which is then dissolved in water after conversion to $Ca(HS)_2$, and the H_2S finally released by precipitation of $CaCO_3$ with CO_2 (11).

In a more "low-tech" vein, it has been suggested that sulfate-reducing bacteria be employed to use gypsum as a substrate to produce SO_2 (12). The only other requirement would be an organic source, which might be municipal sewage sludge, animal manure, etc. The organisms are quite efficient at sulfate reduction, although maintenance of optimum growing conditions, as well as purifying and concentration the SO_2 stream, would be problematic. No feasibility studies have been carried out on a large-scale.

Agricultural Markets for By-product Gypsum

Traditional Markets. These are lmited largely to application to peanuts in the Southeastern U.S. at pegging to supply calcium, and to sodic soils in the Western U.S. to ameliorate soils high in sodium on their exchange complex. Some gypsum is used as a S fertilizer on very sandy soils, and as a filler in blended fertilizers. Of the total 2.5 million t/y applied, about 0.5-0.75 million t are used on peanuts in the southern U.S. In the past much of this has been PG from northern Florida, which has a considerable price advantage over mined gypsum. However, the USEPA ruling on radionuclides in PG has restricting this use, although some PG has low enough radium levels for continued use.

Use on Acidic Subsoils. The use of gypsum on peanuts is specific to that plant in that Ca is poorly translocated to the developing pod, and must be absorbed from the soil directly by the developing fruit. Similarly, however, root tips of all plants require Ca at the growing point, and cannot obtain it from other plant parts via translocation. Thus in subsoils low in Ca (present in the soil solution pore waters or as exchangeable ions on the soil exchange sites), roots will not develop adequately, thereby limiting the soil volume able to be exploited by the plant. Old, highly weathered soils of the Soil Taxonomic orders Ultisols and Oxisols are most likely to exhibit this problem, as soluble Ca sources have weathered and leached away. Exchangeable acids and aluminum (Al^{+3}) then accumulate, causing further root injury due to direct toxicity of particularly Al to growing roots. The result is that on many Ultisols roots are restricted to the topsoil, and consequently suffer repeated drought stress during hot, dry spells.

The use of gypsum, and particularly by-product gypsums, to ameliorate this acid subsoil infertility syndrome, has been reviewed by Shainberg et al. (*3*) and by Sumner (*13*) for soils around the world. Lime is totally ineffective in supplying Ca to subsoils due to its insolubility in limed topsoils; gypsum, however, continually dissolves to move Ca^{+2} with percolating water into subsoil horizons. This process may take some time, depending on the water balance in a particular soil region, and in the Southeastern U.S. requires one or two winter seasons to allow sufficient rainwater to percolate to the Bt horizon. After 5 years of leaching on a Georgia Ultisol receiving 10 t/ha of PG, exchangeable Ca levels had increased throughout the soil profile to levels signficantly above control soil (Figure 1).

Gypsum also has an effect on exchangeable Al levels in subsoil horizons, consistently reducing amounts present after gypsum application (Figure 1). The mechanism of this reduction is not definitely known, but has been ascribed to the role of specific adsorption of sulfate on Fe and Al oxides in generating either displaced OH^- or higher surface charge for Al precipitation or adsorption, or to direct precipitation of Al as hydroxysulfate solid phases such as alunite ($AlOHSO_4$) (*13*).

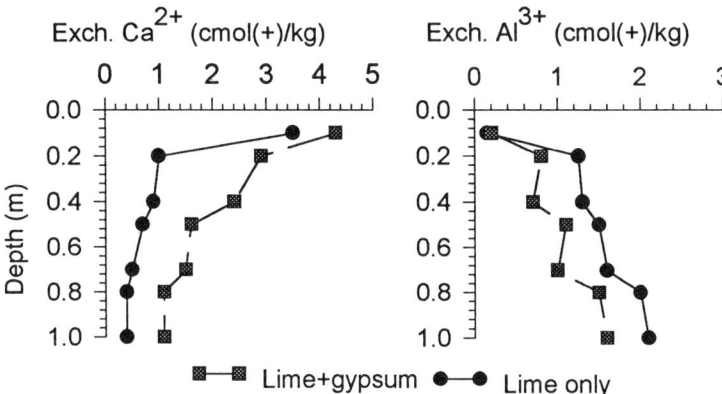

Figure 1. Exchangeable Ca and Al levels in Cecil soil over time after amendment with 5 mt/ha PG (after Shainberg et al., [*3*]).

The net effect of these subsoil chemical changes is enhanced plant vigor and yield on many Ultisols amended with gypsum. Excavated alfalfa root systems from Cecil Ultisol profiles in Georgia have shown that few roots penetrate the Bt horizon, largely due to the unfavorable chemical properties described above; gypsum-amended profiles, however, show considerable rooting to a depth of 1 m several years after amendment. Measurements of root density of maize and apple indicate up to 5 times more roots at 0.75 m depth with gypsum compared to controls (*3*). Water extraction measured from subsoils on gypsum-amended soils also indicate that much more water is removed from these depths than on unamended soils.

Greater water availability related to deeper subsoil rooting is likely to be the major mechanism for yield increases often observed with gypsum amendment. Yield differentials with gypsum are most likely to occur on soils with pronounced acid subsoils (pH<4.7), with normally deep-rooted crops (especially perennial legumes), and during years where rainfall patterns produce periods of drought stress. These conditions are not unusual in the Southeastern U.S. and other areas with semi-tropical soils and weather conditions. Shainberg et al. (*3*) have summarized a range of yield data from the U.S., Africa, Central America, and Australia, which indicate the kinds of responses obtained.

Most of the field studies to date have used mined gypsum or PG; the latter is nearly pure (90%+ gypsum) and rapidly soluble in water. Recent studies with a FDG produced in Georgia have shown that this by-product gives similar results (*4*). At four locations ranging from the Georgia-Tennessee mountains to the Georgia-Florida area, alfalfa yields for the year following application of FDG or a mixed (1:1) FDG-fly ash (FA) material showed a positive response to amendment at 3 of the 4 locations tested (Table II). Two of the sites (unlimed Piedmont and Coastal Plain) were poorly suited for alflafa, yet gave strong yield increases with FDG; the Mountain site gave excellent yields on the control soil, and in excess of 1 t/ha increases with the higher rate (20 mt/ha) of FDG. Only the limed Piedmont site did not respond statistically, although yield increases of 1 t/ha were also observed.

Yield effects of FDG on annual row crops such as corn (maize), soybean, and sorghum are often less pronounced than for perennials, perhaps in part due to their lesser

Table II. Effect of FDG and FDG-fly ash mixtures on alfalfa yield at four locations one year after application

Treatment	Rate (mt/ha)	Alfalfa yield (mt/ha/yr)			
		Piedmont (limed site)	Piedmont (unlimed site)	Coastal Plain	Mountains
Control		13.43	0.14	2.36	11.22
FDG	5	13.67	1.15	2.43	12.33
	10	14.02	2.73	2.82	12.96
	20	14.63	5.84	3.19	13.43
FDG+FA	5	14.32	0.55	2.47	13.07
	10	14.87	0.61	2.77	12.96
	20	14.30	2.51	3.06	13.85
LSD (0.05)		NS	0.98	0.63	1.40

tendency to root deeply into subsoils. Maize, with its high water requirement, often does respond to gypsum (*3, 13*), although increases of 10-20% above controls are often not statistically significant in field experiments (*4*).

The questions of rates and economics of gypsum use on field crops other than peanuts are not simple to answer; most trials show 10 t/ha is a reasonable one-time application rate, which for forages appears to suffice for at least 5, and maybe up to 10 years. This is due to the more-or-less permanent effect of the leached Ca on subsoil exchangeable cations, although re-treatment will clearly be necessary eventually. The conomics might be evaluated by assuming the cost of a 10 t/ha application were roughly $300 ($30/t); if that sum were invested at 7.5% interest paid quarterly, at the end of 5 y it would be worth $430, and after 10 y it would be worth $630 (Table III). Thus, gypsum application should give an average annual yield increase over these time periods at least equal to these annual values ($86 and $63). From Table III it is evident that for a relatively low yielding, low-value crop like corn, large yield increases (30-40%) are required to "pay for" the initial application; for higher-value crops such as soybean, only 20-25% yield increases are needed. For alfalfa, a high-yielding but lower value crop, only very modest increases are required to justify gypsum use. Thus, probably only quite acidic soils should be considered for gypsum application to row crops in order to assure strong yield responses, while the small yield increases needed for alfalfa probably recommend the use of gypsum on most soils growing this crop.

Application to Dispersive Soils. Soils containing as little as 5-10% Na^+ on their exchange complex have been long recognized as dispersive, and display a range of physical problems, including poor structural stability, low infiltration rate under rainfall, poor aeration and root development, and consequent high soil erosion rates when found

Table III. Economics of gypsum use on crops, showing yield increase needed with gypsum application to equal annual return from investment of $300 at 7.5% interest rate

Flat Return Rate: Term (y)	Future value	Ave. annual value
5	$430	$86
10	$630	$63

Gypsum Application:			Average annual yield increase needed, for period--	
Crop	Base yield	Value	5 y, t (% incr.)	10 y, t (% incr.)
corn (grain)	2.2 t	$100/t	0.86 (40%)	0.63 (29%)
soybean	1.0 t	$350/t	0.25 (25%)	0.18 (18%)
alfalfa (hay)	6.0 t	$100/t	0.86 (14%)	0.63 (10%)

on sloping topography (*14*). This dispersion refers to the tendency of soil colloids (< 2 μm diameter) to develop strong repulsive forces due to the inability of Na to adequately shield surface charge on the colloids; lower levels of soluble solution cations similarly allow electric fields to extend further into the solution and cause particle repulsion. This phenomenon also occurs in soils with very low Na levels (<2%) if sufficient charge develops, and soluble cations are dilute in the soil solution. Such conditions are common in topsoils of highly weathered soils, where leaching has removed most of the soluble cations, and liming has increased variable charge (*3*).

Gypsum has been used for remediation of sodic soils for some time, acting largely to replace exchangeable Na with Ca; it also has a marked effect on non-sodic, dispersive soils by supply electrolyte to the soil solution at a relatively controlled rate based on its solubility in water (15 *mM*). Studies of the Ca level needed to flocculate colloids of serveral Georgia soils showed that 2-5 *mM* were adequate (*15*). When gypsum is present at the soil surface during rainfall, this flocculation allows larger water transmission pores to remain open, and infiltration rates to remain high; runoff and attendant erosion is thereby lessened, and sediment delivery to surface water reduced.

Studies in the southern U.S., Israel, and Australia, summarized by Shainberg et al. (*3*), indicate that gypsum applied on the soil surface prior to rainfall signficantly increases infiltration rate, total infiltrated water, and reduces soil erosion. These effects are typically on the order of 2- to 3-fold, but vary somewhat with soil properties in an unpredictable way. After most of the gypsum has dissolved, the electrolyte concentration decreases and the effect largely disappears, often after about 100 mm of rainfall. The cumulative effect of raindrop impact overcomes any remaining flocculating effect, and the surface "seals", with final infiltration rates often dropping to very low levels (2-5 mm/h).

In the Western U.S., gypsum is sometimes injected into feed lines for overhead sprinklers on very dispersive soils to encourage water infiltration. In Israel, some field-scale applications have been made, but it has been difficult to economically justify such applications even where surface sealing is a major problem. By-product gypsums are often very fine particle size, dissolving rapidly at the soil surface, and so are good candidates for such uses; but even at low cost per t applied, yield responses have not been consistently documented. The value of erosion and sediment control may be considerable, however, given local and Federal requirements to protect surface water from sediments washed from cropland and construction sites. Some use has been suggested on roadsides and construction sites, where acid, dispersive soils might be treated at low relative cost, but no trials have demonstrated the potential benefit of such practices.

Use on Hard-setting or Dense Soils. Hard-setting is a term used in Australia to define topsoils that slump and disperse readily when wetted, then dry to a brick-hard consistency. Sumner (*13*) notes that similar behavior occurs to some degree in other soils around the world, although not to the degree of the Australian case. Gypsum readily ameliorates these soils, greatly improving their stucture and water intake rates, and largely preventing the hardening upon drying. Some years ago gypsum was recommended for improving drainage and subsoil properties of Coastal Plain soils of Virginia (*16*), perhaps by a similar flocculation mechanism.

Many highly weathered soils contain clayey argillic (Bt) horizons with high bulk densities that have been suggested to impede root growth. Often poor chemical conditions combine with this physical effect to limit root extension. Sumner (*13*) reviewed several studies showing that both mined gypsum and PG applications caused a lowering of penetration resistance (pressure required to drive a probe into the subsoil) to depths of up to 0.8 m on several Piedmont and Coastal Plain soils. The reductions (up to 50%) occurred on both cropped and fallow plots, and on cropped areas were associated with greater rooting density as measured by excavation. It thus appears that gypsum ameliorates both physical and chemical properties of highly weathered subsoils, thereby resulting in the improved plant performance noted earlier. The exact mechanism by which gypsum is able to effect lowered subsoil strength is not clear, but must be related to its aggregating or flocculating effects on structural elements in the subsoil.

Horticultural Uses. The horticultural industry, including landscape installation and the propagation and production of ornamental as well as vegetable plants and fruits, is a large and economically important agricultural sector in many parts of the country. Many horticultural plants grown on a variety of acid soils have the potential to benefit from gypsum applications, ranging from fruit crops (peaches, apples) to field vegetables to ornamentals installed on residential lots. The high value of horticultural crops and installations suggests that even small improvements in plant performance would easily justify gypsum use.

Both apples and coffee were shown to respond favorably to gypsum application in Brazil (*3*). Unfortunately, there is little further research to demonstate beneficial effects in this country. Anecdotal information suggests that lawn trees (*Acer rubrum*) give superior early growth when treated with gypsum (Smalley, T., University of Georgia Horticulture Dept., Athens, GA); a range of northern species unaccustomed to acid soils should be good candidates for such trials. Certainly more research is due in this area, where a large potential market for by-product gypsum might develop if the expected responses in growth can be documented.

Environmental Issues in By-product Gypsum Use

Obviously no wide-spread use of by-product gypsums will be made unless such uses pose no threat to the public health, in terms of persons exposed to any contaminants that may be present in these by-products.

Miller (*17*) and Alcordo and Rechcigl (*1*) have recently reviewed environmental considerations in the agricultural use of by-product gypsums. For the FDG and acid neutralization materials, some trace metal contamination is possible during their production, and may be locally significant for some materials. The major problem with FDG is that some fly ash may be mixed with the gypsum, either through improper operation of the elecrostatic precipitator, or intentionally if the EP is shut down. Fly ash contains a range of elements of potential concern, particularly As. In the absence of fly ash, FDG typically contains trace contaminants below the level of concern. Only Hg has been cited as a potential question mark, as it readily volatilizes in the furnace of a power plant; few Hg measurements have been made on FDG, and its status as a contaminant is not clear. Such contamination may have an impact on land application only if levels are high enough to deliver significant contaminant to the soil using a typical application rate.

For PG, the issue of radionuclide levels has been extensively treated by Alcordo and Rechcigl (*1*), and will not be discussed in depth here. Phosphogypsum does contain measurable levels of radio-isotopes in the ^{238}U decay series, originally present in the sedimentary phosphate rock ore and carried into the PG during processing (although if P fertilizer is produced, some radioactivity also is present in that product as well). The production of gaseous ^{222}Rn, which is highly toxic, in this decay series makes the prospect of use of PG-derived building products for indoor use problematic. Measurements of Ra release from a test structure made of PG, however, gave acceptable air quality, which was further improved by simply painting the gypsum board with ordinary house paint (*18*). Outdoor use in cement products or replacements, or even in roadbeds, may also be challenged if it is judged that maximum future exposure exceeds tolerances. In the chemical recovery area, radionuclides may be an issue depending on where they partition in the chemical process used; ^{226}Ra, the most common radionuclide, is chemically similar to Ca, and will likely concentrate in the solid Ca phase, perhaps resulting in that product being less marketable.

In agricultural use, USEPA has restricted all use of PG with ^{226}Ra, the most common radionuclide, present at >10 pCi/g. This level, arrived at using risk assessment modeling after a number of years of consideration, is currently being challenged as too conservative; most of the Florida PG stockpiles exceed this level and cannot be moved for any use at present. The risk models used in this determination consider a variety of exposure pathways including direct gamma exposure, ingestion of contaminated food and water from the treated soil, and inhalation of radionuclides from the air, using as the most exposed individual one who lives full-time on an amended site and eats and drinks food and water derived solely from the site. A large number of assumptions are made for transport coefficients, uptake and leaching coefficients, and exposure effects of the different pathways, all of which may be questioned based on our incomplete understanding of radionuclide behavior. Most of the available direct measurements made on water quality, plant uptake, and atmospheric levels of radionuclides, particularly ^{226}Ra but including other daughter products, do not show elevated levels associated with even large applications of PG (*1, 17*). It is clear that this issue must be resolved prior to any wide-spread use of PG; continued studies are underway in parallel with the appeal of the EPA ruling.

Radon removal from PG has been proposed as a solution; the Ra is apparently present in the PG in a mixed Ba-Sr sulfate phase which has a very small (<10 µm) particle size. Moisset (*19*) has suggested an initial clean-up of the PG by hydrocycloning to remove the fine fraction, which reduces activity from 38 pCi/g to about 10 pCi/g. By adding Ba and Sr salts and heating to allow the Ra to migrate to recrystallizing Ba and Sr sulfate centers, further reductions may be possible. The cost of such operations would be considerable.

Conclusions

Between 40 and 50 million t of by-product gypsum is produced annually in the U.S., with perhaps three times this much made overseas. Its disposal or utilization is a major concern for the phosphate industry, which produces most of this amount, and the electric power generating industry, which is making increasing amounts as it finds ways to remove SO_2 from its coal furnace emissions. The domestic gypsum market, currently supplied by mined gypsum, is only about 25 million t/y; therefore, new uses of this material must be found, even if by-product gypsum can penetrate the current market.

On-going research has identified several new areas into which by-product gypsum may move, including new building products, new cement formulations, road bases, and recovery of sulfuric acid. Of these, acid recovery seems the most advanced technically and with the most economic incentive for the industry. Re-conversion of gypsum to H_2SO_4 would, however, probably only affect current gypsum production, as this would likely satisfy the industry need for new process acid; nearly 1 billion t of stockpiled PG remains.

Agricultural uses have the potential to use a considerable amount of this stockpile; if something approaching 25% of the 40 million acres of cropland in the nine Southeastern U.S. states were treated with an average of 0.5 t/y, 5 million t/y could be utilized. Horticultural uses may contribute some additional demand. The amounts of by-product material stockpiled and produced annually obviously dwarf any one use; combinations of new markets will need to be developed, and will be fostered by the availability of this resource, in order to begin to utilize available material.

Literature Cited

(1) Alcordo, I.S.; Rechcigl, J.E. In *Soil Amendments and Environmental Quality*; Rechcigl, J.E., Ed.; Lewis Publishing: Boca Raton, FL, 1995; pp. 365-425.
(2) Ferguson, F. In *Proc. Second Internl. Symp. on Phosphogypsum*; Florida Institute of Phosphate Research: Bartow, FL, 1988, vol. I; pp. 117-131.
(3) Shainberg, I.; Sumner, M.E.; Miller, W.P.; Farina, M.P.W.; Pavan, M.A.; Fey, M.V. *Adv. in Soil Sci.* **1989**, 9, 1-111.
(4) Miller, W.P.; Sumner, M.E.; Radcliffe, D.E.; Hoveland, C.S. *Final Report: Use of by-product flue gas desulfurization gypsum as an ameliorant for Southeastern soils;* US Dept. of Energy: Washington DC., 1997.

(5) Lin, K.T.; Ouyang, C.; Chang.W.F. In *Proc. Second Internl. Symp. on Phosphogypsum*; Florida Institute of Phosphate Research, Bartow, FL, 1988, vol. II; pp. 301-316.
(6) Bhanumathidas, N.; Kalidas, N. In *Proc. Second Internl. Symp. on Phosphogypsum*; Florida Instititute of Phosphate Research: Bartow, FL, 1988, vol I; pp. 277-295.
(7) Kenley, W.C.; Chang W.F. In *Proc. Second Internl. Symp. on Phosphogypsum;* Florida Institute of Phosphate Research, Bartow, FL, 1988, vol. II; pp. 353-360.
(8) Ghafoori, N.; Chang, W.F. In *Proc. Second Internl. Symp. on Phosphogypsum*; Florida Institute of Phosphate Research: Bartow, FL, 1988, vol. II; pp. 215-239.
(9) Wheelock, T.D; Fan, C.W.; Floy, K.R. In *Proc. Second Internl. Symp. on Phosphogypsum;* Florida Institute of Phosphate Research: Bartow, FL, 1988, vol. II; pp. 3-25
(10) Clur, D.A. In *Proc. Second Internl. Symp. on Phosphogypsum*; Florida Institute of Phosphate Research: Bartow, FL, 1988, vol. II; pp. 39-51.
(11) Sliger, A.G. In *Proc. Second Internl. Symp. on Phosphogypsum*; Florida Institute of Phosphate Research: Bartow, FL, 1988, vol. II; pp. 83-108.
(12) Taylor, B.F. In *Proc. Second Internl. Symp. on Phosphogypsum;* Florida Institute of Phosphate Research: Bartow, FL, 1988, vol. II; pp. 159-166.
(13) Sumner, M.E. *Adv. Agron.* **1993,** 51, 1-32.
(14) Shainberg, I.; Letey, J. *Hilgardia,* **1984,** 52, 1-57.
(15) Miller, W.P.; Frenkel, H.; Newman. K.D. *Soil Sci. Soc Am. J.* **1990,** 54, 346-351.
(16) Rinehart, J.C.; Blake, G.R.; Tedrow, J.; Bear, F. *Gypsum for improving drainage of wet soils;* N. Jersey Agric. Exp. Sta.: Bull 772, 1955, 15p.
(17) Miller, W.P. In *Agricultural Utilization of Urban and Industrial By-products*; Karlen, D., Ed. American Society of Agronomy: Madison, WI, 1995, pp. 183-208.
(18) Trefler, M.; Wu, X.; Medora, R. In *Proc. Second Internl. Symp. on Phosphogypsum;* Florida Institute of Phosphate Research: Bartow, FL, 1988, vol. I; p. 25-25.
(19) Moisset, J. In *Proc. Second Internl. Symp. on Phosphogypsum*; Florida Institute of Phosphate Research: Bartow, FL, 1988, vol. I; pp. 303-319.
(20) US Geologic Survey. Mineral data sheet: gypsum. WWW address: http://minerals.er.usgs.gov.minerals/pubs/mcs; 1995.

Chapter 15

Ammonium Sulfate Fertilizer as By-Product in Flue Gas Desulfurization: The Dakota Gasification Company Experience

Daniel L. Wallach

Dakota Gasification Company, P.O. Box 1149, Beulah, ND 58523–1149

Introduction

The Great Plains Synfuels Plant, the nations first commercial coal gasification plant located 8 miles northwest of Beulah, North Dakota, produces high quality substitute natural gas (SNG) from lignite. Taking into account the need for environmental accountability, Great Plains incorporated into its design numerous process units to treat waste water and gas streams. The Plant does not surface discharge treated process wastewater but rather reuses the water within the gasification process and cooling tower or injects low pressure boiler blowdown and water softener regeneration wastewater in deepwells. A Sulfur Removal Unit at Great Plains was used to remove hydrogen sulfide from a waste gas stream prior to the gas being burned in the main boilers. This was to reduce the sulfur dioxide content of the boilers flue gas to permitted levels.

Following start up and initial operation of Great Plains, the Sulfur Removal Unit (started up in 1984) was found to be unable to reduce the hydrogen sulfide content of the waste gas to the point that the sulfur dioxide content of the Plant's flue gas would meet permitted levels. Numerous changes in the Sulfur Removal Unit's operation and design modifications were unsuccessful in improving removal of hydrogen sulfide.

In late 1990, the Dakota Gasification Company (DGC), owner and operator of the Great Plains Synfuels Plant, commissioned a Task Force to review all options and recommend a solution to meet the main stack's sulfur dioxide environmental permit. Several options of pretreating the waste gas to remove sulfur species prior to burning in the Plant's main boilers along with flue gas desulfurization (lime, wet limestone, dual alkali) were investigated. The Task Force found that wet limestone forced oxidation flue gas desulfurization (FGD) for the flue gas from the Plant's main boilers was the best solution.

During this review period by the Task Force, representatives from GE Environmental Systems Inc. (GEESI) approached DGC to discuss their new ammonia scrubbing process. This process was of interest to DGC as it had the potential to

15. WALLACH (NH$_4$)$_2$SO$_4$ Fertilizer from Flue Gas Desulfurization

produce fertilizer grade ammonium sulfate rather than the gypsum that would be produced from a wet limestone FGD. Gypsum would be a waste and require disposal in an approved landfill site. Production of salable ammonium sulfate fit in with DGC's on-going efforts to enhance by-product production at Great Plains to lessen the reliance on the sale of SNG.

Following the successful operation of an ammonium sulfate FGD, DGC and GEESI entered into an agreement to build the first commercial scale Ammonium Sulfate Forced Oxidation FGD system. Not only would DGC benefit from the production of a salable by-product rather than a waste but other industries contemplating means to meet sulfur dioxide environmental requirements would also benefit by seeing that there is an option.

Construction of the Ammonium Sulfate Forced Oxidation FGD was completed in June 1996 with flue gas being introduced on June 29, 1996. The FGD has shown that it can remove sulfur dioxide and produce fertilizer grade ammonium sulfate. Some 16,000 short tons of ammonium sulfate have been produced.

Overview of Great Plains Synfuels Plant

The process of turning lignite into SNG begins by feeding 18,500 tons of sized coal to 14 Lurgi Mark IV gasifiers on a daily basis. In the Lurgi moving bed gasifiers, steam and oxygen are fed to the bottom of the gasifier and distributed by a revolving grate. The steam and oxygen slowly rise through the coal bed, reacting with the coal to produce a raw gas stream.

The raw gas stream which exits each gasifier is first cooled in a waste heat boiler which generates 100 psig saturated steam. After this initial cooling step, two-thirds of the raw gas is sent to additional waste heat recovery and cooling water units where 50 psig and 25 psig saturated steam is generated and the gas is cooled to 95 degrees F. The remaining one-third of the raw gas is sent to shift conversion where the composition of the gas is modified by converting some CO and water to CO_2 and hydrogen. The shifted gas is then cooled to 95 degrees F and combined with the cooled raw gas to make mixed gas.

The mixed gas stream is sent to a Sulfur Removal System where CO_2 and sulfur compounds are removed by a cold methanol wash. The "sweet gas" from the Sulfur Removal System is then sent to Methanation where SNG is produced. Following Methanation, the SNG is compressed, dried and sent to the pipeline.

The Sulfur Removal System also produces a waste gas stream. The waste gas stream is comprised primarily of the CO_2 and sulfur compounds removed from the mixed gas. The waste gas stream is used as gaseous fuel in the plant's three main boilers. While the waste gas stream is low in BTU value, 48 BTU per SCF (HHV), the quantity of waste gas is such that this stream provides approximately 36% of the heat input required by the main boilers. The remainder of the heat input is supplied by a medium BTU gas stream and by two liquid streams. Each of the boilers is capable of generating 500,000 LB/HR of 1150 PSIG superheated steam. The normal operating range for each boiler is 380,000 to 400,000 LB/HR.

Original Sulfur Removal System. The original design of the Great Plains Synfuels Plant (GPSP) called for the waste gas stream to be treated in the Stretford Sulfur

Recovery Unit (SRU) to remove primarily hydrogen sulfide prior to being burned in the main boilers. The Stretford SRU is a vanadium based Redox technology designed to remove low concentrations of hydrogen sulfide in gas streams. The sulfur compounds in the gas stream being treated are absorbed into the Stretford solution where they are ultimately converted to elemental sulfur. Operational problems have plagued the SRU ever since start-up in 1984. Severe plugging and sulfur deposit problems were experienced. The absorbers in this system were originally packed, however, the packing had to be removed due to the severe plugging problems.

When efforts to resolve the operating difficulties with the Stretford SRU had been exhausted, the SRU was switched over to Sulfolin chemistry (another vanadium based Redox technology) in late 1987. This chemistry change significantly reduced the severe plugging and sulfur deposit problems experienced with the Stretford chemistry. The better performance of the now Sulfolin SRU prompted DGC to submit an application with the North Dakota State Health Department (NDSDH) to Amend the Air Pollution Control Permits for the Great Plains Synfuels Plant. The Amended Permit proposed a parallel Sulfolin SRU as Best Available Control Technology (BACT) for the main stack emissions. Subsequent testing of the existing Sulfolin SRU revealed that the emission levels proposed in the Amended Permit could not be achieved even with a parallel Sulfolin SRU of the same size as the existing unit. Other operating problems with the packed absorbers, regeneration of the solution, and recovery of the sulfur through the centrifuges and decanters limited the plant performance. Also at the projected high sulfur levels expected in the coal from the Freedom mine, there was concern that even with Sulfolin operating perfectly the plant's boilers would be not meet the required 0.8 LB SO_2 per MMBTU due to the high sulfur levels in the other fuels.

The main flaw with both the Stretford and the Sulfolin vanadium based Redox technologies appeared to be the composition of the waste gas. The waste gas stream is high in CO_2, approximately 90%. The high CO_2 impacts the operation of the process in ways that had not been anticipated by the process licensers resulting in reduced performance of the oxidizers and the absorbers.

Development of Alternative Sulfur Removal System.

In late 1990 a Task Force was commissioned to review all options and recommend a solution. The Task Force was made up of representatives from DGC and Basin Electric Power Cooperative (BEPC), DGC is a wholly owned subsidiary of BEPC. A main stack assessment of BACT for sulfur dioxide emissions control was commissioned by the Task Force. Several options for treating the waste gas prior to burning it in the plant's main boilers were investigated along with flue gas desulfurization. None of the pretreatment processes that were reviewed during the assessment process were found to be BACT. When the assessment was complete, a wet limestone forced oxidation flue gas desulfurization for the flue gas from the plant's main boilers was found to be BACT. A Revision to the Application to Amend the Air Pollution Control Permit to Construct for the Great Plains Synfuels Plant was filed in late fall of 1990 with the NDSDH. The Revision to the Application was accepted by the NDSDH and a Permit to Construct was granted on March 23, 1993.

Under a Consent Agreement between DGC and the NDSDH, the environmental fix for the main stack SO_2 emissions is to be in operation by March 23,1997. Following the selection of a wet limestone forced oxidation FGD as BACT, DGC formed a Construction Department responsible for implementing the environmental fix for Great Plains Synfuels Plant. An engineering firm, Raytheon Engineers and Constructors, was contracted to prepare the design specifications for the limestone forced oxidation (LSFO) FGD, the bid specifications and the request for proposal.

During the period between the selection of wet limestone forced oxidation FGD as BACT and the granting of the Permit to Construct by the NDSDH, DGC began talking to numerous companies that market FGD systems and visiting numerous plants that had operating FGD systems. It was at this time that GEESI approached DGC to discuss their new ammonia scrubbing process. GEESI presented operating data on their ammonia scrubbing process that had been obtained from a mini pilot plant that was installed and operated in their research facilities in Lebanon, PA. The ammonia process was of interest to DGC as it would produce a salable byproduct, ammonium sulfate, rather than a waste product, gypsum, as does a wet limestone forced oxidation FGD. DGC was at this time developing numerous byproducts from gasification of coal at Great Plains Synfuels Plant.

In late May 1992, DGC and GEESI entered into an agreement to demonstrate the ammonia scrubbing process on a 10,000 ACFM (approximately 3 MW) pilot plant. Under the agreement, the pilot plant would be built at the Great Plains Synfuels Plant and would take a slipstream from one of the plant's main boilers flue gas stream. Due to the uniqueness of the flue gas, DGC required that the pilot plant be constructed and operated to demonstrate the ammonia scrubbing technology for SO_2 control. The commercial scrubber would have to operate at the design conditions listed in Table I. The pilot plant would be equipped with SO_2 addition to the flue gas feed stream to simulate the full range of emissions expected in the flue gas from the plant's main boilers. The pilot plant would also be equipped with vents for treated and untreated flue gas for comparison of opacity.

Table I. Design Conditions

Inlet Flue Gas Conditions	Maximum	Average	Minimum
Temperature, °F	450	400	350
Gas Flow Rate, Lb/Hr.	3,396,900	3,396,900	1,019,100
Gas Flow Rate, ACFM	1,141,476	1,132,100	339,600
N_2, Volume %	60	55	50
O_2, Volume %	8	6	4
CO_2, Volume %	34	31	28
H_2O, Volume %	9	7	5
SO_2, ppmv	4,780	3,710	2,650
SO_3 ppmv	100	78	56
No_x, ppmv	150	60	25
Particulate, Lb/Hr	86	86	0

Pilot Plant - Ammonium Sulfate Scrubbing Process. Construction of the pilot plant was begun in June 1992 with completion in mid October 1992. Start up of the pilot plant began on October 24, 1992. 10,000 ACFM of flue gas from the "S" main boiler at a temperature of approximately 350 degrees F was ducted from the discharge of the boiler's induced draft (ID) fan to a co-current prescrubber followed by a spray tower absorber. Initial operation of the pilot plant was conducted without the prescrubber in which case the hot flue gas was sent directly to the spray tower absorber. Also, the initial operation of the pilot plant was conducted with limestone to demonstrate that the equipment could be adapted back to a conventional wet limestone FGD and to determine the correlation between pilot plant data and commercial limestone FGD systems. During the period when the pilot plant was operated as an ammonia scrubber without the prescrubber, the absorber was operated in both dilute and saturated ammonium sulfate liquor mode. The amount of SO_2 used to spike the flue gas feeding the pilot plant was limited during this initial operation as an ammonia scrubber when the liquor was saturated due to the physical limitation of the equipment to oxidize the liquor.

When the pilot plant was operated with the prescrubber, no removal of SO_2 from the flue gas occurred in the prescrubber. The main purpose of the prescrubber was to separate SO_2 absorption and oxidation from crystallization of ammonium sulfate. Crystallization of ammonium sulfate was accomplished in the prescrubber by cooling and saturating the flue gas. This was done by the three co-current sprays in the spray section of the prescrubber. Saturated liquor from the bottom of the prescrubber was pumped to the three co-current sprays. Make-up to the prescrubber for level control was provided by absorber liquor either directly to the storage section of the prescrubber or by mist eliminator wash in the outlet ducting from the prescrubber to the absorber. The prescrubber was also equipped with a "dentist bowl" at flue gas inlet. The purpose of the "dentist bowl" was to move the wet/dry interface down into the spray section of the prescrubber thereby avoiding any build up at the flue gas inlet.

The absorber in the pilot plant was a conventional spray tower type. It was equipped with eight spray levels, each containing a single spray nozzle of commercial size. Scrubbing liquor was pumped to the sprays by two pumps from the oxidation/recycle tank. The spray section of the absorber was sized to handle SO_2 concentrations of up to 6100 parts per million volume (PPMv). This allowed the pilot plant to simulate all possible SO_2 concentrations that could possibly occur in the flue gas from the plant's three main boilers. The SO_2 concentration of the inlet flue gas was maintained at the desired levels by taking pure liquid SO_2 from a rail car, vaporizing it and injecting the vapor into the flue gas.

A two-stage mist eliminator was installed in the absorber above the spray section for removal of entrained liquor. A standard wash system was provided for the mist eliminator, however, the system was not used during the ammonia scrubber testing. Inspection of the mist eliminator after ammonia scrubber testing was completed showed that there was no build-up. The scrubbed flue gas leaving the absorber was reheated by mixing hot bypass gas. The reheated gas was then sent back to the suction of the "S" main boiler's ID fan by the pilot plant's induced draft fan. During ammonia scrubber mode testing, vaporized ammonia was fed to the absorber tank via the air sparger from an anhydrous ammonia storage trailer and vaporizer.

Ammonium sulfate crystals were removed from the system by a dewatering system consisting of a hydroclone and a centrifuge. Liquor from the prescrubber was pumped to a hydroclone. The overflow from the hydroclone was returned to the prescrubber while the underflow was sent to the centrifuge feed tank. The liquor from the centrifuge feed tank was pumped to the centrifuge for final dewatering. Several different types of centrifuges were tried during testing of the pilot plant. The centrifuge that proved the most successful was a vertical basket type (reminds you of a washing machine). Water recovered from the centrifuge was sent back to the prescrubber. The ammonium sulfate crystals from the centrifuge were stored in 55 gallon drums for later characterization and granulation. Testing done by outside firms showed that the ammonium sulfate "sugar crystals" produced by the pilot plant could be easily compacted to produce the desirable granules.

The pilot plant operated around the clock with only minor interruptions. A series of parametric runs were conducted. Data collected from the Great Plains Synfuels Plant's Honeywell TDC-2000 system was down loaded into a personal computer spreadsheet program. The duration of the parametric runs were generally 2 or 4 hours in length.

The pilot plant was first started with water, with ammonia being injected into the absorber via the oxidation air sparger. The rate of injection was controlled via flow control that was reset by pH control on the absorber to maintain approximately a 5.5 pH. Makeup water (softened water) was added to the absorber to maintain a constant level. Performance data were collected as the concentration of ammonium sulfate in the absorber liquor increased, however, performance data between 25% and 35% was of most interest. In DGC's application, the absorber liquor concentration would be in this range in the commercial unit. The parameters that were studied during the pilot plant testing were liquid-to-gas ratio (L/G), pH of the absorber liquor, inlet SO_2 concentrations, sulfite ion concentration in the solution, and absorber gas velocity. As the potential commercial Ammonia Scrubber would be forced oxidation, varying the air rate was important to determine the oxidation rate. Varying the sulfite ion concentration in the scrubbing solution was done to determine the effect on SO_2 removal efficiency.

At an SO_2 concentration of 2000 PPMv, a pH of 5.8 and a zero sulfite level in the liquor the removal efficiency was 95% with 4 spray levels. The efficiency increased to 98% with 6 spray levels. Under comparable conditions using limestone in the pilot plant for SO_2 removal the efficiency with 4 spray levels was only 90% while the efficiency with 6 spray levels was 95%. The effect of pH on removal efficiency was found to be rather modest. Increasing pH from 5.2 to 5.8 only increased the efficiency by 1 to 2 percentage points.

The inlet SO_2 concentration and number of operating spray levels show a strong effect on performance. Removal efficiencies of 93% were obtained at a SO_2 of 6100 PPMv with 6 spray levels and a pH of 5.5. The 6100 PPMv level is the highest that could be expected in the flue gas from the plant's main boilers. A 24 HR duration test was conducted at 3710 PPMv SO_2 level. The pilot plant was able to meet 93% removal efficiency for every hourly average during the test.

The ammonia concentration in the gas leaving the absorber was monitored under all conditions using Drager tubes. In addition, the opacity of the gas before and after scrubbing was monitored using opacity meters. This was accomplished by the installation of two side-by side stub stacks, one for the untreated gas and one for the gas that had been treated and reheated. The ammonia slip under most conditions was zero. Maximum ammonia slip of approximately 3 PPM was measured when sulfite levels in the absorber liquor were 11,000 PPM with a pH of 5.8. The bypass stack opacity measured 3-4%. The opacity in the reheated treated gas was always below the opacity of the untreated gas until the pH of the absorber liquor went above 5.9.

The ammonium sulfate crystals produced during the pilot plant testing averaged approximately 300 microns in size. Trial runs were made by two companies, one in the United States and one in Germany, to granulate the ammonium sulfate by compaction. Both firms were able to compact the material and produce ammonium sulfate granules which were compatible with other granulated fertilizers. The granulated ammonium sulfate product consisted of 2-3 mm sized granules with an abrasion index of 2-3%. The abrasion index indicates the attrition rate (amount of dust produced due to handling) for the ammonium sulfate granules, which is important due to the bulk handling of the granules. The granulated ammonium sulfate product produced from the pilot plant testing fully met commercial fertilizer quality requirements.

Following successful completion of the ammonia scrubber pilot plant testing, DGC continued the process of requesting bids on a LSFO FGD. An alternate bid from GEESI for an ammonium sulfate forced oxidation (ASFO) FGD was accepted by DGC for review. Following review of the LSFO and ASFO FGD bids and considering the benefits of producing a valuable byproduct rather that a waste, DGC and GEESI entered into a contract for an ASFO FGD in March 1994.

Commercial ASFO FGD System. Design of the Absorber and Dewatering Areas of the ASFO FGD is very similar to a wet limestone forced oxidation FGD. One major departure that the ASFO FGD has is the Compaction Area where ammonium sulfate crystals are turned into fertilizer grade product.

Flue gas from the three Riley boilers combine in a single duct before going to the Absorber Area (Figures 1 and 2). The flue gas is first processed in the Prescrubber where it is contacted with a solution of ammonium sulfate circulated through two spray headers. The purpose of the Prescrubber is the same as the one in the pilot plant; it separates absorption of SO_2 from crystallization of ammonium sulfate. The flue gas evaporates water from the Prescrubber's ammonium sulfate solution causing crystallization of ammonium sulfate product. The pH of the circulating ammonium sulfate solution is < 2.0 thus no SO_2 is removed. The evaporated water is replaced by subsaturated ammonium sulfate solution from the Absorber. This solution is added to the Prescrubber through the mist eliminator wash system. Saturated flue gas leaving the Prescrubber passes through a single stage mist eliminator. Unlike the Pilot Plant's Prescrubber, the commercial plant's Prescrubber is countercurrent in design.

The Absorber is a conventional countercurrent scrubber where flue gas is contacted by a circulating solution of subsaturated ammonium sulfate. SO_2 in the flue gas is removed, reacted with ammonia and oxidized to form ammonium sulfate. Liquid ammonia is first vaporized then added to the oxidation air to maintain a pH of < 6.0 in the Absorber's recycle solution. Oxidation air is injected at the bottom of the

Absorber through nine sparge headers to oxidize ammonium sulfite into ammonium sulfate. Fresh water make-up is added to the Absorber under level control to replace the ammonium sulfate solution that is sent to the Prescrubber. Scrubbed flue gas leaving the Absorber passes through two stages of mist eliminator to remove any entrained droplets. These two stages of mist eliminator are not equipped with a wash system as the solution in the Absorber is subsaturated.

Saturated ammonium sulfate solution is pumped from the Prescrubber to the hydroclone cluster in the Dewatering Area (Figures 3 and 4). The hydroclones (six operating with two spare) increase the total suspended solids (TSS) of the ammonium sulfate solution from approximately 10 to 50 percent. The underflow from the hydroclones is directed to the centrifuge feed tank while the overflow is directed to the centrate tank for return to the Prescrubber. Both underflow and overflow are directed to the centrate tank after the TSS of the ammonium sulfate solution is reduced to 6 percent.

The ammonium sulfate slurry in the centrifuge feed tank is pumped through a circulation loop back to the tank. Four vertical basket centrifuges take feed from the circulation loop. The centrifuges dewater the solids in the slurry to > 97 weight percent. Solids from the centrifuges are conveyed to a rotary drum dryer where the ammonium sulfate is dried to less than 1 weight percent moisture. The dried ammonium sulfate is conveyed from the dryer to a storage silo where it is stored prior to compaction. In the event that there are operating problems in the Compaction Area, the dried ammonium sulfate can be diverted to an intermediate storage dome where it would await further processing.

The sugar-like ammonium sulfate crystals are converted into granules in the Compaction Area (Figure 5). Raw ammonium sulfate feed from the storage silo along with fines generated in the Compaction Area are conveyed to two compactors. The compactors compress the material into cigar shaped "flakes" of ammonium sulfate. The "flakes" are first sized by a "lumpbuster" under each compactor and then conveyed to sizing mills where granules that are 4 to 5 millimeters in size are produced. The compacted and sized granules are first screened by two primary screens to remove any dust. The dust is recirculated as feed to the compactors. The on-size material from the two primary screens is conditioned in the final product dryer and then cooled in the cooling drum. The material is then again screened in a double deck screen to remove oversized and undersized material from the final product. The final product is conveyed to a 90 day storage dome. From the dome, the ammonium sulfate granules, sold under the trade name of DAKSUL 45®, is loaded into either rail cars or trucks. All of the ammonium sulfate is sold in bulk.

Construction of the ASFO FGD began in the summer of 1994 with the erection of a new acid brick lined main stack. Detail design of the ASFO scrubber was also begun at this time. One main departure that the commercial FGD made from the Pilot Plant was to construct the Prescrubber and Absorber out of fiberglass reinforced plastic (FRP). Due to the low pH and chloride content in the Prescrubber slurry, this vessel would have to had been constructed from lined carbon steel, exotic metal or FRP. FRP was chosen for the Prescrubber due to the corrosive nature of the slurry and to reduce cost. The Absorber was constructed from FRP to also reduce cost. Material of construction for piping is either rubber lined carbon steel or FRP (for certain small

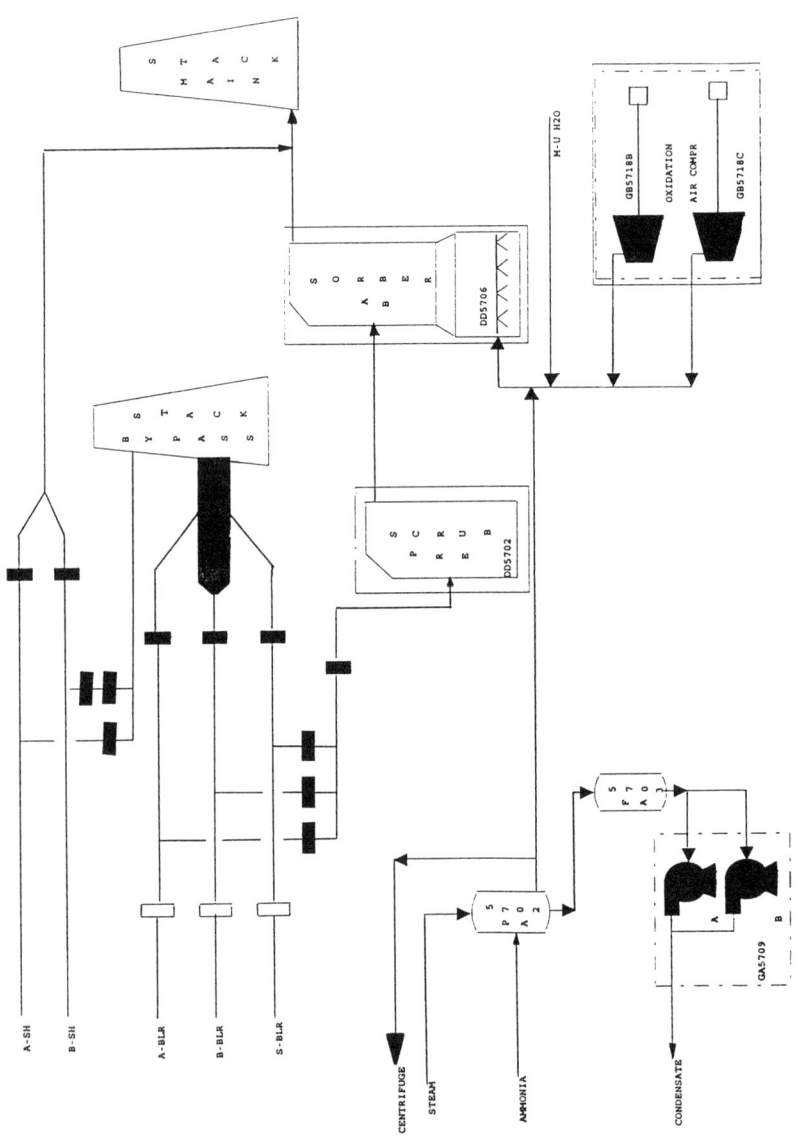

Figure 1. Absorber Area - Gas Streams

15. WALLACH (NH$_4$)$_2$SO$_4$ *Fertilizer from Flue Gas Desulfurization* 249

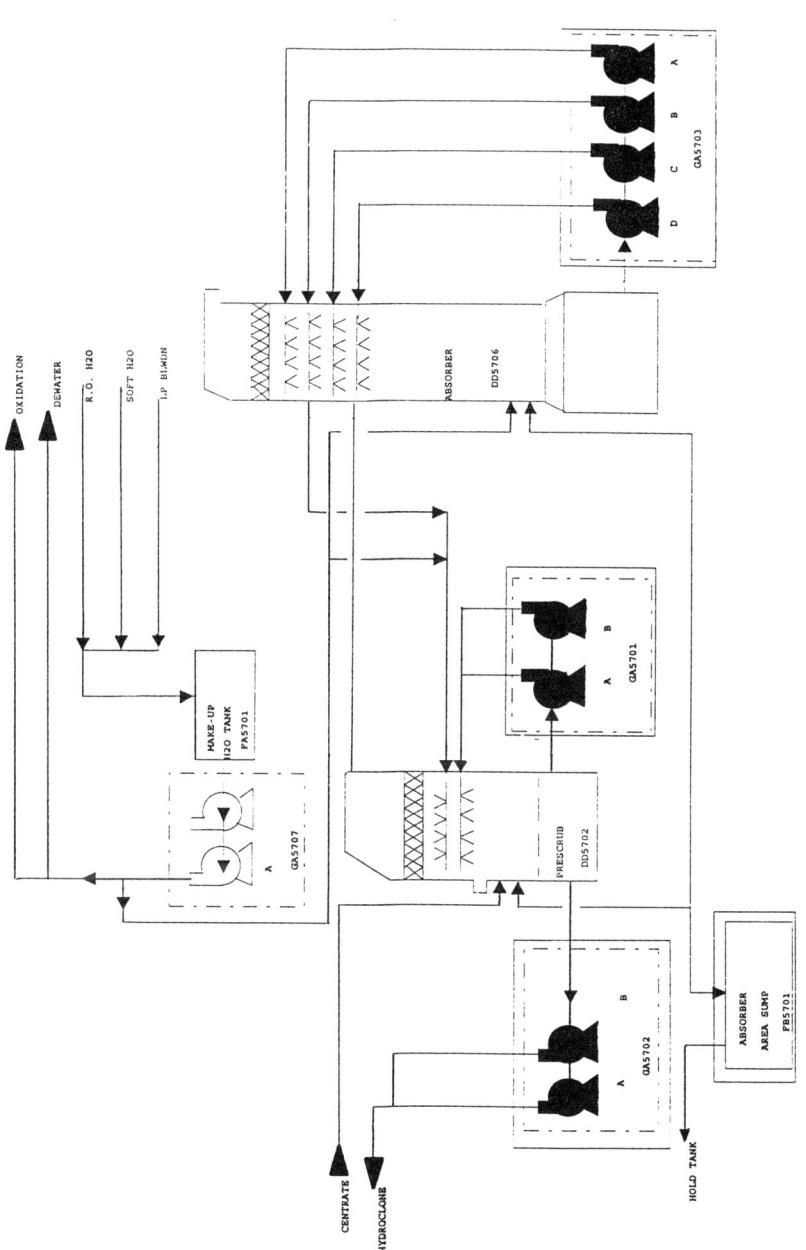

Figure 2. Absorber Area - Liquid Streams

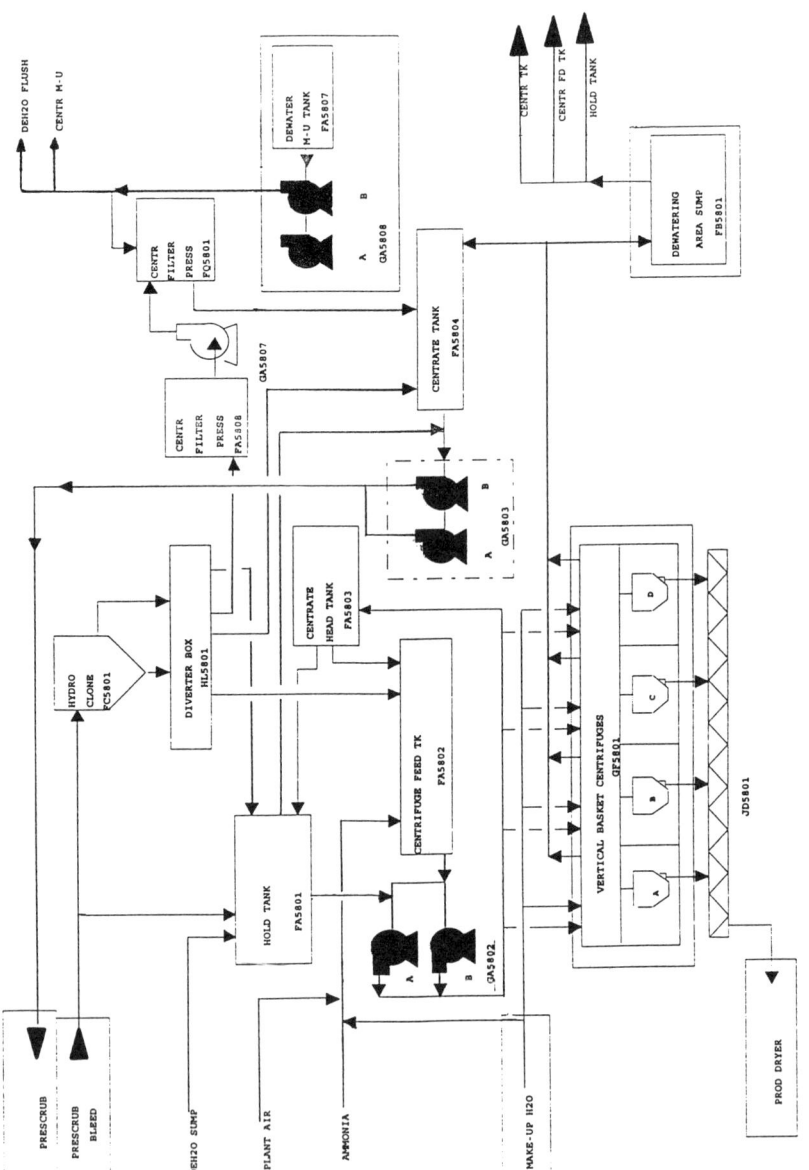

Figure 3. Dewatering Area - Liquid Streams

15. WALLACH (NH$_4$)$_2$SO$_4$ *Fertilizer from Flue Gas Desulfurization*

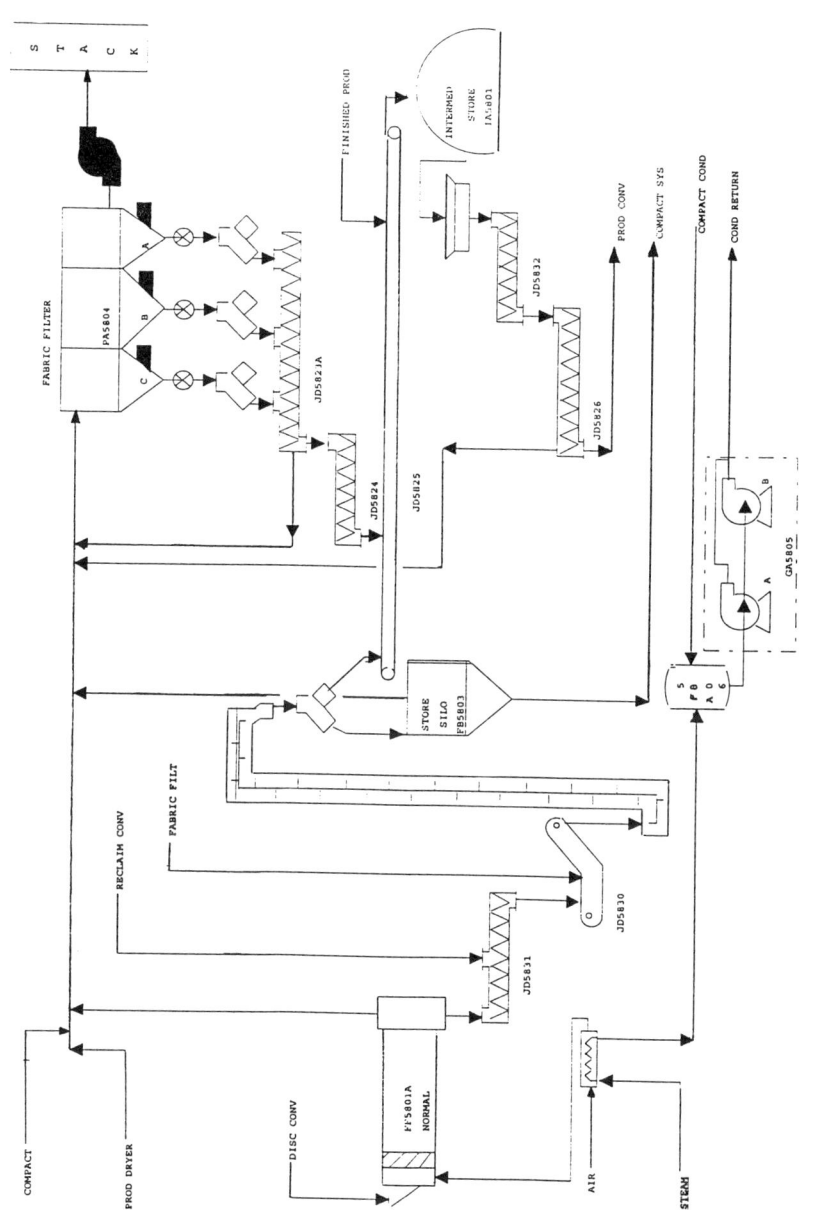

Figure 4. Dewatering and Compaction Areas - Raw Product Drying

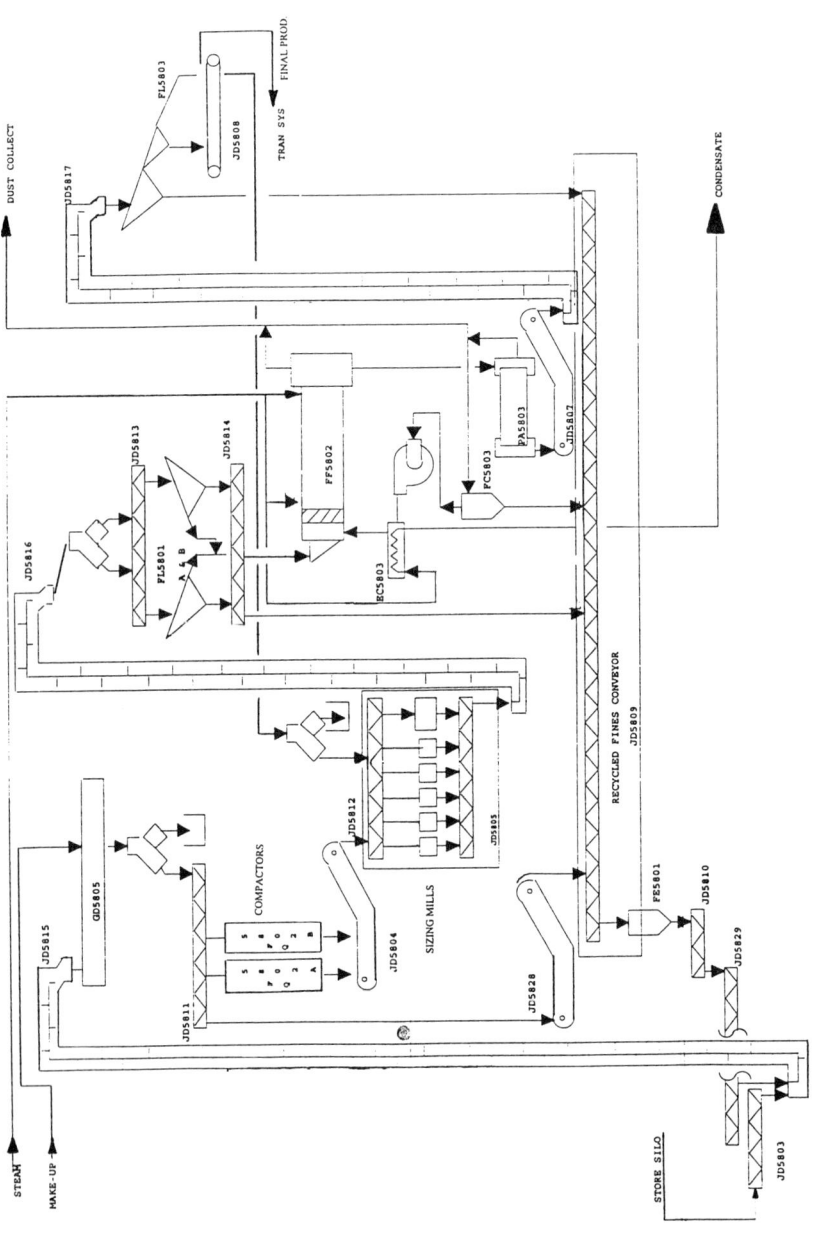

Figure 5. Compaction Area

pumps provided by Warman. The absorber recycle pumps have metal impellers while the rest of the pumps use rubber coated impellers.

Initial Operation of ASFO FGD. Construction of the ASFO FGD stood at over 90% complete by the end of June 1996. Early in June 1996, start up activities were begun. The FGD was filled with water and circulation begun to test the various systems. On June 29, 1996 the first flue gas was introduced into the FGD. There were the normal initial operating difficulties experienced during the early days of start up. Several days were required to gain control of the chemistry in the Prescrubber and Absorber. Commissioning activities also began in the Dewatering and Compaction Areas to be ready when the first ammonium sulfate crystals were formed. On July 4, 1996 the first ammonium sulfate crystals were noted in the Prescrubber slurry. Following the formation of ammonium sulfate crystals, the underflow of the operating hydroclones was diverted to the centrifuge feed tank. Centrifuging and drying of the ammonium sulfate crystals began on July 5, 1996. The dried ammonium sulfate crystals was diverted to the intermediate storage dome at first until their purity could be determined. Laboratory analysis showed that the crystals met purity requirements for fertilizer grade ammonium sulfate. Several attempts were then made to compact the ammonium sulfate crystals. Numerous problems with plugging and burn out of the motor on the fixed speed compactor delayed compaction and granulation of the crystals until July 7, 1996. Numerous trips in the Compaction Area were experienced during the next several days. During these trips, dried ammonium sulfate crystals were diverted to the intermediate storage dome.

On July 9, 1996 flue gas was taken out of the FGD and diverted to the by-pass stack as it had become evident that certain modifications would be required in the Dewatering and Compaction Areas. Major modifications accomplished during the outage was to move the motor from the variable speed compactor to the fixed speed compactor, add "flites" to certain screw conveyors, replace the apexes and vortex finders in the hydroclones and modify the spray headers in the centrifuges. After all modifications were completed, only the Compaction Area was restarted on July 20, 1996. The ammonium sulfate crystals that had been sent to the intermediate storage dome were reclaimed and compacted to produce granulated product, DAKSUL 45®. Granulated product which did not meet the fertilizer grade size guide number (SGN) was also reclaimed and recompacted.

The Absorber Area of the FGD was brought back on-line on July 24, 1996. Troubles with the side entry agitators in the Prescrubber required that flue gas be taken out of the FGD on July 26, 1996. The problems with the agitators were a broken shaft on one and seal leaks on the others. The shafts, seals, and blades from the "A" and "C" Prescrubber agitators were packaged up and shipped back to the manufacturer for failure analysis. Numerous discussions between DGC and GEESI were held to review the acceptability of side entry agitators in the Prescrubber and to explore alternative means of agitation. The FGD remained down from July 26, 1996 through August 5, 1996. Several attempts were made to operate the FGD on August 6, 7, and 9, however, problems were encountered keeping solids in the Prescrubber in suspension. On August 9, 1996 the decision was made to shut down the FGD and proceed with design and installation of an eductor system in the Prescrubber in lieu of the side entry agitators.

Flue gas was reintroduced into the FGD on August 27, 1997. The FGD has operated almost continuously since then with minor outages to repair equipment or make modifications in the Compaction Area. During most of this time when the FGD was operated flue gas from only two of the three Riley boilers was sent to the Absorber Area. This was primarily due to the Compaction Area being unable to process all of the ammonium sulfate that would have been produced by scrubbing the SO_2 in the flue gas from all three boilers. Recent revisions have allowed the FGD to treat the flue gas from all three Riley boilers. Since startup, approximately 16,000 tons of DAKSUL 45® have been produced. The quality of this product is shown on Table II.

TABLE II - DAKSUL 45® PRODUCED	
	Typical Analysis
% Nitrogen	21
% Sulfur	24
% Moisture	0 - 0.10
% Water Insolubles	0 - 0.10
Size Guide Number (SGN)	230 - 240
Uniformity Index (UI)	30 - 40
% Attrition	1.0 - 2.5

The ability of the commercial ASFO FGD to scrub SO_2 from the Riley boilers flue gas has been found to be as successful as the Pilot Plant. With three of the four absorber spray levels in operation and at a pH of 5.6, a removal efficiency of 93% can be obtained at an inlet SO_2 concentration of 2700 ppmv.

The ammonium sulfate product, DAKSUL 45®, from the FGD meets all fertilizer grade requirements and has been well received.

There remain minor problems to improve reliability; however, the ASFO FGD has proven to be the best solution to DGC's environmental difficulties.

Chapter 16

By-Product Usage in Fertilizer Micronutrients

James M. Wyatt

Frit Industries, Inc., P.O. Box 1589, Ozark, AL 36361-1589

Nutrients supplied by fertilizers are divided into three categories: primary, secondary, and micronutrients. Neither the primary nutrients (nitrogen, phosphate, and potassium) nor the secondary nutrients (calcium, magnesium and sulfur) are the subject of this paper, but many by-products are utilized beneficially in the fertilizer industry to provide these nutrients.

Micronutrients refer to the remaining essential plant nutrients that must be present for plant growth, but are required only in very small or micro quantities. Nutrients that are classified as micronutrients are boron (B), cobalt (Co), chlorine (Cl), copper (Cu), iron (Fe), manganese (Mn), molybdenum (Mo), vanadium (V), and zinc (Zn). The metallic forms of most of these metals are not available to plants. Chemical forms that are used are frits, oxides, sulfates, nitrates, carbonates, and chelates. The physical forms in which micronutrients are applied are liquids, suspensions, powders, and granules. Analytical methods followed by fertilizer control officials are those specified in the fertilizer section of the <u>Official Methods of Analysis of The Association of Official Analytical Chemists</u> (AOAC).

Micronutrients are typically applied in accordance with recommendations received from agronomic consultants, who consider as soil test results, the crop and cultivar being grown, tissue analysis of the growing plant, yield levels, and soil conditions in a given geographical area. Micronutrient deficiencies occur most frequently in sandy soils, high pH soils, and older, highly-weathered soils from which the nutrients have been leached.

Since a plant's root system must intercept soil nutrients, it is important for soil nutrients to be maintained at a high level to prevent deficiencies in crops, especially when high yields are being sought.

The agronomic importance of the level of water solubility of the micronutrient compound has long been in dispute within the fertilizer industry as has the assertion that a lower application rate of chelated micronutrients will give the same crop response as a much higher rate of oxides or sulfates. Placement rather than the source can reduce the amount of product used. Banding or foliar application is more effective than broadcasting micronutrients thus lower rates may be applied, but broadcasting utilizes less expensive materials and takes far less time and equipment.

With granular products, the size of the individual particle is important. The larger the particle, the fewer particles there will be per unit of area. Granular product (made up of very fine individual particles) with a diameter between two and four millimeters is the most commonly used due to that being the typical size of the nitrogen, phosphate, and potassium sources. Powdered micronutrients may be incorporated into the fertilizer granule during the manufacturing process. This method results in better distribution in the soil compared to a physical blend of micronutrients and other fertilizer materials, but all fields upon which the fertilizer is applied receive the same micronutrient application. Physical blends can be produced in small lots, thus offering a method to include only the micronutrients needed by a particular field. Coating powders onto granules in a blender is another method of manufacturing. However, the handling of dry, dusty powders is not acceptable to most blenders, so this is not a very widespread practice.

Liquid fertilizers may be true solutions or suspensions. Liquid fertilizer can be applied uniformly and allow the amount of micronutrients applied to a field to vary. Reactions between micronutrients and other fertilizer components can occur and present problems.

In some instances micronutrients are applied alone, as in a foliar application of a liquid micronutrient. This application may be to correct a deficiency which becomes apparent during the growing season or may be due to a high pH or some other soil condition which converts soil-applied micronutrients to an unavailable form.

16. WYATT   *By-Product Usage in Fertilizer Micronutrients*

During the early periods of our country's development, it was common for very little fertilizer to be used and the land to be farmed until the soils were depleted of nutrients and then the farmer move and leave the land fallow. In the 1800s and early twentieth century, farmers began to use bone meal, tankage (animal rendering wastes), manure, fish meal, guano, cotton hull ashes, and rock superphosphate as fertilizers. Micronutrients were included in fertilizers due to their presence in these materials. As an example, one study found that manure contained an average of 20 ppm B, 201 ppm Mn, 1 ppm Co, 16 ppm Cu, 96 ppm Zn, and 2 ppm Mo (Russell). According to The Fertilizer Handbook, bone meal contains 20,000 ppm Zn, normal superphosphate contains 10,0000 ppm Zn, and cotton hull ashes contain 40,000 ppm Cu, 60,000 ppm Mn, and 70,000 ppm Zn. All of these materials were used in large volumes but contained relatively low levels of nutrients. These materials were commonly used as late as the 1960s. Plant varieties commonly used, row width, and plant populations were such that the nutrient demands placed on the soils and the depletion of micronutrient levels in the soils were far less than today. Using corn as an example, in 1940 corn plant populations were 10-12,000 per acre. By the 1950s the plant populations had increased to 16,000 (Aldrich), and today we see 24,000 plants per acre as the norm (NASS). Prior to 1935, the average corn yield had not changed since records were first kept in 1880. However, from 1935 to 1960 average yields increased by 140% (25 Bu/acre to 60 Bu/acre) and from 1960 until the present, average yields have increased from 60 Bu/acre to 110 Bu/acre or another 80% (NASS). Since 1980 little change has occurred in average yields due to the emphasis placed on minimizing inputs and the excess grain in world stockpiles.

Micronutrient deficiencies have been reported in the United States since the 1920s (Cunningham) but were not addressed by the fertilizer or farming sectors until the 1960s. The first national survey on micronutrient deficiencies was done in 1960 by the Council on Fertilizer Application and was followed by a series of surveys which culminated in a report by the Soil Testing Committee of the Soil Science Society of America. Their report listed specific micronutrient needs by crop and gave recommendations for corrective measures (Cunningham).

Micronutrient usage has been growing steadily since the 1960s. The biggest increase was in the 1973-1980 period when marginal land was brought into production due to high grain prices and easy farm credit. New cropping practices were adopted that emphasized maximum yield. As fertilizer rates increased, other factors limiting maximum yield were revealed, such as insufficient micronutrient levels in the soil. During this period, micronutrient use increased and new micronutrient sources from by-products increased dramatically. Today, most micronutrients are produced from by-product materials.

In 1983, the growth in fertilizer usage stopped when one-third of U.S. corn and wheat acreage was removed from production when the government's payment in kind (PIK) program was implemented. This program gave surplus grain in government storage to farmers in exchange for their taking land out of production. Since then and until recently, government programs have continued to emphasize reduction in both planted acreage and crop production. The new credo for the 80s was maximum economic yield, which in the 90s became maximum sustainable yield. With the depletion of world grain reserves this summer and increases in grain prices, a new emphasis will probably be placed on increased production, but with farmers and the government taking a more cautious approach than in the 1970s.

While the above mentioned changes in cropping practices were occurring and greater attention was being paid to soil deficiencies, governmental regulations were being put in place to eliminate air pollution. Pollution from automobile exhaust and the smokestacks of the nation's power plants, steel mills, copper, zinc, brass and bronze smelters, and foundries had served as major sources of some secondary nutrients and micronutients. As the air was cleaned up, particulate from pollution no longer provided nutrients. Smokestacks can still provide secondary nutrients and micronutrients, but in a different form--baghouse dust, scrubber sludge, or other types of material from an air pollution control system. There are, however, many limits on by-product use because of their physical form, high freight costs for low analysis products, lack of a way for an individual farmer to apply these materials, and federal regulations. These factors present insurmountable barriers to direct use of these materials.

The federal regulations having the most impact on recycling were introduced as part of the Resource Conservation and Recovery Act of 1976 (RCRA). The stated purpose of the act was to encourage the recovery and recycling of wastes and to reduce the volume of materials being landfilled or improperly disposed. Regulations implementing RCRA are found in 40 CFR, Parts 260-272. Hazardous waste categories were created for materials posing special hazards. Regulations were promulgated controlling the use, transportation, storage, and disposal of certain wastes. Some materials were included by name (listed waste), others were included because of a characteristic (characteristic waste) such as corrosivity or flammability. A tracking system for hazardous wastes was created, requiring that wastes be manifested when transported and that any facility generating or accepting hazardous wastes had to be permitted by EPA to assure the safe storage, handling, and disposal of the waste. Restrictions on how these materials could be handled were established. The greatest impact on micronutrient raw materials came from regulations making any material captured in a baghouse or electrostatic precipitator a sludge (since this equipment serves the same purpose as a scrubber and a scrubber sludge is by "definition" hazardous; therefore, baghouse or precipitator dust must also be a hazardous waste). Under RCRA the disposal of certain hazardous wastes by placing them on the land or burying in unregulated landfills was prohibited. This prohibition includes any product "derived from" a listed hazardous waste, regardless of any benefits that may derive from its use (40 CFR 261.3). In the case of fertilizers, a determination was made that a listed waste, or a material "derived from" one of these wastes by being utilized in a fertilizer was being placed on the land and this use constituted disposal and was prohibited under the land ban. This obviously has had a "chilling" effect on the use of many baghouse dusts and scrubber sludges in fertilizers due to record keeping requirements, the cost of mandatory insurance coverage for hazardous waste activity, and all other costs associated with compliance with an ever-growing number of burdensome regulations. Prior to 1980 when the regulations first were put into place, large quantities of baghouse dust and scrubber sludge were utilized as micronutrient raw materials. Some wastes containing high levels of zinc are still utilized today but in decreasing amounts. Many of these materials contain unwanted lead, cadmium, nickel, chromium, and arsenic as impurities that render the materials characteristically hazardous. Whether any danger is posed by processing and/or applying a micronutrient containing lead and cadmium is of some debate. A material can be tested to determine if it should be considered a hazardous waste by subjecting it to the Toxicity Characteristic Leaching Procedure (TCLP). If the level of certain designated metals are in excess of specified limits, the material is considered a hazardous waste. There is an on-going debate between regulators and generators as to what is a product, co-product, by-product, or waste. If a material is not a waste, it cannot be a hazardous waste regardless of its characteristics.

While some of the listed and characteristic wastes were being buried or "dumped," others were being beneficially recycled, bringing significant economic value. As with most regulatory enforcement, every material covered by the regulations was treated the same with few exceptions. In the case of electric arc furnace dust(K061), a listed hazardous waste, a temporary exemption from RCRA was granted for its use in fertilizer as long as it undergoes a chemical change, can be used without further processing, and is registered for sale as a fertilizer (50 CFR 666). This exemption was granted as a result of evidence submitted to the EPA showing that without K061 there would not be an adequate supply of zinc for use as a micronutrient, additionally the lead and cadmium present in K061-derived products would cause no significant change in soil concentrations when applied at normal rates. It is possible to have a material removed from RCRA authority if it can be proven that it poses no risk or exhibits none of the hazardous characteristics for which it was initially listed. This is not a simple or easy task.

Many companies who in the past sold their baghouse dust or scrubber sludge for use in fertilizers are now paying huge amounts to have these materials recycled or disposed of in a licensed treatment facility, recovery facility, or permitted landfills. Some treatment processes turn materials into insoluble briquettes, which are then buried, but in most cases these materials are stabilized by mixing with cement or other binders and simply buried in a regulated landfill. In cases where the metals are being recovered in a process declared to be the best demonstrated available technology (BDAT), they can be sold even for use as fertilizers. In one BDAT process the recovered material has the harmless impurity (iron) removed, but the heavy metals are still present (albeit in a more concentrated form). This material can then be utilized as a fertilizer with no restrictions. Some processes do remove the lead and cadmium at a high cost. The resulting product is often too expensive for use as a fertilizer micronutrient.

A discussion of each of the micronutrients and the by-products from which they can be derived follows. An attempt will be made to explain the available types of materials and what physical or chemical characteristic might limit a product's use.

Boron. Boron is critical to the germination of pollen grains and to cell wall formation. Boron deficiency may appear in vegetable, fruit, and grain crops. Typical symptoms of boron deficiency in corn are ears that are poorly filled or twisted in appearance. In sugar beets, broccoli, and cauliflower the deficiency may appear as a hollow area in the stem, while in peanuts it may appear as a hollow heart, a void in the center of the nut. Deficiencies of boron were reported in 44 states in a 1966 survey by Burgess (Cunningham). A soil is considered deficient if it contains less than .3 ppm and the typical application is 1-1.5 pounds of B per acre. Boron usage has been fairly constant for the past thirty years with estimates of about 5,000 tons of elemental B being utilized as fertilizer borate. Estimates of boron consumption have been the most reliable of all of the micronutrients. There are two domestic producers of borate products. All imports can be traced by examining import records maintained by the U.S. Customs Service. In the past twenty years, imports of borate products from Turkey and South American countries have increased, but their primary use has been for industrial applications. The chemical forms can vary, but sodium borate is the most commonly used form both for industrial and agricultural applications. There has been an increase in the use of sodium-calcium borate (ulexite) in the past few years due to the availability of low cost materials from South America as well as high quality ulexite from Turkey. Calcium borate (colemanite) has also been utilized in fertilizers for many years, but its use has been limited due to supply problems as well as its relatively high cost.

The water solubility levels of various boron sources are of concern due to the potential for water leaching boron out of the plant's root zone in coarse textured or sandy soils, where most boron deficiencies are found.

Boron toxicity problems can arise when excessive levels of boron are applied, and for this reason, some users prefer a boron source that is not all immediately soluble, such as

ulexite or colemanite. Boron slags are not readily soluble
and are used to supply boron to crops grown in soils that
are easily leached.
    Few by-products are available for use as boron sources.
Some baghouse dusts from glass bead production and slags
created by the fluxing of metal recovery smelters contain
boron, as well as other metals, and are utilized on a
limited basis.

Chlorine. Chlorine is rarely deficient in the soil due to
its presence in muriate of potash (potassium chloride),
which is the most commonly used potassium fertilizer. There
is currently no practical reason for by-product chlorine
sources to be used as a fertilizer. As with several other
micronutrients, there is concern about chlorine toxicity or
leaf burn from its application on certain crops. For this
reason potassium sulfate and potassium nitrate are used in
greenhouses and on leaf crops.

Cobalt. Cobalt is needed by legume root nodules to fix
atmospheric nitrogen. Like molybdenum, it is used by the
plant and in fertilizers in extremely small quantities due
to cost and the fact that it can be coated onto seed or be
foliar applied if it is needed. Due to an unusual soil
condition, relatively large quantities of cobalt are used in
New Zealand and Australia. The usage of cobalt there
fluctuates with the price of cobalt and sheep.
    By-product cobalt can be found in spent catalyst or
from off-grade cobalt oxide and sulfate products.

Copper. Copper is needed for chlorophyll formation in the
plant. Where copper is used as a fungicide, the need for its
use as a micronutrient fertilizer is reduced. Copper
deficiencies were reported in 14 states in the 1966 survey
by Burgess(Cunningham). In arriving at the total amount of
copper used as a fertilizer, one must be careful to take
this dual use into account. The 1984 USDA report estimated
fertilizer use at 1,100 metric tons of elemental copper
while the 1980 estimate was 1,500 tons (Mortvedt). Frit's
estimate for copper use in 1995 is 1,100 tons on an
elemental basis. The major reason for the decline in the
use of copper is its cost versus commodity prices,
especially wheat.

Copper used for fertilizers must be extractable by hydrochloric acid if it to be guaranteed as a plant nutrient. Metallic copper and some of the copper contained in brass slags and skimmings is not hydrochloric acid extractable. However, this copper can be recovered by nitric acid extraction to produce copper nitrate or by heating to produce copper oxides. Copper oxide and copper sulfate are the primary sources of fertilizer copper. All domestically produced copper sulfate is now produced from copper scrap.

Useable by-product copper can be found in various petroleum refining catalysts, copper recovery operations, smelter effluent streams, furnace slags, scales from copper wire drawing, copper plating wastes, and baghouse dust from copper tube manufacturing.

Iron. The consumption of iron has increased substantially in the last decade with the growth in usage of "specialized" turf fertilizers that contain iron. Burgess reported in 1966 that 25 states had iron deficiencies in their soils(Cunningham). Iron serves as an oxygen carrier in the plant and is an integral part of the chlorophyll molecule. Iron will cause turf grasses to become darker green in color, which is associated with a healthy lawn. In 1984, iron use was estimated at 5,900 metric tons on a 100% iron basis (Mortvedt). Frit's estimate for 1995 is 6,400 tons. Iron deficiencies can occur due to an imbalance of iron with molybdenum, copper, manganese, and phosphate, or from a high soil pH(Olsen). Iron sulfate from the neutralization of wastes from the cleaning of steel coils and sheets prior to galvanizing and from titanium oxide production is widespread in use. Iron oxide and slags from copper smelting, iron baghouse dusts, and mill scale from primary steel production are also used in large quantities to provide an inexpensive source of iron for fertilizers.

Manganese. Manganese functions as part of the enzyme system of plants and aids in chlorophyll production. Deficiencies are typically found in wet, cold soils or soils containing high levels of organic matter. Thirty states reported manganese deficiencies in 1966(Cunningham). The use of manganese in fertilizers has remained fairly constant from 1984 to 1995. The 1984 USDA estimate of 15,100 metric tons of manganese (Mn) (Mortvedt) is considerably higher than

Frit's estimate of 13,100 tons for both 1984 and 1985. Sources of manganese are baghouse dusts, scrubber sludges, and slags produced during the production of ferromanganese, silica manganese, or electrolytic manganese dioxide, and the fine particles produced when these manganese alloys are crushed and screened to size. Manganese ores (MnO_2) can be used to produce manganese fertilizer products, but they must first be reduced to manganous oxide (MnO++) to be available for plant uptake. Manganese presents agronomic problems due to its tendency to oxidize in the soil and revert to manganese dioxide or some other unavailable form of manganese.

Molybdenum. Molybdenum usage in fertilizers is quite limited although 30 states reported deficiencies in 1966(Cunningham). Molybdenum is required by the symbiotic bacteria found on roots of legumes that enable the plant to fix elemental nitrogen. Its use is limited by the small amount required and, due to the cost, it is often coated onto the seed prior to planting. Little molybdenum is applied to the soil or foliage, due to its unique role in plant nutrition. It is difficult to estimate the usage of molybdenum as fertilizer but in a typical micronutrient fertilizer mixture, the molybdenum guarantee will be .005% and in a mixed fertilizer, the guarantee will be .0005%. While there is little used in micronutrient fertilizers, by-product molybdenum can be found in spent catalyst and off-grade molybdic oxide and sulfate.

Vanadium. Vanadium use is rare, and there is no consensus among agronomists that supplemental applications of vanadium are needed, although some tests performed at Texas A&M have shown that vanadium applications benefit grain and cotton crops under drought conditions(Wendt).

Zinc. Zinc is vital in the formation of chlorophyll and synthesis of carbohydrates. It is the most widely used micronutrient, with 43 states reporting deficiencies. Zinc is routinely applied to corn, wheat, rice, and other grain crops. Zinc deficiencies are more widespread than any other micronutrient and usually appear in sandy or high pH soils. High soil phosphate levels can also induce zinc deficiency.

In 1984 the U.S. Department of Agriculture reported 37,300 metric tons of elemental zinc was used as a fertilizer micronutrient(Mortvedt). Frit Industries estimates that 29,500 metric tons of zinc were applied in 1995. Sources of zinc raw materials are: plating wastes, baghouse dust from smelters, electric arc furnace dust from steel production, zinc dross and zinc skimmings from galvanizing operations, zinc-copper baghouse dust from brass and bronze smelters, boiler ash, and various streams from industrial processes or water treatment systems. Most zinc materials contain lead and, in some cases, cadmium as an impurity due to their occurrence in the same geological deposits(Lagerwerff) or the improved flowability of molten zinc which has lead included as part of the ingot being melted. Prime western zinc is an example of this.

Zinc containing by-products having a low bulk density or dust present problems for users due to OSHA regulations restricting worker exposure to dusts--especially those containing lead and cadmium. Mechanical conveying systems used in the fertilizer industry are typically bucket elevators and belt conveyors. Neither fine dry materials nor sludges are handled safely by these conveyances. One very important limitation to a material's utilization is its content of very fine metallic zinc which will oxidize when exposed to a source of moisture, generating extreme heat.

Micronutrient usage has leveled off in recent years. The source of the fertilizer micronutrient raw materials will continue to be by-products. Fertilizer micronutrients already consume, with a great benefit to society, many by-product materials that have no other economic use. If further regulation prohibits this use, it would drive more materials into landfills or into non-economic, subsidized recovery programs. The production of all copper, iron, and zinc material is already heavily dependent on recycled materials. Even if purified products are used as micronutrients, their source will continue to be by-products.

As agricultural practices continue to evolve, the trend will probably be towards more selective application of fertilizers and micronutrients. Selective application of fertilizers does not mean less fertilizer or micronutrients will be used. Environmentally sensitive or marginally productive land will probably not be farmed; instead the more productive and/or less environmentally sensitive land will be more heavily fertilized to feed the world's population. For this reason, micronutrients will continue to be needed to optimize production.

Bibliography

Alsrich, Samual R., Scott, Walter O., Leng, Earl R., Modern Corn Production, A&L Publications, Champagne, IL 1976.

Brown, A. Lloyd, Bureau, Richard G., Giger, David R., "Flue Dusts As Zinc Fertilizers," California Agriculture, August, 1977.

Cunningham, H.G., Micronutrients In Agriculture, "Trends in the Use of Micronutrients", Soil Science Society of America, Madison, WI, 1982.

Farm Chemical Handbook '95.Vol. 81., Meister Publishing Company, Willoughby, OH, 44094, 1995.

The Fertilizer Handbook, The Fertilizer Institute, Washington, DC, 1982.

Foth, Henry D. Fundamentals of Soil Science, John Wiley & Sons, New York, 1978.

Hignett, T.P.,Fertilizer Manual, Developments in Plant and Soil Sciences, Volume 15, Martinus Nijhoff, Dr. W. Junk Publishers, Dordrecht, The Netherlands, 1985.

Lagerwerff, J.V., "Lead, Mercury, and Cadmium as Environmental Contaminants", Micronutrients in Agriculture, Soil Science Society of America, Madison, WI, 1972

Lepp, N.W., Effect of Heavy Metal Pollution on Plants, Volume 2, Metals in The Environment, Applied Science Publishers Inc., Englewood, NJ 1981.

Mortvedt, J.J. 1987. "Cadmium Levels in Soils and Plants From Long Term Soil Fertility Experiments in The United States of America," J. Environ. Qual.,16:137-142.

Mortvedt, J.J. 1985. "Plant Uptake of Heavy Metals in Zinc Fertilizers Made From Industrial Byproducts," J. Environ. Qual., 14:424-427.

Mortvedt, J.J., "Use of Industrial By-Products Containing Heavy Metal Contaminants in Agriculture," Residues and Effluents- Processing and Environmental Considerations, The Minerals, Metals & Materials Society, 1991.

National Agricultural Statistical Service "Track Records, United States Crop Production,", Beltsville, MD, 20705-2351, 1996.

Nicholas, D.J.D., Egan, Adrian R., Trace Elements in Soil Plant-Animal Systems, Academic Press, New York, 1975.

Nriagu, Jerome O., Zinc in The Environment, Part 1: Ecological Cycling, John Wiley & Sons, New York, 1980.

Olsen,S.R., Micronutrients in Agriculture, "Micronutrient Interactions", Soil Science Society of America, Madison, WI, 1972.

Page, A.L., Logan, T.J., Ryan, J.A., Land Application of Sludge, Lewis Publishers, Inc., Chelsia, MI, 1987.

Russell, E.W., Soil Conditions and Plant Growth, Longman Group, Ltd., New York, 1973.

Sittag, Marshall, Metal and Inorganic Waste Reclaiming Encyclopedia, Noyes Data Corporation, Park Ridge, NJ, 1980.

Wendt, Charles, "TAES Research", The Cross Section, High Plains Underground Water Conservation District No.1, Lubbock, TX, May, 1991.

Weiss, Stanley A., Manganese, The Other Uses, Metal Bulletin Books Ltd., Garden City Press Limited, Letchworth, Hertfordshire, Great Britain, 1977.

Wolnik,Karen, Fricke, Fred L., Braude, George L., Meyer, Milton W., Satzger, R. Duane, Bonnin, Evelyn. "Elements in Major Raw Agricultural Crops in the United States. 1. Cadmium and Lead in Lettuce, Peanuts, Potatoes, Soybeans, Sweet Corn, and Wheat," J. Agric. Food Chem., Vol.31, No. 6, 1983.

Chapter 17

Iron Humate Production, Processing, Properties, and Usage

B. E. Rehberg and G. L. Smith

IMC Vigoro, 150 Third Street Southwest, Winter Haven, FL 33880

Iron Humate is a controlled release iron product generated by the water treatment industry during the production of potable water from humate rich river water sources. Use of pure iron salts to precipitate these complex color agents provides a new raw material that offers more environmentally sound use options than humates precipitated with aluminium or contaminated iron salts. The raw iron humate filter cake can be processed in different ways to produce a range of controlled release iron products. Dried iron humate is the slowest release iron source but exhibits fragile physical characteristics. Reaction of wet or dried humate with different acid-base systems and process equipment improves it's physical quality and modifies the release properties of the iron. This integrated approach of linking water purification to the manufacture of multi-use final products hold promise for future environmental management projects.

Traditionally, colored surface water was clarified with aluminum sulfate (alum) by potable water treatment facilities. This treatment coagulated the colored humates and created an aluminum sludge which, after dewatering, was discarded. In the past, the sludge was dumped back into the water source, to the detriment of water quality. Current disposal options include lagoon storage, landfills and discharge into sanitary sewers. Inadequacies in all the disposal techniques has led to increased interest in land application of water treatment sludge.

Iron salts have also been used to coagulate humates but, too often produced a sludge high in heavy metals due to impurities. Another problem encountered with this potable water treatment method was it generally resulted in a sludge with unnecessarily high concentrations of hydrous metal oxides. These metal hydroxides are strong absorbents of metals and phosphorus and can make these essential nutrients unavailable for plant use. Additionally, excessive iron hydroxide levels impede the humate dewater process which translates into increased processing and handling costs. Such problems do not afford materials well suited to land application.

Integrated Approach to Water Clarification and Slow Release Iron Production.

Existing water treatment processes focus primarily on producing high quality drinking water with little regard for the quality of the resulting sludge. The goal of the integrated water treatment process is to produce both high quality water and raw materials for industry. One example of just such a system is seen in a patent issued for the production of iron humates from a potable water color clarification process (*1*). There are three critical factors required to assure the goals of clear water and effective iron humate can be met: 1) pure ferric sulfate salt, 2) proper iron-humate stoichiometry, and 3) effective pH control.

Production of the Iron Humate Raw Material. The coagulant salt (ferric sulfate) is manufactured specifically for water treatment from a pure grade of iron ore and virgin sulfuric acid. Raw water is tested at the treatment facility to determine the optimum pH conditions (3.8-5.5) and the proper ferric salt to humate stoichiometry to provide clear water and preferred iron humate. Once these factors are determined, the raw water is adjusted to the prescribed pH. Downstream of the acidification zone the pure iron salt coagulant is added at a rate previously established to maximize clarification and minimize iron hydroxide formation. This solution is sent to a flash mixer for a minimum of 15 seconds, then to a flocculating mixer for 3 to 10 minutes depending upon the temperature of the water. After the reaction is completed, the iron humate is allowed to settle, separate, and is dewatered to about 20% solids. This filter cake is air dried to 55-65% solids, classified and represents the base form of the raw material iron humate.

Chemical Composition of Raw Material Iron Humate. Elemental analysis for the principal components are 16.5 to 19.5% Fe, 35% to 40% organic carbon, and 35% to 45% moisture. The physical character of the iron humate is quite fragile. Particle crush strength values run less than 1-2 lb of force. Additional elements present in this raw material are listed below (Table I).

Table I. Analysis of Iron Humate for Primary, Secondary, Minors and Metals

Component	Wt%	Component	ppm	Component	ppm
Phosphorous	1.0-1.5	Magnesium	180-350	Tin	12.0
Potassium	0.5-1.0	Manganese	150-280	Lead	4.5
Sulfur as SO_4	2.5-3.0	Barium	50-120	Titanium	4.4
Calcium	1.5-2.5	Copper	25 - 85	Selenium	0.14
Chloride	0.2-0.3	Zinc	15 - 50	Mercury	<3.0
Aluminum	0.5-0.8	Nickel	5 - 10	Cadmium	<1.7
Silica	0.1-0.2	Chromium	7 - 10	Antimony	<5.0

The heavy metal content of this water treatment raw material falls well below the EPA 503 sludge regulations for land application (*2*).

Agronomic Effectiveness of Raw Material Iron Humate [0-0-0-18(Fe)]. Crop yield increases were most dramatically seen with its use on citrus. A study in Arcadia, Florida on Hamlin oranges indicated the effectiveness of iron humate. During this three year study, iron humate treated plots (88g Fe/tree) were compared against identical plots treated with chelated iron (EDDHA 22g Fe/tree). Control plots received fertilizer only. Results showed a clear trend of higher pounds solids (15%-20%) from trees treated with iron humate over chelates or fertilizer control (*3*). A similar study on Flame grapefruit was conducted in Fort Pierce, Florida and showed iron humate treatments affording fruit and juice yields equal to EDDHA chelated iron treatments (*3*). In both these studies the soils were sandy with pH readings in the range of 7.5 -8.0.

Other potential areas of agricultural application for straight iron humate raw material are in vegetable and horticultural production, as well as supplements in animal feeds. Initial studies with tomatoes and bell peppers in sandy soils were encouraging but required application rates of 500 to 750 lb/acre. Feeding studies with pigs showed the iron humate to be a very palatable source of complex iron for these animals. More testing is required in both areas before specific economic assessments can be made.

Reactions with Iron Humate for Nutrient and Physical Character Enhancement

Although iron humate had a number of admirable agronomic successes, it would not support more than a few market applications in its raw physical and chemical form. As noted above, this raw material is friable, which prevents its use in most blending operations. Raw iron humate releases iron very slowly so it will not fit a number of "short duration" crop applications. The product is very insoluble in water and has a neutral citrate solubility of only 30%. This analysis is a laboratory indicator of the limited availability of the iron in the complex organic structure. As a consequence of these features the raw iron humate provided its best results when; 1) it was used in sandy or porous media, 2) the weather was warm and wet, 3) it was incorporated rather than top-dressed and 4) it was used on perennial woody plants. To improve the value of the iron humate, improvements in its chemical availability, and physical characteristics were pursued.

Acid - Base Reactions Using Sulfuric Acid. To improve physical characteristics and increase iron solubility, the iron humate was mixed with gypsum, basic reactants, and granulated using sulfuric acid. The basic materials consisted of carbonates or oxides of several divalent metals such as calcium, magnesium, manganese, zinc, or copper. In general, these granulated products were round and hard with crush strengths over 2 lb of force. The acidulation processes altered the water solubility of iron up to four-fold, and doubled or triple its availability. This increase in available iron was thought to occur by ion displacement and/or humate complex denaturing.

Composition of Granulated Solubilized Dolomitic Iron Humate. Granulation of 40% iron humate, 40% of a mixture containing dolomite, lime, magnesia, gypsum, and 20% sulfuric acid gave a physically improved, solubilized dolomitic iron humate. This improved material [0-0-0-8(Fe)] was produced using common raw materials, and standard acid granulation equipment. Elemental analysis of the principal components of this product is listed below (Table II).

Table II. Analysis of Granulated Solubilized Dolomitic Iron Humate

Component	Wt%	Component	Wt%	Component	Wt%
Total Iron	8.0	Magnesium	5.0	Calcium	7.0
Soluble Fe	1.2	Soluble Mg	2.0		
Available Fe	4.8				
Sulfur as SO_4	7.7				

Agronomic Effectiveness of Solubilized Dolomitic Iron Humate [0-0-0-8(Fe)]. The combination of more available iron with extra calcium, magnesium and sulfur afforded positive results for row crops on sandy soils. Independent research results on this dolomitic iron humate (8% Fe) versus iron chelate (10% Fe) on tomatoes and bell peppers were notable (unpublished data). At a rate of 250 lb/acre the solubilized iron humate gave a 2.5% increase over chelates of the number of 25 lb cartons/acre for tomatoes and a 45% increase in the number of 28 lb cartons/acre for peppers. Yields over the fertilizer control were 6.5% and 65% respectively. In comparison, the raw material iron humate (18% Fe) was not "as effective" even at rates up to 750 lb/acre on either crop (Table III). Control plots received fertilizer only.

Table III. Effect of Iron Source and Rate on Yields of Tomato and Pepper

Treatment	Rate lb/acre	Tomatoes 25lb Cartons/acre	Peppers 28lb Cartons/acre
Dolom IronH	250	2062	1439
Iron Humate	750	1994	893
Chelate Iron	5.0	2014	973
Fertilizer Only	---	1932	857

Chemical Composition of Soluble Nitrogen Mineralized Iron Humate. Reactive granulation of 40% iron humate, 46% of a mixture containing gypsum, magnesia, urea, manganous oxide, ammonium sulfate, zinc sulfate, iron sulfate and 14% sulfuric acid produced an attrition resistant, activated iron humate with added minerals (*4*). This new product [7-0-0-8(Fe)] was designed for use on turf for improving color. Turf type applications require higher iron availability than that achieved in the dolomitic humate product while keeping water soluble iron to a minimum. This combination of iron features maximize turf color yet minimize staining potential. Analysis of the product components is listed below (Table IV). Additional products made in this same series were 4-0-0-18(Fe), 10-0-0-7(Fe), 10-0-4-7(Fe) and 8-0-8-8(Fe). All the products contained the elements calcium, magnesium, manganese, zinc and sulfur as sulfate. Acceptable crush strength values of 3 lb force were obtained on this product set.

Table IV. Analysis of Soluble Nitrogen Iron Humate Plus Minors

Component	Wt%	Component	Wt%	Component	Wt%
Total Nitrogen	7.4	Zinc	0.2	Calcium	5.2
Total Iron	9.4	Magnesium	3.3	Manganese	3.0
Soluble Fe	0.1	Soluble Mg	2.6	Soluble Mn	2.4
Available Fe	8.9				
Sulfur as SO_4	13.6				

Agronomic Effectiveness of Soluble Nitrogen Iron Humate Plus [7-0-0-8(Fe)]. Adding nitrogen and manganese to the iron humate raw material helps the uptake and effectiveness of the iron in a turfgrass environment. Applications of the 7-0-0-8(Fe) to Penncross Creeping Bentgrass at rates of 0.4 lb Fe/6wk and 0.8 lb Fe/6wk gave excellent color and quality readings (unpublished data). Application of the iron humate raw material [0-0-0-18(Fe)] at rates of 1 lb Fe/15wk to 2 lb Fe/15wk under these conditions showed no improvement over the non-iron check plots. Five months of quality ratings and product comparisons are reported below (Table V).

Table V. Quality Ratings of Penncross Bentgrass putting Greens

Treatment	Rate/M	12/6	1/13	1/26	2/16	3/7	3/28	4/18	5/8
[0-0-0-18(Fe)]	1 lbFe/15wk	5.2[a]	5.5	5.5	5.6	6.5	6.0	7.4	6.8
[0-0-0-18(Fe)]	2 lbFe/15wk	5.4	5.4	5.5	5.6	6.2	6.1	7.2	6.7
[7-0-0-8(Fe)]	0.4 lbFe/6wk	7.1	6.2	6.6	6.6	6.7	6.4	7.7	7.0
[7-0-0-8(Fe)]	0.8 lbFe/6wk	7.4	6.6	7.0	7.4	7.0	6.9	7.6	7.3
FeSO$_4$ spray	0.13 lbFe/3wk	5.9	5.6	6.6	6.2	6.7	6.1	7.5	6.8
Check plots	- - - -	5.4	5.5	5.6	5.9	6.4	5.7	7.0	6.9

[a]Visual turfgrass quality: 1 to 9; 9 = best; 5 = minimum acceptable

Acid - Base Reactions Using Phosphoric Acid. To improve physical characteristics and increase iron availability, the iron humate raw material can be reacted with and matrixed into slow release fertilizer salts like magnesium ammonium phosphate and manganese potassium phosphate (5). Reactant materials consist of oxides of divalent metals (calcium, magnesium, manganese, zinc, or copper), diammonium phosphate (DAP), phosphoric acid (purity requirement dependent on use as fertilizer or animal feed additive), potassium hydroxide and iron humate. For these acid-base reactions to succeed, high intensity mixer-granulator equipment and process technology is required. The increase in available iron is thought to occur by ion displacement and/or phosphate-iron-humate complexing.

Chemical Composition of Manganese Potassium Iron-Phospho-Humate. Reactive granulation of 35% iron humate, 40% of a mixture containing manganous oxide, potassium hydroxide solution (50%) and 25% black phosphoric acid (53%) afforded a slow release iron-potassium product with high attrition resistance.(5). This new slow release product [0-16-10-10(Fe)-13(Mn)] was designed for use on high sand-low CEC soils to improve the available levels of potassium, iron and manganese. This product application requires use of the latest technologies available to target and meet the product design criteria. Elemental analysis of the principal features of this product is listed below (Table VI).

Table VI. Analysis of Slow Release Manganese Potassium Iron H Phosphate

Component	Wt%	Component	Wt%	Component	Wt%
Total P$_2$O$_5$	17.4	Available P$_2$O$_5$	9.9		
Total Iron	9.9	Manganese	13.4	Potassium	11.5
Soluble Fe	0.1	Soluble Mn	0.7	Soluble K$_2$O	3.1
Available Fe	8.9	Available Mn	5.9		

Additional products made in this same series were the new matrix formulations 0-14-10-14(Fe)-5(Mg), and 0-16-10-10(Fe)-7(Mn)-4(Mg). These products, containing the controlled release elements of magnesium and manganese, are derived from the salts of metal potassium phosphate. Excellent crush strength values of 3 to 4 lb force were obtained with these granules.

Agronomic Effectiveness of Manganese Potassium Iron-Phospho-Humate [0-16-10-10(Fe)-13(Mn)]. Matrixing the dry iron humate raw material into a manganese potassium phosphate salt is an effective method for the manufacture of products with slow release character for all the listed elements. To maintain efficacious levels of iron, manganese and potassium in high-sand-low-CEC soil environments, use of controlled release nutrients is beneficial. Proper linkage of product technology to agronomic need does provide benefit. Applications of the 0-16-10-10(Fe)-13(Mn) to creeping bentgrass at rates of 1.0 lb Fe/9wk gave only marginal color and quality readings (unpublished data). However, mixing this product 1:1 with a controlled release nitrogen source provided good results at application rates as low as 0.25 lb Fe/9wk. Use of the iron humate raw material [0-0-0-18(Fe)] at rates of 1 lb Fe/15wk to 2 lb Fe/15wk under these conditions showed no improvement over the non-iron check plots. Five months of quality ratings and product comparisons are reported below (Table VII).

Table VII. Quality Ratings of Penncross Bentgrass putting Greens

Treatment	Rate/M	12/6	1/13	1/26	2/16	3/7	3/28	4/18	5/8
[0-0-0-18(Fe)]	1 lbFe/15wk	5.2[a]	5.5	5.5	5.6	6.5	6.0	7.4	6.8
[0-0-0-18(Fe)]	2 lbFe/15wk	5.4	5.4	5.5	5.6	6.2	6.1	7.2	6.7
[0-16-10-10(Fe)]	1 lbFe/9wk	5.7	5.5	5.9	6.4	6.4	5.6	7.3	6.3
[15-8-5-5(Fe)]	0.25 lbFe/6wk	6.2	6.5	6.4	7.1	6.6	6.2	7.4	7.2
FeSO4 spray	0.13 lbFe/3wk	5.9	5.6	6.6	6.2	6.7	6.1	7.5	6.8
Check plots	- - -	5.4	5.5	5.6	5.9	6.4	5.7	7.0	6.9

[a]Visual turfgrass quality: 1 to 9; 9 = best; 5 = minimum acceptable

Chemical Composition of Magnesium Ammonium Iron-Phospho-Humate. Reactive granulation of 60% iron humate, 25% of a mixture containing magnesium oxide, diammonium phosphate (DAP) and 15% low fluoride phosphoric acid (53%) made a slow release ammonium-iron product with high attrition resistance.(5). This new slow release product [3-17-0-12(Fe)-5(Mg)] was designed specifically as an available iron source for feeding animals. Such a product application requires use of the best quality raw materials and latest technologies to target and meet animal feed product specification criteria. Analysis of the composition of this product is listed below (Table VIII).

Table VIII. Analysis of Slow Release Magnesium Ammonium Iron H Phosphate

Component	Wt%	Component	Wt%	Component	Wt%
Total P_2O_5	16.6	Available P_2O_5	12.1		
Total Iron	14.4	Magnesium	5.6	Total Nitrogen	3.2
Soluble Fe	0.01	Soluble Mg	0.5	Soluble Nitrogen	1.0
Available Fe	12.0	Available Mg	4.9		

Additional products made in this series were new matrix formulations 4-22-0-10(Fe)-4(Mn)-6(Mg)-2(Zn)-0.5(Cu), and 4-22-10-10(Fe)-7(Mn)-3(Mg). The products containing the controlled release elements of nitrogen, phosphorus, copper, zinc, magnesium, and manganese, are derived from the salts of metal ammonium phosphates. Acceptable crush strength values of 3 lb force were obtained with these granules.

Animal Feed Effectiveness of Magnesium Ammonium Iron-Phospho-Humate [3-17-0-14(Fe)]. To improve iron uptake in feed animals the water solubility of the humate product must be low and its availability high. The physical/chemical interaction of the magnesium ammonium phosphate - iron humate matrix provided the controlled release characteristics that are necessary in this application. The [3-17-0-14(Fe)] matrix nutrient granules and two commercially available iron supplements were administered to six litters of pigs to provide a comparison of their utilization characteristics. One third of the pigs were orally fed 75 grams/litter/day of the matrix nutrient granules, another third were fed 75 grams/litter/day of a commercial oral iron product, and the last third were injected intravenously with 100 mg iron/day from a commercial iron dextran material. Blood samples were taken from each pig at seven different time intervals throughout the 30 day testing period with hemoglobin determinations made on each. The results of the test showed that this from of modified iron humate provided excellent availability and consistency for the pigs over its 30 day duration. Results of the test are shown below (Table IX).

Table IX. Hemoglobin Levels of Pigs Treated with Iron Supplement

Days after Treatment	Hemoglobin Level, G/DL		
	Modified Iron Humate	Comm Oral Iron	Comm Injection Iron
1	10.42	10.41	9.36
3	7.72	7.60	8.44
5	9.00	7.10	9.13
10	11.25	9.84	9.44
15	11.40	10.42	10.26
22	13.75	11.22	9.14
30	11.59	10.91	7.50

Conclusions on Uses of By-Products and Waste in Agriculture.

Generating wastes in large quantities is easy, but manufacturing a raw material suitable for agricultural or other uses takes considerably more planning and research. The production of iron humate is small when compared to bio-solid production in the United States but the concepts, principals, planning, and research required for implementation are similar. For effective utilization of waste materials to occur, targeted end use products have to be identified, defined, and the waste raw material engineered to meet those targets. Planing needs to happen at the beginning of a waste resolution project, not the end. This project was, and continues to be, an attempt to meet the requirements of producing an economical agricultural raw material while resolving a potable water treatment disposal problem.

Literature Cited

1. Hjersted, L.N.; U.S. Patent 5,213,692 (1993).
2. Walker, J.; Knight, L.; Stein, L. *A Plain English Guide to the EPA Part 503 Biosolids Rule*; EPA/832-R-93-003; EPA, Office of Waste Water Management: Washington, D.C., 1994; pp. 29.
3. *Iron Humate Studies Show Results at Reduced Cost*; Garner, F.C., Ed.; Florida Grower and Rancher; Meister Publishing Company: Willoughby, OH, 1996, Vol.89 No. 9, pp. 11.
4. Moore, W.P.; Wilson, B.B.; Rehberg B.E. *Hard Particle Iron Humate Matrix Fertilizers*; (paper presented at the 212th ACS National Meeting, Orlando, FL, Aug 1996.
5. Moore, W.P.; U. S. Patent 5,354,350 (1994).

Indexes

Author Index

Barker, Allen V., 120,140
Black, E. D., 204
Camberato, J. J., 185
Edwards, J. H., 163
Forste, Jane B., 50
Galloway, Dale F., 22
Hall, M. B., 91
Huntley, E. E., 120
Lindsay, Billie J., 63
Logan, Terry J., 63
Mikkelsen, Robert L., 110
Miller, W. P., 226
Mitchell, C. C., 204
Muchovej, Rosa M. C., 1
Pacovsky, R. S., 1

Rehberg, B. E., 268
Rubin, Alan B., 28
Sims, J. Thomas, 72
Smith, G. L., 268
Someshwar, A. V., 185
Southworth, Robert M., 28
Stratton, M. L., 120
Sumner, M. E., 226
Titko, Steve, 63
Van Horn, H. H., 91
Vance, E. D., 185
Walker, John M., 22,28
Wallach, Daniel L., 240
Wyatt, James M., 255

Affiliation Index

Agricultural Research Service, 1,163
Auburn University, 204
Clemson University, 185
Dakota Gasification Company, 240
Florida Certified Organic Growers & Consumers, Inc., 120
Frit Industries, Inc., 255
IMC Vigoro, 268
National Council of the Paper Industry for Air and Stream Improvement, 185
North Carolina State University, 110

Ohio State University, 63
RDE, Inc., 22
Scotts Company, 63
University of Delaware, 72
University of Florida, 1,91,120
University of Georgia, 226
University of Massachusetts, 120,140
U.S. Department of Agriculture, 1,163
U.S. Environmental Protection Agency, 22,28
Wheelabrator Water Technologies Inc., 50

Subject Index

Acid–base reactions, iron humate, 270–272
Acidic subsoils, use of byproduct gypsums, 232–234,235f
Agricultural issues
  cattle manure management, 91–107
  poultry waste management, 72–88
  swine waste management, 110–118
Agricultural markets
  byproduct gypsums, 232–236
  regulations, 67
Alkaline treatment, biosolids, 58–59,60t
Alkalinity, composts, 154f,155–156
Aluminum, content from waste paper, 175–176
Ammonia emissions, cattle manure, 97–98
Ammonia scrubber, applications, 14
Ammonium sulfate fertilizer, byproduct in flue gas desulfurization, 240–254
Ammonium sulfate forced oxidation flue gas desulfurization
  commercial system, 246–253
  construction, 241
  description, 240–241
  development, 242–243
  initial operation, 253–254
  pilot plant, 244–246
Anaerobic treatment, cattle manure, 100–101
Anaerobically digested biosolids, nutrient composition, 51
Animal products, organic fertilizers, 123–127
Animal tissues, use as fertilizer, 126–127
Animal wastes
  applications, 3–4
  contamination, 10
  elemental concentration, 9
  environmentally sound management, 72–73
  properties, 26
  quantities, 8
  storage, 9–10

Basalt rock dusts, use as fertilizers, 132

Biodegradable mulch for vegetable crops, use of organic wastes, 173
Bioenergy, poultry waste, 87–88
Biological control agents, use of organic wastes, 174
Biosolid(s)
  alkaline treatment, 58–59,60t
  applications, 61
  bagged products, 69,71
  composting, 57
  definition, 69,71
  dewatering, 54
  digestion, 53–54
  drying, 59,61
  metals limits, 56
  nitrogen-based agronomic application rates, 52–53
  organic matter and inorganic nutrients, 50
  organic nitrogen, 51
  paper manufacturing, 8
  properties for beneficial uses, 51–53
  regulatory requirements, 54,56
  treated, 63
  U.S. Environmental Protection Agency policies of disposal, 50–51
  wastewater treatment, 6–8
Blended fertilizers, use, 134
Boiler wood ash
  applications, 13
  description, 205
  disposal methods, 205,206f
  land applications, 204–223
  sources, 205,206f
Bone meal, use as fertilizer, 125–126
Boron
  applications, 14
  byproduct use, 261–262
Byproduct(s)
  inorganic, *See* Inorganic byproducts
  management objectives, 22–23
  micronutrients, 14
  minimization, 24
  policies, 29–37
  recycling on land, 28

Byproduct(s)—*Continued*
  regulation changes due to
    intensification, 24
  regulatory requirements, 3
  rules, 29–37
  strategy for holistic management, 38–44
  subsidization, 24
  tailoring, 15
  U.S. Environmental Protection Agency
    regulations, 15–16
  use as feedstock for other processes, 25
  value, 22
  watershed degradation, 23–24
  zero waste goal, 24–25
  See also Byproduct gypsums
Byproduct gypsums
  agricultural markets, 232–236
  applications, 14
  demand, 229–230
  engineering markets, 230–232
  environmental issues, 237–238
  production, 227–228
  sources, 227–229
Byproduct use in fertilizer micronutrients
  boron, 261–262
  chlorine, 262
  cobalt, 262
  copper, 262–263
  iron, 263
  manganese, 263–264
  molybdenum, 264
  vanadium, 264
  zinc, 264–266

Calcium, role in compost, 152–153
Carbon, importance in composting, 143
Carbon mineralization, description, 177–178
Cattle, wastes, 8–10
Cattle manure management
  advantages and disadvantages, 104–107
  air quality issues
    ammonia emissions, 97–98
    methane emissions, 98,99f
    odor control, 96–97
  emission regulation, 92
  environmental issues, 91–107
  nutrient management to ensure water quality, 92–96

Cattle manure management—*Continued*
  partitioning and processing alternatives
    anaerobic treatment, 100–101
    export of nutrients off farm, 102–103
    solids separation, 101–102
    processing of urban wastes, 105–106
    resource value estimation, 104–105
Chlorinated pesticides, occurrence in composts, 156–157
Chlorine, byproduct use, 262
Cobalt, byproduct use, 262
Colored surface water, clarification, 268
Combination ash, description, 205
Compost(s)
  alkalinity, 154f,155–156
  applications, 11
  calcium, 152–153
  demand, 142,143t
  function, 144
  magnesium, 153
  micronutrients, 153
  nitrogen
    carbon:nitrogen ratios, 144–148,150f
    nitrogenous constituents, 148–151,154f
    organic compounds, 156–157
  phosphorus, 151–152
  potassium, 152
  production, 142
  quantities, 10
  soluble salts, 156
  sulfur, 152
  trace elements, 153,155
  use as fertilizers, 129
Composting
  applications, 140,141t
  biosolids, 57
  definition, 140
  description, 143
  importance of carbon, 143
  organic wastes, 164
  poultry waste, 87
  swine waste, 118
Copper, byproduct use, 262–263

Desulfurization, ammonium sulfate forced oxidation, 240–254
Dewatering, biosolids, 54

INDEX

Digestion, biosolids, 53–54
Digestive tract, role in waste production, 25–26
Dioxin
content in waste treatment sludges, 191,192*t*
occurrence in composts, 157
Dried blood, use as fertilizer, 126
Drying, biosolids, 59,61

Electric arc furnace dust, recycling, 260

Fertilizer
ammonium sulfate byproduct in flue gas desulfurization, 240–254
history of use, 257
horticultural, use of processed biosolids, 63–71
manufacture, 86–87
micronutrients, 255–266
organic, *See* Organic fertilizers
specialty, 12–13
substitute, 50–61
uniform standards, 68–69,70*t*
Flue gas desulfurization, ammonium sulfate forced oxidation, 240–254
Food wastes, use as fertilizers, 129–130
Furan, content in waste treatment sludges, 191,192*t*

Glacial dust, use as fertilizer, 132
Granite dust, use as fertilizer, 131–132
Granular fertilizers, description, 256
Granulated solubilized dolomitic iron humate, composition, 270–271
Great Plains synfuel plant
ammonia scrubbing unit, 240
description, 241
development of alternative sulfur removal system, 242–254
original sulfur removal system, 241–242
substitute natural gas production, 240
sulfur removal unit, 240
Green manures and cover crops, use as fertilizers, 128–129
Greensand, use as fertilizer, 132

Guano, use as fertilizer, 124
Gypsums
applications, 226
byproduct, *See* Byproduct gypsums
production as byproduct, 226
use as fertilizer, 132–133

Hazardous waste categories, 259–260
Heat-dried biosolid pellets, source of nutrients for pasture, 52
Horticultural fertilizers, use of processed biosolids, 63–71
Humification, description, 143
Humus, 143–144
Hydrogen sulfide, removal from waste gas at Great Plains synfuels plant, 240–254

Inorganic byproducts and wastes
ammonia scrubber, 14
boiler wood ash, 13
byproduct gypsums, 14
Iron, byproduct use, 263
Iron humate
acid–base reactions
phosphoric acid, 272–274
sulfuric acid, 270–272
production, 269–270
use as specialty fertilizer, 13

Kelp, use as fertilizer, 130

Land application
boiler wood ash in southeastern United States
environmental concerns, 206,212,214
logistics of application, 206
plant nutrients, 208,209*t*,211–212,213*t*
regulation, 214,216–217,222
use as lime, 207–208,209*t*,210*f*
variability of ash, 214,215*f*
description, 50
organic wastes, 164–174
paper manufacturing residuals, 185–200
potential environmental problems from organic wastes, 175–177
swine waste, 118
Langbeinite, use as fertilizer, 133–134
Limes, use as fertilizers, 133

Limestone, use as fertilizer, 133
Liquid fertilizers, description, 256

Magnesium, role in compost, 153
Magnesium ammonium iron phosphohumate, chemical composition, 273–274
Manganese, byproduct use, 263–264
Manganese potassium iron phosphohumate, chemical composition, 272–273
Manure resources, environmental effects, 91–92
Metal(s)
 content in waste treatment sludges, 187–191
 health effects, 6
 land application of boiler wood ash, 211–212,213t
Metal limits, biosolids, 56
Metal loading, buildup from land applications of organic wastes, 175
Methane emissions, cattle manure, 98,99f
Microbial population shifts, description, 178
Microbiomass, dynamics, 177–178
Micronutrients
 application methods, 255
 deficiencies, 257
 description, 255
 examples, 14,255
 fertilizer, byproduct use, 255–266
 land application of boiler wood ash, 211–212,213t
 role in compost, 153
 role of particle size, 256
Mineral sources, use as fertilizers, 130–134
Mixed ash, description, 205
Molybdenum, byproduct use, 264
Municipal sewage sludge, policy on beneficial use, 33–34
Municipal solid waste, 164–165

Nitrate, buildup from land applications of organic wastes, 175
Nitrogen
 cattle manure, 91–107
 challenge for poultry waste management, 76–77,79–80f

Nitrogen—*Continued*
 role in compost, 144–151,154
 swine waste, 115–118
Nitrogen mineralization–immobilization, use of organic wastes, 171–172
Nutrient(s)
 availability, 122
 categories, 255
 content in waste treatment sludges, 187–191

Organic amendments, definition, 121
Organic compounds
 application, 163–178
 composts, 156–157
 content in waste treatment sludges, 187–191
Organic fertilizers
 animal products, 122–127
 definition, 121
 groups, 122
 mineral sources, 130–134
 nutrients, 121–122,127
 plant products, 127–130
Organic wastes
 advantages, 5–6
 composting, 164
 environmental issues, 175–177
 land application, 164–165
 metal content, 167–168
 nutrient content, 167
 organic components, 166
 physical nature and size, 165–166
 role of soil microbial dynamics, 177–178

Paper manufacturing, biosolid waste, 8
Paper manufacturing residuals
 amount generated, 185–186
 causticizing residuals, 194–198
 disposal methods, 187
 effects of land application on water quality and wildlife, 198–199
 examples, 185
 regulation of land application, 199–200
 waste treatment sludges, 187–194
Pelletized paper, characterization, 165
Phosphate fertilizer industry, source of byproduct gypsum, 227

Phosphoric acid, acid–base reactions
with iron humate, 272–274
Phosphorus
buildup from land applications
of organic wastes, 175,176t
cattle manure, 91–107
challenge for poultry waste management,
81–83
land application of boiler wood ash, 211
role in compost, 151–152
swine waste, 115,118
Phosphorus mineralization–immobilization,
use of organic wastes, 172
Plant available nitrogen, calculation, 51–52
Plant nutrients, boiler wood ash, 208–213
Plant products, use as fertilizers, 128–130
Polychlorinated biphenyls
manufacturing, processing, distribution in
commerce, and use prohibitions, 30–33
occurrence in composts, 156–157
Polycyclic aromatic hydrocarbons,
occurrence in composts, 156–157
Potable water treatment facilities, sludge
disposal, 268
Potash, source, 211
Potassium
land application of boiler wood ash, 211
role in compost, 152
swine waste, 115
Potassium-containing ores, use as
fertilizers, 133–134
Poultry
use as fertilizer, 123–124
wastes, 8–10
Poultry waste management
animal feeding programs, 87
bioenergy, 87–88
composting, 87
environmental issues, 72–88
fertilizer manufacture, 86–87
land management programs, 84–86
Processed biosolids, characteristics and
standards in manufacture and marketing
of horticultural fertilizers and soil
blends, 63–71

Quicklime, use as fertilizer, 133

Recyclable byproducts, types, 28
Regulation
agricultural use of byproducts and
wastes, 28–44
biosolid disposal, 50–51
biosolids, 54,56
boiler wood ash, 214
byproducts and wastes, 3
land application of boiler wood ash,
204–223
Resource Conservation and Recovery
Act of 1976, 199,258–259
Section 405(d) of Clean Water Act, 51
waste-derived soil additives and
manufactured soil products, 71

Secondary nutrients, examples, 255
Seed meals, use as fertilizers, 127
Sewage biosolids, use as fertilizers, 125
Sewage sludge, standards for use or
disposal, 34–37
Sludge, from paper manufacturing, 8
Soil
additives
horticultural markets, 64–65,66t
uniform standards, 68–69,70t
blends, use of processed biosolids, 63–71
microbial dynamics, 177–178
nutrient imbalances, 176–177
organic matter content, 173–174
Solid separation, cattle manure, 101–102
Solid waste disposal facilities and
practices, criteria for classification, 29–32
Solubilized dolomitic iron humate,
agronomic effectiveness, 271
Soluble nitrogen iron humate, chemical
composition, 271
Soluble salts, composts, 156
Soybean production, enhancement using
organic wastes, 173
Specialty fertilizers, 12–13
Sulfur, compost, 152
Sulfur dioxide removal
exhaust gases of coal-fired power plants,
source of byproduct gypsum, 227–228
waste gas at Great Plains synfuels plant,
240–254

Sulfuric acid, acid–base reactions with
    iron humate, 270–272
Swine waste management
    challenges, 111
    chemical treatment, 118
    composting, 118
    environmental issues, 110–118
    factors affecting nutrient release,
        117–118
    land applications, 118
    limitations, 116–117
    manure-derived nutrients, 111–112
    manure production, 111,112*t*
    methods, 112–115
    nutrient composition, 115–116
    nutrition, 111

Tailoring, byproducts and wastes, 15
Tobacco stems, use as fertilizers, 128
Toxicity characteristic leaching
    procedure, description, 259–260
Trace elements, role in compost, 153,155

Uncomposted waste paper, 12,163–178
Urban wastes, cattle manure, 105–106

Vanadium, byproduct use, 264

Wallboard waste, source of byproduct
    gypsums, 229
Waste(s)
    agricultural use, 28–44
    costs of disposal, 204–205
    inorganic, *See* Inorganic wastes
    intensification problems, 23
    management objectives, 22–23
    minimization, 24
    need for management, 2–3

Waste(s)—*Continued*
    policies, 29–37
    subsidization, 24
    tailoring, 15
    use as feedstock for other processes, 25
    value, 22
    watershed degradation, 23–24
    zero waste goal, 24–25
    See also Organic wastes
Waste paper
    benefits from land application, 169–174
    land application, 164–165
    method and time of application, 168–169
    organic components, 165
    physical nature and size, 165–166
    uncomposted, 12,163–178
Wastewater treatment
    biosolid waste, 6–8
    product, 50
Water quality, role of paper
    manufacturing residuals, 198–199
Water runoff, reduction using organic
    wastes, 170
Water solubility of micronutrient
    compound, 256
Watershed degradation, wastes, 23–24
Weed suppression, use of organic
    wastes, 173
Wind erosion, reduction using organic
    wastes, 170–171
Wood ashes, 130,205

Xenobiotic organic substances, use in
    composts, 156

Yard waste, 165

Zinc, byproduct use, 264–266

Highlights from ACS Books

Desk Reference of Functional Polymers: Syntheses and Applications
Reza Arshady, Editor
832 pages, clothbound, ISBN 0-8412-3469-8

Chemical Engineering for Chemists
Richard G. Griskey
352 pages, clothbound, ISBN 0-8412-2215-0

Controlled Drug Delivery: Challenges and Strategies
Kinam Park, Editor
720 pages, clothbound, ISBN 0-8412-3470-1

Chemistry Today and Tomorrow: The Central, Useful, and Creative Science
Ronald Breslow
144 pages, paperbound, ISBN 0-8412-3460-4

Eilhard Mitscherlich: Prince of Prussian Chemistry
Hans-Werner Schutt
Co-published with the Chemical Heritage Foundation
256 pages, clothbound, ISBN 0-8412-3345-4

Chiral Separations: Applications and Technology
Satinder Ahuja, Editor
368 pages, clothbound, ISBN 0-8412-3407-8

Molecular Diversity and Combinatorial Chemistry: Libraries and Drug Discovery
Irwin M. Chaiken and Kim D. Janda, Editors
336 pages, clothbound, ISBN 0-8412-3450-7

A Lifetime of Synergy with Theory and Experiment
Andrew Streitwieser, Jr.
320 pages, clothbound, ISBN 0-8412-1836-6

Chemical Research Faculties, An International Directory
1,300 pages, clothbound, ISBN 0-8412-3301-2

For further information contact:
American Chemical Society
Customer Service and Sales
1155 Sixteenth Street, NW
Washington, DC 20036
Telephone 800-227-9919
202-776-8100 (outside U.S.)
The ACS Publications Catalog is available on the Internet at
http://pubs.acs.org/books

Bestsellers from ACS Books

The ACS Style Guide: A Manual for Authors and Editors
Edited by Janet S. Dodd
264 pp; clothbound ISBN 0–8412–0917–0; paperback ISBN 0–8412–0943–X

Writing the Laboratory Notebook
By Howard M. Kanare
145 pp; clothbound ISBN 0–8412–0906–5; paperback ISBN 0–8412–0933–2

Career Transitions for Chemists
By Dorothy P. Rodmann, Donald D. Bly, Frederick H. Owens, and Anne-Claire Anderson
240 pp; clothbound ISBN 0–8412–3052–8; paperback ISBN 0–8412–3038–2

Chemical Activities (student and teacher editions)
By Christie L. Borgford and Lee R. Summerlin
330 pp; spiralbound ISBN 0–8412–1417–4; teacher edition, ISBN 0–8412–1416–6

Chemical Demonstrations: A Sourcebook for Teachers, Volumes 1 and 2, Second Edition
Volume 1 by Lee R. Summerlin and James L. Ealy, Jr.
198 pp; spiralbound ISBN 0–8412–1481–6
Volume 2 by Lee R. Summerlin, Christie L. Borgford, and Julie B. Ealy
234 pp; spiralbound ISBN 0–8412–1535–9

From Caveman to Chemist
By Hugh W. Salzberg
300 pp; clothbound ISBN 0–8412–1786–6; paperback ISBN 0–8412–1787–4

The Internet: A Guide for Chemists
Edited by Steven M. Bachrach
360 pp; clothbound ISBN 0–8412–3223–7; paperback ISBN 0–8412–3224–5

Laboratory Waste Management: A Guidebook
ACS Task Force on Laboratory Waste Management
250 pp; clothbound ISBN 0–8412–2735–7; paperback ISBN 0–8412–2849–3

Reagent Chemicals, Eighth Edition
700 pp; clothbound ISBN 0–8412–2502–8

Good Laboratory Practice Standards: Applications for Field and Laboratory Studies
Edited by Willa Y. Garner, Maureen S. Barge, and James P. Ussary
571 pp; clothbound ISBN 0–8412–2192–8

For further information contact:
American Chemical Society
1155 Sixteenth Street, NW ♦ Washington, DC 20036
Telephone 800–227–9919 ♦ 202–776–8100 (outside U.S.)
The ACS Publications Catalog is available on the Internet at
http://pubs.acs.org/books